Gary L. Peterson

$29.95 TR

D0164506

Introduction to High Energy Physics

2nd Edition, Revised, Enlarged, and Reset

Introduction to
High Energy Physics

2nd Edition,
Revised, Enlarged, and Reset

DONALD H. PERKINS
University of Oxford
Oxford, England

1982

ADDISON-WESLEY PUBLISHING COMPANY
Advanced Book Program
Reading, Massachusetts

London · Amsterdam · Don Mills, Ontario · Sydney · Tokyo

First printing, 1982
Second printing, 1983

Library of Congress Cataloging in Publication Data

Perkins, Donald H.
  Introduction to high energy physics.

  Bibliography: p.
  Includes index.
  1. Particles (Nuclear physics) I. Title.
QC793.2.P47   1982   539.7′21   82-3952
ISBN 0-201-05757-3   AACR2

Copyright © 1982 Addison-Wesley Publishing Company, Inc.
Published simultaneously in Canada.

All rights reserved. No part of this publication may be reproduced, stored in a retrieval system, or transmitted, in any form or by any means, electronic, mechanical, photocopying, recording, or otherwise, without the prior written permission of the publisher, Addison-Wesley Publishing Company, Inc., Advanced Book Program, Reading, Massachusetts 01867, USA

Printed in the United States of America

To My Family

For their patience, forbearance, and encouragement

Contents

Preface

The main intention behind this book has been to present the more important aspects of the field of high-energy physics, or particle physics, at an elementary level. The content is based on courses of lectures given to undergraduates in Oxford specializing in nuclear physics, but the book would also serve as an introductory text for first-year graduate students in experimental high-energy physics. I have tried to make the coverage as broad as possible while keeping the text to a reasonable length.

Since the first edition was written twelve years ago, high-energy physics has undergone many revolutionary developments, and both the volume and range of the subject has increased many times. This has meant a substantial rewriting of the text and a modest expansion in length. The interrelation between different aspects of the subject is now so strong that the division of the material under the various chapter headings has perforce been rather arbitrary.

The first chapter presents basic introductory ideas, the historical development, and a brief overview of the subject; the second and third chapters deal with experimental methods, conservation laws, and invariance principles – just as in the first edition. The following chapters deal in turn with the strong interactions between hadrons, the quark model of hadron constitution, weak interactions, and lepton-hadron scattering. The final chapter deals with the nature of the fundamental interactions between the lepton and quark constituents of matter, and the possible unification of these independent, elementary forces. During the last few years, the astrophysical and cosmological implications of results and ideas from high-energy physics have become important and indeed vital to our understanding of the development of the universe. I have tried to convey some of the flavor of this connection, since it will clearly help to shape the trends in high-energy physics in the foreseeable future.

As in the first edition, the interplay between experiment and theory has been emphasized, and some discussion given of key experiments in the field. Long theoretical treatments have been avoided, and for much of the mathematical detail the student is referred to Appendices or other texts. Some knowledge of elementary quantum mechanics is assumed, but generally the material has been presented from the empirical viewpoint, with a minimum of formalism and using an intuitive approach. Physics is about numbers, and I have taken the view that it was more important that a student should know how to calculate a cross-section or a decay rate, in order of magnitude, than how to derive a complicated formula (usually based on assumptions of questionable validity) without any real idea on how to confront it with experiment. In the same spirit, I have included a list of (mostly numerical) problems for each chapter, together with worked solutions at the end of the book.

The material of the text has been compounded from advanced texts, review articles, original papers, and particularly the proceedings of summer schools. A short bibliography at the end of each chapter is given for further reading. A table of physical constants and a short list of particle states and properties is included.

ACKNOWLEDGMENTS

For permission to reproduce various photographs, figures and diagrams, I am indebted to the authors cited in the text and to the following laboratories and publishers: Brookhaven National Laboratory, Long Island, New York; CERN Information Services, Geneva; Rutherford and Appleton Laboratories, Chilton, Didcot, England; Stanford Linear Accelerator Laboratory, Stanford, Calif.; DESY Laboratory, Hamburg; Max Planck Institute, Munich; Annual Reviews Inc., Palo Alto, Calif.; the American Institute of Physics, New York, publishers of *The Physical Review, Physical Review Letters*, and *Reviews of Modern Physics*; The Italian Physical Society, Bologna, publishers of *Il Nuovo Cimento*; North Holland Publishing Co., Amsterdam, publishers of *Physics Letters, Nuclear Physics*, and *Physics Reports*; the Institute of Physics, Bristol, England, publishers of *Reports on Progress in Physics*; Pergamon Press Ltd., Oxford.

Many people helped me with suggestions and advice during the preparation of the text, and were kind enough to point out errors in the first edition. I owe a great debt to Chris Llewellyn-Smith (Oxford) and Chris Quigg (Fermilab) for their careful reading of the manuscript, detailed criticisms and comments, and their many suggestions for improvements. Several people were very kind in supplying me with original photographs and diagrams, and I should like especially to thank the staff of the CERN Information Services; Fred Combley, of the University of Sheffield; and Bärbel Lücke, Sabine Platz and Rolf Felst, of the DESY laboratory, Hamburg.

Finally, I wish to record my special thanks to Irmegarde Smith for her preparation of the line drawings, Cyril Band and his colleagues for the photographic work, and Suzanne Motyka for her careful typing of the manuscript.

DONALD H. PERKINS

Now the smallest Particles of Matter may cohere by the strongest Attractions, and compose bigger Particles of weaker Virtue. . . . There are therefore Agents in Nature able to make the Particles of Bodies stick together by very strong Attractions. And it is the Business of experimental Philosophy to find them out.

Newton, *Optics* (1680)

Les Philosophes qui font des systèmes sur la secrète construction de l'univers, sont comme nos voyageurs qui vont a Constantinople, et qui parlent du Sérail: Ils n'en ont vu que les dehors, et ils prétendent savoir ce que fait le Sultan avec ses Favorites.

Voltaire, *Pensées Philosophiques* (1766)

Chapter 1

History and Basic Concepts

1.1 INTRODUCTION

High-energy physics deals basically with the study of the ultimate constituents of matter and the nature of the interactions between them. Experimental research in this field of science is carried out with giant particle accelerators and their associated detection equipment. High energies are necessary for two reasons; first, in order to localize the investigations to the very small scales of distance associated with the elementary constituents, one requires radiation of the smallest possible wavelength and highest possible energy; secondly, many of the fundamental constituents have large masses and require correspondingly high energies for their creation and study.

Forty or fifty years ago, only a few "elementary" particles – the proton and neutron, the electron and neutrino, together with the electromagnetic field quantum (the photon) – were known. The universe as we know it today appears indeed to be composed almost entirely of these particles. However, attempts to understand the details of the nuclear force between protons and neutrons, as well as to follow up the pioneering discoveries of new, unstable particles observed in the cosmic rays, led to the construction of ever larger accelerators and to the observation of many hundreds of new particle states. These so-called *hadrons* (or strongly interacting particles) are very unstable under terrestrial conditions but are otherwise, as far as we can tell, just as fundamental as the familiar neutron and proton. The hadron states can be grouped into families or multiplets, and the neutron and proton occupy undistinguished places in one of these multiplets. Their predominance in our universe today appears to be an accident of the conservation laws and of the ambient conditions. Physicists have long believed however that a deeper understanding of the nature of neutrons and protons could only emerge from a comprehensive picture of particle states as a whole. Our present model of the constituents of matter is that they consist of fundamental pointlike spin $\frac{1}{2}$ fermions – the *quarks*, with fractional electric charges ($+\frac{2}{3}e$ and $-\frac{1}{3}e$) and the *leptons*, like the electron and neutrino, carrying integral electric charges. Neutrons and protons are built from quarks, three at a time. These constituents can interact by exchange of various fundamental

1

bosons (integral spin particles) which are the carriers or quanta of four distinct types of fundamental interaction or field. *Gravity* is familiar to everyone, yet on the scales of mass and distance involved in particle physics, it is by far the least important of the four. Apart from gravity, *electromagnetic* interactions account for most extranuclear phenomena in physics (because electromagnetic forces have the longest range) and lead to the bound states of atoms and molecules. *Weak* interactions are exemplified by the extremely slow process of radioactive β-decay of nuclei. *Strong* interactions are postulated to hold together the quarks in a proton, and their residual effects apparently account for the interactions between neutrons and protons, that is, for the nuclear binding force.

There are many unusual, even bizarre, aspects to this picture. The fractionally charged quarks have not been convincingly observed as free particles, and they may well be permanently confined in hadrons. Quarks come in a variety of types or *flavors* (at least five are known) as do the leptons (three types of charged and of neutral leptons). We neither understand the mechanism of confinement, nor the real reason for the "Xerox copies" of quark and lepton flavors, when the universe, on the basis of what we see today, seems to be constructed predominantly from just two types of quark and one neutral and one charged lepton.

The multiplicity of quark and lepton flavors is paralleled by the existence of the four types of fundamental interaction. Here, some real progress has been made. There are good grounds for supposing that some, perhaps all, the interactions are *unified*, that is, different aspects of one single interaction. For example, the weak and electromagnetic interactions appear to have the same intrinsic coupling of fermion constituents to the respective mediating bosons. Compared with electromagnetism (mediated by the massless photon field with infinite range), the weakness of the weak interactions is ascribed to their short-range nature (they are mediated by massive bosons $W^{\pm}, Z^0$, whose mass is believed to be nearly 100 proton masses). At high enough energies, well above such boson mass scales, electromagnetic and weak interactions are expected to have the same actual strength; all the low-energy experiments support this extrapolation.

Why the high-energy symmetry is badly broken at low energy, and the respective bosons have such widely differing masses, is still an unsolved problem. The important point however is that the strengths of the different interactions are not fixed once and for all; they depend on energy scales. At high energies, weak interactions are observed to grow stronger, and strong interactions appear to grow weaker. Possibly these two interactions also merge at some unattainably high energy ($\sim 10^{14}$ GeV), and there quark-lepton transitions might become commonplace.

1.2 HISTORICAL DEVELOPMENT

High-energy physics had its foundation in experiments with cosmic rays, the only available source of very energetic particles up to the early 1950s. The

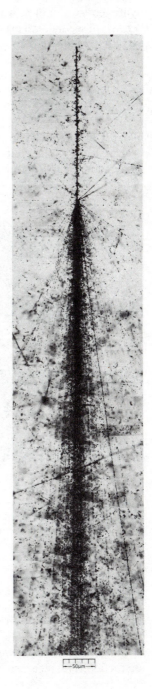

Fig. 1.1 Collision of a primary cosmic-ray iron nucleus, of energy 5000 GeV per nucleon, with a nucleus in nuclear photographic emulsion, carried by balloon in the stratosphere. Both nuclei are fragmented, and in addition about 200 new particles (mostly pions) are created. The pions decay in flight in the stratosphere, producing leptons (muons and neutrinos). The charged muons form the bulk of the cosmic-ray flux (~ 1 per cm^2 per minute) at sea level.

primary cosmic rays consist principally of high-energy nuclei of atoms, from hydrogen to uranium, which permeate the galaxy. The radiation at sea level consists of secondary particles produced when the primary nuclei impinge on the earth's atmosphere. An example of such a collision is shown in Fig. 1.1. Among the important discoveries in this field was the observation (Fig. 1.2) in a cloud chamber of the *positron* (e^+) by Anderson and by Blackett and Occhialini in 1933. The existence of the positron, as the antiparticle to the electron (e^-), with the same mass but opposite charge and sign of magnetic moment, had been predicted by Dirac in 1928. Shortly afterwards, the *muon* ($\mu^\pm$), a charged member of the lepton family, with 206 times the mass of the electron, was first observed in cloud-chamber experiments by Street and Stevenson (1937) and Anderson and Neddermeyer (1938).

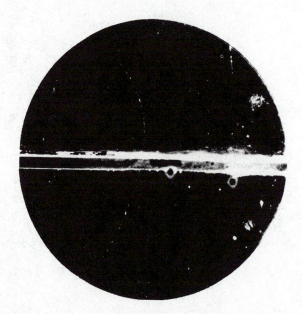

Fig. 1.2 Discovery of the positron by Anderson in 1932, in a cloud chamber. The particle's momentum and sign of charge are inferred from its direction and curvature in the applied magnetic field. It loses energy in the central lead plate, and was therefore moving upwards and positive. The droplet density along the track implies a velocity $v \simeq c$ and thus a small mass.

At first, the muon was incorrectly identified with the strongly interacting quantum of the nuclear field, which had been postulated by Yukawa in 1935. Yukawa's key idea was to attribute the finite range ($\sim 10^{-15}$ m) of nuclear forces to the existence of a massive field quantum or carrier of integral spin, the exchange of which between proton and neutron would account for their interaction. This idea is so important that it is worth describing it in a little detail.

To create a particle of mass m, associated with a static field, implies violation of energy conservation unless it is restricted within a time limited by the uncertainty principle, $\Delta t \leqslant \hbar/mc^2$ (where $\hbar = h/2\pi$). In this time, the particle could cover at most a distance $R = c\,\Delta t < \hbar/mc$. Thus, the range of the field is given by the Compton wavelength of the associated quantum. We can make this argument more quantitative with the help of the relativistic relation between the total energy E and momentum p of a free particle of mass m,

$$E^2 = p^2c^2 + m^2c^4. \tag{1.1}$$

The differential equation describing the wave amplitude ψ of such a free particle is obtained by substituting in (1.1) the quantum-mechanical operators

$$E_{\mathrm{op}} = i\hbar\frac{\partial}{\partial t}, \qquad p_{\mathrm{op}} = -i\hbar\nabla = -i\hbar\frac{\partial}{\partial \mathbf{r}}$$

which give the Klein-Gordon wave equation

$$\nabla^2\psi - \frac{m^2c^2}{\hbar^2}\psi - \frac{1}{c^2}\frac{\partial^2\psi}{\partial t^2} = 0 \tag{1.2}$$

describing the propagation in free space of spinless particles of mass m. If we set $m = 0$, (1.2) becomes the familiar wave equation describing the propagation of an electromagnetic wave, with ψ interpreted either as the potential at a point in space and time, or as the wave amplitude of the associated free, massless photons. We are interested here not so much in the propagation of particle waves as in static potentials. Dropping the time-dependent term in (1.2) therefore, the resulting equation for the static potential U has the spherically symmetric form

$$\nabla^2 U(r) = \frac{1}{r^2}\frac{\partial}{\partial r}\left(r^2\frac{\partial U}{\partial r}\right) = \frac{m^2c^2}{\hbar^2}U(r)$$

for values of $r > 0$ from a point source at the origin, $r = 0$. The solution is

$$U(r) = \frac{g}{r}e^{-r/R}, \tag{1.3}$$

where

$$R = \frac{\hbar}{mc}. \tag{1.4}$$

Here the quantity g is a constant of integration identified with the strength of the point source. The analogous equation in electromagnetism is $\nabla^2 U(r) = 0$ for $r > 0$, with solution $U = Q/r$, where Q is the charge at the origin. Thus, g in the Yukawa theory plays the same role as charge in electrostatics, and is a measure of the "strong nuclear charge".

Yukawa's idea of linking the finite range of the nuclear field with a massive mediating boson has turned out to be, as we shall see, a fundamental concept in the theory of fields. The Fourier transform of the potential (1.3) will yield the dependence of the amplitude for a scattering process on the momentum q transferred by the mediating boson. This so-called propagator term has the form (see Chapter 7, Eq. (7.28))

$$f(q) = \int U(r)e^{iqr}r^2 \, dr$$

yielding a dependence for the scattered intensity, or amplitude squared, on q as follows

$$ff^* = \frac{1}{(q^2 + 1/R^2)^2} = \frac{1}{(q^2 + m^2)^2}, \tag{1.5}$$

using (1.4) and with units $\hbar = c = 1$. This result also holds relativistically, if we treat q as the four-momentum transfer. For $m = 0$, for example, the photon propagator introduces a $1/q^4$ dependence of the cross-section for scattering between two charged particles, and is the basis of the famous Rutherford scattering formula.

Returning to the cosmic-ray muons, experiments were then made with counters by Conversi *et al.* (1947), showing that their interpretation as Yukawa quanta was untenable. With the aid of electromagnets, they were able to observe separately the fate of the positive and negative muons coming to rest in blocks of various absorbers. Positive muons were observed to decay (to a positron and neutrinos) with lifetime 2 μsec. Negative particles, if they were the Yukawa quanta, would be expected to fall into Bohr orbits about an atomic nucleus, and thence suffer rapid nuclear absorption (in 10^{-12} sec). On the contrary, upon stopping in light materials like carbon, they also were observed to decay. This was a surprising and unexpected result. It could be estimated, on the basis of the energies and radii of the Bohr orbits of the negative muon in the nuclear Coulomb field, that the muon in the first atomic S-state would spend $\sim 10^{-3}$ of its time inside nuclear matter, that is, some 10^{-9} sec, in which time it would cover ~ 1 cm, or 10^{13} times the internucleon distance R_0. Clearly, such a particle could not have strong nuclear interaction. Indeed, the muons are leptons having only weak and electromagnetic coupling.

In the same year (1947) Marshak and Bethe suggested that *two* particles might be involved, the lighter muons being decay products of heavier mesons, which were to be identified with the Yukawa quanta. Using special photographic emulsions exposed at mountain altitudes, Lattes *et al.* (1947) in fact found events (see Fig. 1.3) corresponding to decay of a *pion* (π) into a muon and a neutral particle (neutrino):

$$\pi^+ \rightarrow \mu^+ + \nu_\mu, \tag{1.6}$$

this decay proceeding through the weak interaction, and having a lifetime of 10^{-8} sec.

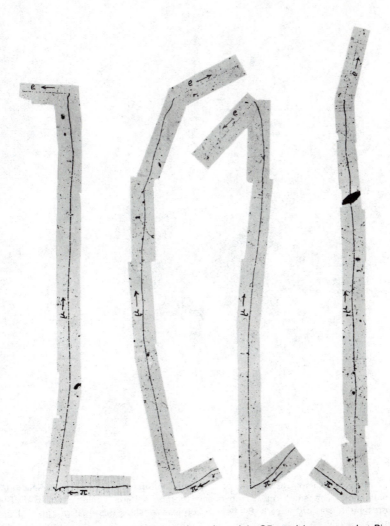

Fig. 1.3 Examples of the decay sequence $\pi^+ \to \mu^+ \to e^+$ in G5 emulsion exposed at Pic du Midi. The constancy of range ($\simeq 600 \ \mu$m) of the muon implies two-body decay at rest of the pion: $\pi^+ \to \mu^+ + \nu_\mu$. The first examples of pion decay were observed by Lattes, Muirhead, Occhialini, and Powell in 1947. The electron emitted in muon decay, $\mu^+ \to e^+ + \nu_e + \bar{\nu}_\mu$, was not observed in the early experiments employing less sensitive emulsions. (Photograph courtesy University of Bristol).

Events were also observed (Perkins 1947, Occhialini and Powell 1947), which were interpreted as nuclear capture of negative pions (π^-) when they came to rest in emulsion, the mass energy released leading to disintegration of the capturing nucleus (Fig. 1.4). Thus, pions had strong interactions with nuclei, and were created by energetic cosmic-ray primary particles (mostly protons) undergoing

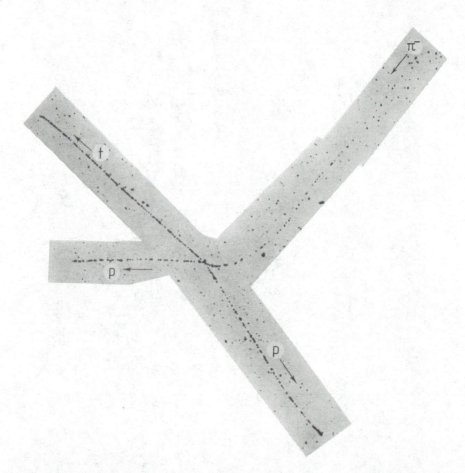

Fig. 1.4 First example of capture of a negative pion (π^-) by a nucleus, leading to its disintegration. The emulsion was exposed in high-altitude aircraft from RAF Benson, Oxfordshire. It was of the early B1 type, much less sensitive than that of Figs. 1.1 and 1.3 (Perkins 1947).

nuclear collisions in traversing the atmosphere. The pions usually decay in flight in the stratosphere, producing the daughter, noninteracting muons which constitute the bulk of the cosmic ray particles at sea level and which had been observed in the previous experiments. Later work established that the muons in turn decay according to the scheme $\mu^+ \to e^+ + \nu_e + \bar{\nu}_\mu$. The muon decay is also shown in Fig. 1.3.

In addition to the positive and negative pions, neutral pions decaying by the electromagnetic process

$$\pi^0 \to 2\gamma \tag{1.7}$$

were also shown to exist, both in cosmic-ray experiments (Carlson *et al.* 1950) and at the newly commissioned 184-in. synchrocyclotron at Berkeley (Bjorklund *et al.* 1950). These experiments showed the neutral and charged pions to be produced in nuclear interactions with similar cross-sections and to have similar masses, as expected for charged and neutral counterparts of the same particle. The detection of neutral as well as charged pions was of great importance, since as Kemmer had emphasized some ten years previously, they were necessary in the Yukawa theory to account for the observed near-equivalence of the proton-proton and neutron-proton potentials.

In the same year (1947) in which the pion was discovered, the cloud-chamber observations of Rochester and Butler on penetrating cosmic-ray showers revealed the existence of still-heavier unstable particles. In one picture they observed a "V-event" ascribed to a neutral particle decaying into two charged particles; another picture showed a heavy charged particle decaying in flight in the gas of the chamber (Fig. 1.5). At almost the same time, the Bristol group reported the decay at rest, in emulsion, of a heavy charged meson into three charged secondaries, which were soon to be identified as pions.

By 1953, although the picture was still confused, the existence had been established of unstable particles heavier than nucleons, collectively called hyperons: notably the Λ-hyperon, decaying in the mode $\Lambda \to p + \pi^-$; the Σ-hyperon, decaying according to $\Sigma^+ \to p + \pi^0$ or $\Sigma^+ \to n + \pi^+$; the cascade or Ξ-hyperon, giving a "double V-event", $\Xi^- \to \pi^- + \Lambda$, $\Lambda \to p + \pi^-$. In the same year, artificial production of such particles was observed in a hydrogen diffusion cloud chamber using a 1.4-GeV/c negative-pion beam from the newly commissioned proton synchrotron (Cosmotron) at the Brookhaven National Laboratory. Apart from the hyperons, a variety of decay modes had been established for particles, both neutral and charged, of mass between that of the pion and the proton — generically labeled K-mesons or kaons. Again, the experiments were able to establish that one was dealing with charged and neutral counterparts of one particle, but one with a variety of decay modes.

However, the greatest problem posed by kaons and hyperons was their copious production (some 10% of the pion production) in view of their long lifetimes, in the range 10^{-8} to 10^{-10} sec. Suppose a Λ-hyperon is produced in a reaction like

$$\pi^- + p \to \Lambda + \pi^0. \tag{1.8}$$

Then, by the principle of microscopic reversibility, the inverse process,

$$\Lambda \to \pi^0 + \pi^- + p \to \pi^- + p \tag{1.9}$$

can occur. The first stage in Eq. (1.9) violates the conservation of energy; it is supposed that it can occur virtually, i.e. on such a short time scale as to be allowed by the uncertainty principle $\Delta E = \hbar/\Delta t$. The typical time involved in the production reaction is of the order of the range of the strong interaction (10^{-13} cm) divided by the velocity of light ($\sim 10^{-23}$ sec). Why then was the decay

Fig. 1.5 First observations of V-events in a cloud chamber, by Rochester and Butler (1947). The upper picture is of a "neutral V-event", consisting of a wide-angle fork occurring in the gas a few millimeters below the horizontal plate. Subsequent analysis suggests that it was due to the decay $K^0 \rightarrow \pi^+ + \pi^-$. The lower picture is of a "charged V-event", seen as a fork near the right-hand top corner of the picture. The secondary traverses the 3-cm lead plate without interaction. The measured momenta are in fact consistent with the decay scheme $K^+ \rightarrow \mu^+ + \nu$, or what is now called the $K_{\mu 2}$ decay mode of the charged kaon. (Courtesy Pergamon Press.)

process not as rapid? The corresponding problem in the case of the slowly decaying, but apparently copiously produced, muon had been solved, as we saw, with the discovery of the pion.

The hypothesis of *associated production* introduced by Pais (1952) solved the problem elegantly; the kaons and hyperons must be created (and destroyed) in pairs. Then the production process would appear as

$$\pi^+ + n \rightarrow \Lambda + K^+. \tag{1.10}$$

The inverse strong decay process must then be

$$\Lambda \rightarrow K^- + n + \pi^+ \rightarrow K^- + p. \tag{1.11}$$

However, $\Lambda \rightarrow K^- + p$ as a real decay process is forbidden by the conservation of energy, since $m_\Lambda < m_K + m_p$. Gell-Mann (1953) and Nishijima (1955) formalized the concept of associated production, by introducing an additive quantum number, called *strangeness* (S). The Λ-hyperon and K^- were assigned $S = -1$, and the K^+, $S = +1$. Conservation of strangeness in a strong

Fig. 1.6 An example of associated production, due to the interaction at A of a 4-GeV/c negative pion in a hydrogen bubble chamber: $\pi^- + p \rightarrow \Lambda + K^0$. The Λ-hyperon decays at B according to $\Lambda \rightarrow p + \pi^-$, and the K^0-meson at C according to $K^0 \rightarrow \pi^+ + \pi^-$. (Courtesy CERN.)

interaction then led to the associated production hypothesis (1.10). Violation of strangeness conservation, as in Eq. (1.9), implies that the decay process must have the long time scale typical of weak interactions. Indeed, the so-called "τ-θ paradox" – the observation that the K-meson could decay to either two or three pions – led directly to the suggestion by Lee and Yang (1956) that the weak interactions were not invariant under the operation of space inversion (parity violation). This is discussed in Chapter 6. An example of the process of associated production is shown in Fig. 1.6.

These early cosmic-ray experiments have been described at some length, in order to explain how some of the nomenclature originated, and because they were of the greatest importance in stimulating the construction of high-energy, high-intensity accelerators, which alone made possible the precision measurements required to put the subject on a quantitative footing.

1.3 CLASSIFICATION OF PARTICLES—FERMIONS AND BOSONS

One of the most fundamental concepts underlying our analysis of the interactions of particles and fields is the spin-statistics theorem (Pauli 1940), connecting the statistics obeyed by a particle with its spin angular momentum. Particles with half-integral spin ($\frac{1}{2}\hbar, \frac{3}{2}\hbar, \ldots$) obey Fermi-Dirac statistics and are thus called fermions, while those with integral spin ($0, \hbar, 2\hbar, \ldots$) obey Bose-Einstein statistics and are called bosons.

The statistics obeyed by a particle determines the symmetry of the wavefunction ψ describing a pair of identical particles, say 1 and 2, under interchange. If the particles are identical, then the square of the wavefunction, $|\psi|^2$, giving the probability of particle 1 at one coordinate and particle 2 at another, will be unaltered by the interchange $1 \leftrightarrow 2$. Thus

$$\psi \overset{1 \leftrightarrow 2}{\rightarrow} \pm \psi.$$

The following rule holds:

Identical bosons: $\psi \overset{1 \leftrightarrow 2}{\rightarrow} + \psi$ symmetric,

(1.12)

Identical fermions: $\psi \overset{1 \leftrightarrow 2}{\rightarrow} - \psi$ antisymmetric.

In order to make use of this rule, the total wavefunction of the pair can be expressed as a product of functions depending on spatial coordinates and spin orientation:

$$\psi = \alpha(\text{space})\, \beta(\text{spin}). \tag{1.13}$$

The spatial part, α, will describe any orbital motion of one particle about the other, and can be represented by a spherical harmonic function $Y_l^m(\theta, \phi)$, as described in Chapter 3.3. Interchange of the space coordinates of particles 1 and 2 (leaving spin alone) is equivalent to the replacement $\theta \rightarrow \pi - \theta$, $\phi \rightarrow \phi + \pi$, and introduces a factor $(-1)^l$ multiplying α. Thus, if l is even (odd), the function α is symmetric (antisymmetric) under interchange. As also indicated in Chapter 3, the spin function β may be symmetric (spins parallel) or antisymmetric (spins antiparallel) under interchange. Equation (1.12) implies that, for identical bosons, α and β must be both symmetric or both antisymmetric; while for fermions, a symmetric α implies an antisymmetric β and vice versa.

As an example, consider the decay of the neutral ρ-meson of spin $J = 1$ into two neutral pions: $\rho^0 \to 2\pi^0$. Both pions are uncharged and spinless. Thus, since β is necessarily symmetric, the rule for identical bosons means α must be symmetric, and thus the two pions can exist only in a state with even total angular momentum J. Hence the decay $\rho^0 \to 2\pi^0$ is forbidden by angular-momentum conservation and Bose symmetry. Decay into charged (non-identical) pions does take place, however: $\rho^0 \to \pi^+\pi^-$, $\rho^\pm \to \pi^\pm\pi^0$. In this case, we can extend the meaning of "identical particles" (i.e. treat all pions of whatever charge as identical bosons) if we include in (1.13) a charge wavefunction (formally called an isospin wavefunction) γ:

$$\psi = \alpha(\text{space})\,\beta(\text{spin})\,\gamma(\text{charge or isospin}). \tag{1.14}$$

The existence of the decay $\rho^\pm \to \pi^\pm\pi^0$ then implies that, since β is symmetric and α antisymmetric, γ must also be antisymmetric. This approach is very useful in relating branching ratios for the decay and production of different charged states in particle reactions.

The Pauli principle is a well-known application of the antisymmetry of the wavefunction of two identical fermions under interchange. Suppose two identical particles are in the same quantum state, so that ψ is necessarily symmetric. This violates the rule that two identical fermions must have ψ antisymmetric. Hence two fermions cannot exist in the same quantum state – the Pauli principle. On the other hand, there is no restriction on the number of bosons (photons, for example) which may exist in the same quantum state. An example of this is the laser.

1.4 PARTICLES AND ANTIPARTICLES

The relativistic wave equation proposed in 1928 by Dirac was able to account for the intrinsic angular momentum, or spin quantum number, of the electron, which had previously been postulated by Uhlenbeck and Goudsmit in order to account for the Zeeman effect in atomic physics. In the Dirac theory free electrons are described by 4-component wavefunctions, corresponding to two spin substates, $J_z = \pm\frac{1}{2}\hbar$, each of positive or of negative energy. The negative-energy states are interpreted in terms of an antiparticle, the positron (see Appendix B). The demonstration of the existence of positrons has already been described. The existence of antiparticles is a general property of both fermions and bosons, the antiparticle having the same mass as the particle, but opposite charge and magnetic moment. The antiproton was discovered in experiments at Berkeley by Chamberlain *et al.* in 1955 (see Chapter 2). Antideuterons and various antihyperons have since been observed.

Fermions and antifermions can only be created or destroyed in pairs. For example, a γ-ray, in the presence of a nucleus (to conserve momentum), can "materialize" into an electron-positron pair (see Fig. 1.7), and an e^+e^- bound

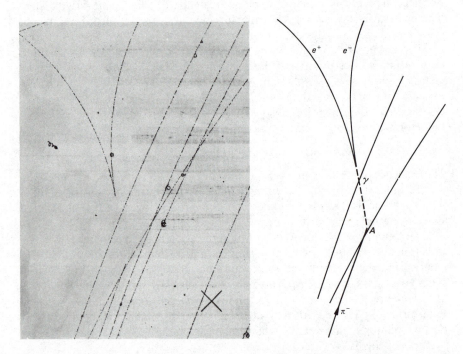

Fig. 1.7 Conversion of a photon into an electron-positron pair in a bubble chamber. The photon originated from the decay $\pi^0 \rightarrow 2\gamma$. Since the neutral pion has a short lifetime (10^{-16} sec), the pair appears to point straight at the interaction vertex, *A*, corresponding to the charge-exchange reaction $\pi^- + p \rightarrow \pi^0 + n$.

state, called positronium, annihilates into two or three γ-rays. Theoretically, particle and antiparticle states are connected by the process of particle-antiparticle conjugation. Fermion number is conserved if each fermion is assigned a fermion number $+ 1$ and each antifermion $- 1$. So the process of particle-antiparticle conjugation for fermions gives an antifermion with opposite charge, magnetic moment, and fermion number, but identical mass and spin angular momentum. There is no law of number conservation for bosons, and particle-antiparticle conjugation has the same effect as charge conjugation for them.

1.5 BASIC FERMION CONSTITUENTS: QUARKS AND LEPTONS

In the remainder of this chapter, we shall discuss briefly the nature of the fundamental fermion constituents of matter—the quarks and leptons—and of the interaction fields (the fundamental boson exchanges) occurring between them. This is meant as an introduction and overview, in order to indicate the scope of the subject and the underlying physical ideas. It will necessarily introduce new and unfamiliar quantum numbers and new concepts, which will be simply stated without theoretical reasons or experimental justification. These will emerge in detail in the following chapters. Present evidence indicates that matter is built from two types of fundamental fermion, called quarks and leptons, which are structureless and pointlike on a scale of 10^{-17} m.

Quarks carry fractional electric charges, of $+\frac{2}{3}|e|$ and $-\frac{1}{3}|e|$. They occur in several varieties or *flavors*, distinguished by the assignment of internal quantum numbers, and are labelled u, d, s, c, b as in Table 1.1. The u- and d-quarks are the lightest and have approximately the same mass (within 1 MeV or so). As indicated in Section 1.6, protons and neutrons are considered to be built from u- and d-quarks, and consequently the near-equality of proton and neutron masses implies the same equality for u- and d-quarks. They have similar strong interactions with other quarks, and differ only in their electric charge and hence electromagnetic interactions. Historically, the equality of strong interactions of

TABLE 1.1 Quarks

Quarks		Antiquarks	
$Q/|e| = +\frac{2}{3}$	$u, c, \ldots$	$Q/|e| = -\frac{2}{3}$	$\bar{u}, \bar{c}, \ldots$
$Q/|e| = -\frac{1}{3}$	d, s, b	$Q/|e| = +\frac{1}{3}$	$\bar{d}, \bar{s}, \bar{b}$

$u =$ "up" quark $\Big\}$ $I = \frac{1}{2}$ doublet	$m_u \simeq m_d \simeq 350\ \mathrm{MeV}/c^2$
$d =$ "down" quark	$m_s \simeq 550\ \mathrm{MeV}/c^2$
$s =$ "strange" ($S = -1$),	$m_c \simeq 1800\ \mathrm{MeV}/c^2$
$c =$ "charmed" ($C = +1$)	$m_b \simeq 4500\ \mathrm{MeV}/c^2$
$b =$ "bottom" ($B = -1$)	

TABLE 1.2 Leptons

Leptons				Antileptons			
$Q/|e| = -1$	e^-	μ^-	τ^-	$Q/|e| = +1$	e^+	μ^+	τ^+
$Q/|e| = 0$	v_e	v_μ	v_τ	$Q/|e| = 0$	$\bar{v}_e$	$\bar{v}_\mu$	$\bar{v}_\tau$

$$m_e = 0.511\ \mathrm{MeV}/c^2$$
$$m_\mu = 105.6\ \mathrm{MeV}/c^2$$
$$m_\tau = 1870\ \mathrm{MeV}/c^2$$

the u- and d-quark constituents appeared as the hypothesis of isospin invariance of interactions between hadrons composed of u- and d-quarks. For this reason, u- and d-quarks are sometimes grouped as an isospin doublet ($I = \frac{1}{2}$, with the third component $I_3 = +\frac{1}{2}$ for u and $-\frac{1}{2}$ for d). The s-quark has strangeness number -1, the c-quark charm number $+1$, and the b-quark (the heaviest observed so far) the bottom quantum number -1. These quantum numbers are discussed more fully in Chapter 5. The quarks are confined in hadrons, and do not appear to exist as free particles. The masses indicated in Table 1.1 should be taken as indicative only. Corresponding to each quark is an antiquark, with the opposite sign of charge, magnetic moment, strangeness, etc. As indicated in (1.15) below, *baryons* consist of three quarks, and *mesons* of a quark-antiquark pair.

The *leptons* carry integral electric charges, 0 or $\pm|e|$, and three types of each are known (see Table 1.2). The neutral leptons are called neutrinos, and have very small or zero rest-mass. The electron is familiar to everyone, the muon has already been mentioned, and the τ-lepton was first observed in accelerator experiments in 1974. The leptons appear in doublets, the neutrinos being assigned a subscript corresponding to the charged member. Charged leptons are distinguished from antileptons by the sign of charge. Neutrinos are longitudinally spin-polarized with $J_z = -\frac{1}{2}$ ("left-handed") where z is the direction of the velocity vector, while antineutrinos have $J_z = +\frac{1}{2}$ ("right-handed").

The charged leptons have electromagnetic and weak interactions, while the neutrinos are distinguished by having only weak interactions with other particles. Quarks, in addition to weak and electromagnetic interactions, are subject to the strong (specifically quark-quark) interactions. While the strong interactions lead to quark composites (hadrons), only loosely bound and unstable combinations of charged leptons occur (for example positronium e^+e^-, bound by the Coulomb interactions).

The conservation rules for fermions apply of course to quarks and leptons. In particular, a lepton number L_e, L_μ, L_τ of $+1$ is given to each type of lepton, and -1 to each antilepton. Examples are

electron pair production by photon:

$$
\begin{array}{cccc}
 & \gamma \rightarrow & e^+ & + \; e^-, \\
L_e & 0 & -1 & +1
\end{array}
$$

pion decay:

$$
\begin{array}{cccc}
 & \pi^+ \rightarrow & \mu^+ & + \; \nu_\mu, \\
L_\mu & 0 & -1 & +1
\end{array}
$$

muon decay:

$$
\begin{array}{ccccc}
 & \mu^+ \rightarrow & e^+ & + \; \nu_e & + \; \bar{\nu}_\mu, \\
L_\mu & -1 & 0 & 0 & -1 \\
L_e & 0 & -1 & +1 & 0
\end{array}
$$

tau decay:

$$\tau^+ \;\rightarrow\; \bar{v}_\tau \;+\; \pi^+,$$
$$L_\tau \quad -1 \quad\;\; -1 \quad\;\; 0$$

while the decay

$$\mu^+ \rightarrow e^+ + \gamma$$
$$L_\mu \quad -1 \quad\;\; 0 \quad\;\; 0$$
$$L_e \quad\;\; 0 \quad\;\; -1 \quad\;\; 0$$

is forbidden by lepton-number conservation. The limit to the branching ratio for this muon decay mode is $< 10^{-9}$.

While the total quark number is conserved in all interactions, the number of quarks of a given flavor is absolutely conserved only in strong interactions (equivalent to conservation of I_3, strangeness, and similar quantum numbers). In weak decay processes, the quark flavor may change ($\Delta S = 1$, $\Delta C = 1$, etc.).

The laws of fermion-number conservation were formulated for baryons by Stueckelberg (1938) and Wigner (1949) and for leptons by Konopinski and Mahmoud (1953). They are no longer considered so sacrosanct as they once were. They do however appear to be valid to a considerable degree of accuracy. For example, the present lower limit on the lifetime of the proton is more than 10^{30} years. The difficulty with an absolute conservation law for baryons is that, in analogy with charge conservation (linked to the existence of the electromagnetic field), there should be a new long-range field coupled to baryons, which should result in a difference in the apparent gravitational force on bodies with the same inertial mass but different baryon number. The Eötvos experiments indicate no such effects, and hence such a field would have to be very much weaker than gravity (Lee and Yang 1955).

1.6 HADRONS—COMPOSITES OF QUARKS AND ANTIQUARKS

As mentioned above, just two types of quark combination are required to account for the observed strongly interacting particles, or hadrons:

$$\text{baryons} = QQQ,$$
$$\text{mesons} \;= Q\bar{Q}. \tag{1.15}$$

Since quarks have half-integral spin, it follows that the baryons are characterized by half-integral, and the mesons by integral, spin. Examples are:

Baryons	Mesons
$uud = p$ (proton)	$u\bar{d} = \pi^+$ (pion)
$udd = n$ (neutron)	$\bar{u}s = K^-$ (kaon)
$uds = \Lambda$	$c\bar{c} = \psi$-meson

As expected from our discussion of bosons and fermions, the conservation rule for quarks is reflected in the conservation of baryon number, while there is

TABLE 1.3 Many-body interactions of composite particles

Structure	Interaction	Force	Interaction in terms of constituents
Molecular and atomic	Molecule-molecule, atom-atom	Van der Waals, electron exchange	Residual EM interaction
Nuclear	Nucleus-nucleus	Nuclear forces (e.g. $\alpha + {}^{14}N \rightarrow p + {}^{17}O$)	Residual quark-quark interactions
Hadron	Hadron-hadron	"Strong" forces (e.g. $p + p \rightarrow \pi^{+} + n + p$)	

no conservation rule for mesons. Several hundred hadron states are known. Figure 5.1 p. 181 shows the mass spectrum of baryons with masses up to 2500 MeV/c^2. Most of them are broad states, decaying by strong interactions with very short lifetimes ($\sim 10^{-23}$ sec) and large widths ($\Gamma \sim 100$ MeV). A few decay by weak interactions, with lifetimes varying from 10^{-10} sec for the hyperons to 15 minutes for the free neutron. The common decay product of all these states is the proton, which can be regarded as the stable ground state of the many excited baryon states. Similarly, there are many dozens of meson states. All are unstable by decay through the strong, electromagnetic, or weak interactions, and give leptons or photons as the ultimate decay products.

Hadrons, being composite particles, have extended structure – in contrast with the apparently pointlike nature of the leptons and quarks. For example, the rms charge radius of the proton is 0.8 fm, that of the pion is 0.6 fm. By the same token, the baryons have large anomalous magnetic moments. Composite particles, be they hadrons, atomic nuclei, atoms, or molecules, have interactions. These are in the nature of relatively complicated residual many-body interactions between the constituents, as indicated in Table 1.3.

1.7 INTERACTIONS AND FIELDS IN PARTICLE PHYSICS

Classically, interaction at a distance is commonly described in terms of a potential or field due to one particle acting on another. In quantum theory, it is viewed in terms of the exchange of specific quanta (bosons) associated with the particular type of interaction.

That these two descriptions are equivalent on a macroscopic scale may be illustrated by considering the electrostatic field between two point charges, Q_1 and Q_2. In the classical case

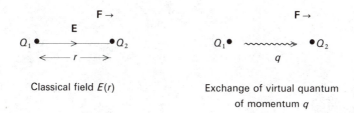

Classical field $E(r)$ Exchange of virtual quantum
of momentum q

the force $\mathbf{F}$ on Q_2 in the diagram is ascribed to the field $E(r)$ due to Q_1; $\mathbf{F} = \mathbf{E}(r)Q_2 = \hat{r}Q_1Q_2/r^2$. Quantum-mechanically, the force between the charges is ascribed to exchange of virtual photons of momentum q, the change of momentum of the charge as it emits or absorbs a photon producing the force. As explained previously, the photons are said to be virtual, as they exist only for a time limited by the uncertainty principle. This links the linear dimension of the system, that is the uncertainty in position of the photon, with its momentum:

$$qr \simeq \hbar.$$

Each photon exchanged involves a momentum transfer q over a period $t = r/c$, or a force $dq/dt = \hbar c/r^2$. The number of photons emitted and absorbed by either charge is assumed to be proportional to the product of the charges, so that one obtains the Coulomb law $F = Q_1Q_2/r^2$ as in the classical case. The quantum concept of continual emission and absorption of virtual photons by the charge source is no more or less fictitious than the classical concept of a field surrounding the source. Neither field nor virtual quanta are directly observable; it is the force that is the measured quantity. Since, however, it is observed that propagating electromagnetic fields are actually quantized in the form of free photons, the quantum description of virtual photon exchange in the static case is appropriate for discussion of interactions on a microscopic scale.

The representation of interactions of particles with quantized fields is frequently visualized with the aid of *Feynman diagrams*, which are associated with formal rules for assigning vertex coupling, propagator terms, etc. in order to compute matrix elements. However, in this text such diagrams will be used mainly as a pictorial representation of interactions via quantum exchange.

1.8 ELECTROMAGNETIC INTERACTIONS

The coupling constant specifying the strength of the interaction between charged particles and photons is the dimensionless fine-structure constant

$$\alpha = \frac{e^2}{4\pi\hbar c} = \frac{1}{137.0360}, \tag{1.16}$$

so called because it determines the magnitude of the fine structure (spin-orbit splitting) in atomic spectra. The quantity α enters in the matrix element for the

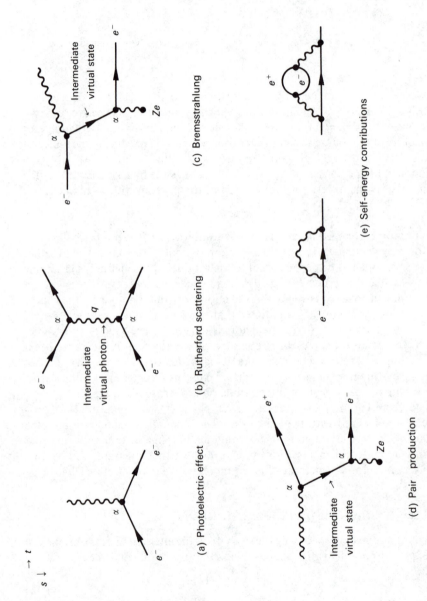

(a) Photoelectric effect

(b) Rutherford scattering

(c) Bremsstrahlung

(d) Pair production

(e) Self-energy contributions

Fig. 1.8 Feynman diagrams for electromagnetic processes.

process under consideration, which after squaring gives the decay probability or cross section.*

Figure 1.8 shows diagrams depicting various electromagnetic processes. Time t flows horizontally, and space s vertically. The arrows indicate the direction of motion of particles entering or leaving vertices. Incoming electrons (momentum $\mathbf{p}$) can always be replaced by outgoing antiparticles (positrons) of momentum $-\mathbf{p}$, without changing the matrix element.

Figure 1.8(a) shows the simplest process of absorption (or emission) of a photon by an electron. This can only take place for an electron bound in an atom, to ensure momentum conservation. The photon couples to the electron with amplitude $\sqrt{\alpha}$ (or e, in units $\hbar = c = 1$), so that the photoelectric cross-section, or matrix element squared, is proportional to α (or e^2). Since α occurs to the first power, this is called a first-order process.

Figure 1.8(b) indicates the second-order process of Coulomb scattering between two electrons, via exchange of a single virtual photon of momentum q, coupling at two vertices. The virtual photon introduces a so-called propagator term $1/q^2$ in the matrix element (see Eq. (1.5)), which is therefore proportional to $\sqrt{\alpha}\sqrt{\alpha}/q^2$. The scattering cross-section is $d\sigma/dq^2 \propto \alpha^2/q^4$ (the Rutherford scattering formula).

Figure 1.8(c) shows the emission of a real photon by an electron which has undergone acceleration in the electric field of a nucleus, charge Ze (the process of bremsstrahlung). A virtual photon must be exchanged with the nucleus, to conserve momentum, and the cross-section is of order α^3 and proportional to $Z^2\alpha^3$. Note that an intermediate virtual electron state is involved, since an electron cannot emit a real photon and conserve energy and momentum without reducing its rest mass (it is said to go "off mass shell").

Finally, Fig. 1.8(d) shows the process of e^+e^- pair creation by a photon in the field of a nucleus, also of magnitude α^3. Diagrams (c) and (d) are closely related, and the latter can be obtained from the former by replacing an incoming electron line with an outgoing positron line.

The field theory employed to compute the cross-sections for such electromagnetic processes is called quantum electrodynamics (QED) and is discussed in Chapter 8. One very important property of QED is that of *renormalizability*. As shown in Fig. 1.8(e), a single electron can emit and reabsorb virtual photons (or pairs), and such "self-energy" terms contribute to the mass (and charge) of the electron: indeed, they give divergent integrals, and the theoretically calculated "bare" mass m_0 or charge e_0 become infinite. Divergent terms of this type are present in all QED calculations, for example of the processes in Fig. 1.8(a)–(d). However, it is found possible to dump all the divergences into m_0 or e_0, and then redefine the mass and charge, replacing them by their physical values e, m (which are determined by experiment). This process

---

* See Section 1.15 for discussion of units employed.

is called renormalization. The result is that QED calculations, if expressed in terms of the physical quantities e and m, always give finite (and incredibly exact) values for cross-sections, decay rates, and so forth.

A second vital property of electromagnetic interactions is that of gauge invariance. In electrostatics, for example, the interaction energy which can be measured experimentally depends only on changes in the static potential and not its absolute magnitude, and is therefore invariant under arbitrary changes in the potential scale or gauge. In quantum mechanics, the phase of a fermion field (e.g. the wavefunction of an electron) is likewise arbitrary, and one could require the freedom to choose the phases of all fermion fields at all points in space-time in any way one pleases, without changing the physics. This local gauge invariance leads to conserved currents and to conservation of electric charge.

Mention has been made of renormalizability and gauge invariance because they are closely related: many particle theories in the past have possessed neither property and have run into insurmountable difficulties. The astounding success of QED, allowing exact calculations of electromagnetic processes to all orders in α, has been such that it is nowadays generally believed that all theories of fundamental fields should be renormalizable gauge theories.

1.9 GRAVITATIONAL INTERACTIONS

Gravity is not an important effect in particle physics, but we mention it briefly for completeness. It is described in terms of the Newtonian constant K, with the force between two equal point masses M given by KM^2/r^2, where r is their separation. By comparing with the electrostatic force between singly charged particles, e^2/r^2, the quantity KM^2/hc is seen to be dimensionless. For example, if M is taken as the proton mass, then

$$KM^2/4\pi hc = 4.6 \times 10^{-40}, \tag{1.17}$$

compared with

$$e^2/4\pi hc = 1/137.$$

Thus for the mass scales common in high-energy physics, the gravitational coupling is negligible in comparison with the electromagnetic or other fundamental interactions. Gravitational quantum effects may be important only when the gravitational energy of a system becomes comparable with the total mass energy, that is for masses M and distances l such that $KM^2/l \simeq Mc^2$, corresponding to a length $l = KM/c^2$. Quantum gravitational effects therefore become significant for particles or systems so massive and pointlike as to have a Compton wavelength of order l. This limit is associated with the Planck mass, given by $\hbar/Mc = l = KM/c^2$, or $KM^2/\hbar c = 1$, that is $Mc^2 = 10^{19}$ GeV. For such a mass value, the dimensionless gravitational coupling (1.17) is seen to be of order unity. Such a mass or energy scale is larger by a factor of 10^5 than that of any fundamental particle which has so far been postulated, so that gravitational effects can be ignored in our subsequent discussions. This is fortunate, since a successful theory of quantum gravity does not presently exist and we would not know how to incorporate this interaction with the others.

A quantum, called the *graviton*, is postulated as the mediator of the gravitational interaction. As the gravitational field has infinite range, the graviton mass is zero. The graviton has spin $J = 2$. This arises because, whereas both negative and positive electric charges — and hence electric dipole oscillators — exist, negative mass does not. The simplest radiator of gravitons is therefore an oscillating mass quadrupole.

1.10 WEAK INTERACTIONS

Historically, the first weak interactions to be observed were those of nuclear β-decay. Denoting a nucleus with N neutrons and Z protons by $A(Z, N)$, the prototype reactions are

$$A(Z, N) \rightarrow A(Z - 1, N + 1) + e^+ + v_e \quad \text{(positron emission)},$$

$$A(Z, N) \rightarrow A(Z + 1, N - 1) + e^- + \bar{v}_e \quad \text{(electron emission)}, \quad (1.18)$$

$$A(Z, N) + e_K^- \rightarrow A(Z - 1, N + 1) + v_e \quad \text{(K electron capture)}.$$

These are all examples of the decay of bound protons or neutrons, where the change in nuclear binding provides enough energy for the creation of the leptons:

$$p \rightarrow n + e^+ + v_e,$$
$$n \rightarrow p + e^- + \bar{v}_e, \quad (1.19)$$

together with the inverse reactions

$$e^- + p \rightarrow n + v_e,$$
$$\bar{v}_e + p \rightarrow e^+ + n. \quad (1.20)$$

The last process of antineutrino absorption was first observed using the high antineutrino flux from a nuclear reactor (Reines and Cowan 1959).

Originally, the weak interaction was described by Fermi (1934) in terms of a four-fermion contact interaction with strength given by the Fermi constant

$$G = 1.02 \times 10^{-5} \hbar c (\hbar/Mc)^2, \quad (1.21)$$

where M is the proton mass. G has dimensions of energy × volume. Instead of Fermi's contact interaction, it is more useful to consider weak interactions as mediated by a weak-interaction quantum acting between fermion *currents*, with the fermions endowed with a "weak charge" g in analogy with electric charge (see Fig. 1.9(a) and (b)). The reactions (1.19) and (1.20) are charge-changing, so the quantum $W^\pm$ must also be charged. Such reactions are called, somewhat misleadingly, "charged current" reactions, because the exchanged particle is charged; the electromagnetic scattering will then be a "neutral current"

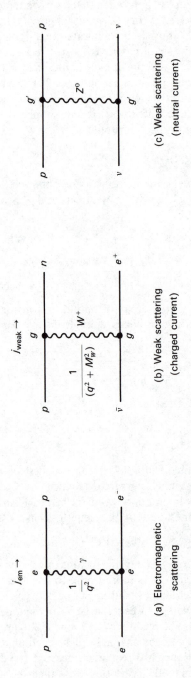

Fig. 1.9   Electromagnetic and weak interactions mediated by virtual boson exchange.

reaction. The boson propagator in Fig. 1.9(b) contributes a term $1/(q^2 + M_W^2)$ to the reaction amplitude as in Eq. (1.5). Thus the matrix element in Fig. 1.9(a) is e^2/q^2, and in Fig. 1.9(b), it is $g^2/(q^2 + M_W^2)$. For small momentum transfers, $q^2 \ll M_W^2$, the latter becomes g^2/M_W^2, so that the Fermi constant can be expressed as

$$G = (g^2/4\pi)(\hbar/M_W c)^2, \tag{1.22}$$

and the dimensionless number characterizing the coupling of the $W^\pm$ to the fermions is, from (1.21),

$$g^2/4\pi\hbar c = 1.02 \times 10^{-5}(M_W/M)^2. \tag{1.23}$$

The point of introducing g instead of the Fermi constant G became more apparent in 1973, when the phenomenon of neutral weak currents was discovered in high-energy neutrino experiments. These correspond to Fig. 1.9(c), in which the elastic scattering process $v + p \to v + p$ is mediated by a neutral boson, called Z^0. Since that time, a formidable amount of experimental evidence has accumulated, consistent with the hypothesis – advanced by Weinberg, Salam, and Glashow – that electromagnetic and weak interactions have the same intrinsic coupling of bosons to leptons, i.e. $g \simeq e$; there is a unified "electroweak interaction". The equality of coupling, together with (1.16) and (1.23), as well as some additional numerical factors discussed in Chapter 8, gives $M_W \simeq 80 \, \text{GeV}/c^2$ and $M_{Z^0} \simeq 90 \, \text{GeV}/c^2$. These very massive bosons of the weak interactions have not yet been observed, but new generations of accelerators are expected to produce them in the near future.

In summary, therefore, the weak interactions are considered to have the same intrinsic coupling as electromagnetic, and at short distances (or, equivalently, large momentum transfers $q > M_W, M_{Z^0}$) are expected to give equal scattering cross-sections. At low energy, the apparent weakness of the weak interaction is ascribed to the very short range associated with a massive propagator in the Yukawa picture. Thus $R_{\text{weak}} \sim \hbar/M_W c \sim 10^{-18}$ m, while the range of the Coulomb interaction, with $m_y = 0$, is infinite.

In the above discussion, weak interactions have been described as between leptons and nucleons. More fundamentally, they occur between lepton and quark constituents; for example

$$v_\mu + d \to \mu^- + u,$$

$$\bar{v}_e + u \to e^+ + d$$

describe charged-current reactions, while

$$v_\mu + u \to v_\mu + u,$$

$$\bar{v}_\mu + e \to \bar{v}_\mu + e$$

describe neutral-current reactions. Neutral current interactions between the quarks themselves are masked completely by their much more powerful strong interaction, and only charge- and flavor-changing processes, forbidden for the strong and electromagnetic interactions, can be detected: for example, the $\Delta S = 1$ weak decay of the Λ-hyperon is written in terms of quarks as

$$
\begin{array}{ccc}
\Lambda & \rightarrow & p & + \pi^-, \\
s & \rightarrow & u & + \bar{u}d. \\
(+u+d) & & (+u+d) &
\end{array}
$$

Since weak interactions are now thought to be unified with electromagnetic, they enjoy the same properties of renormalizability and gauge invariance – while the old Fermi theory of contact interaction is found to diverge at high energy or to high orders in G. The gauge transformations of the unified theory are more general than those of electromagnetism alone, and are discussed in Chapter 8.

1.11   STRONG INTERACTIONS (BETWEEN QUARKS)

The successful theories of electromagnetic and weak interactions hinge on the idea that the reactions between quark and lepton (or lepton and lepton) constituents are dominated by single quantum exchange. This depends on the coupling being small, i.e. $\alpha = e^2/\hbar c \ll 1$, so that higher-order contributions, for example double quantum exchange, may be only a small correction, of relative magnitude α, to single exchange. So the higher-order terms can be treated by perturbation theory. In classical terms, the Coulomb scattering of an electron by a proton is treated as a single act of scattering even though the proton charge is spread over an appreciable volume.

The term strong interactions is applied to quark-quark forces, responsible for the binding of the quarks in hadrons, and on the face of it they are so strong that the idea of single quantum exchange seems quite inappropriate. However, in 1973 it was shown by Gross and Wilczek, and by Politzer that a class of renormalizable, gauge-invariant theories could exist which were "asymptotically free", because the effective coupling at sufficiently small distances or large momentum transfers could become small; only at large distances comparable to the hadron size (1 fm) would the coupling become very strong – and presumably result in quark confinement. This theory seems to be supported by experiments, to be discussed in Chapters 7 and 8. Deep inelastic scattering of high energy leptons by nucleon targets, carried out over the last decade, showed that the complicated process of multiple hadron production – of which Fig. 7.1 shows an example – could be very simply interpreted in terms of the elastic scattering of the lepton by a quasi-free, pointlike constituent or *parton* inside the nucleon. It was found that the partons could be identified with the quarks and thus, at large momentum transfers or short distances, the quarks were behaving almost as free particles. In other words, the interquark interaction in these conditions is very weak, and can in fact be treated by perturbation theory.

The boson mediating these quark interactions, called the *gluon*, is a neutral, flavorless, massless vector particle (spin 1). A strong "color" charge,

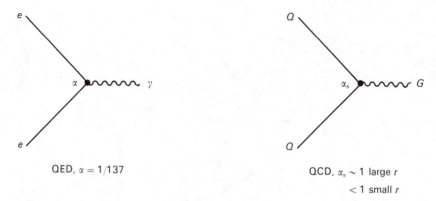

Fig. 1.10 Basic fermion-boson couplings in (a) quantum electrodynamics, (b) quantum chromodynamics.

analogous to electric charge, is carried by both quarks and gluons. Instead of two types of charge, called positive and negative in the electrical case, there are three colors (say red, blue, and green) and three "anticolors". Quarks carry a color, and antiquarks an anticolor, while gluons carry one color and one anticolor. Depending on the colors involved, quark-quark interactions, via gluon exchange, may be attractive or repulsive. Baryons are positively bound states of three quarks with zero net color (that is, one red, one blue, and one green quark). Mesons are bound states of quark and antiquark with zero net color (for example, one red quark and one anti-red antiquark). The quantum field theory of quark interactions is called quantum chromodynamics (QCD) in analogy with quantum electrodynamics (QED) – see Fig. 1.10. The main difference lies in the fact that in QCD there is a strong gluon-gluon interaction. A field of this type, in which the quanta themselves act as a source of field, is said to be non-Abelian. Gravity is also non-Abelian, since gravitons carry energy and momentum and are therefore a source of gravitational field. QED, on the other hand, involves uncharged photons and is Abelian. The non-Abelian nature of QCD is the crucial property which makes the coupling α_s decrease at small distance, since the gluon-gluon interaction "spreads out" the strong color charge.

At large distances, the quark-quark couplings become large and uncalculable, and presumably give rise to quark confinement. An analogy may be useful. Figure 1.11 depicts a proton consisting of three quarks held together by gluon exchanges, which we can imagine as replaced by elastic bands. If these are not stretched, the coupling is fairly weak. However, if one attempts to knock out a quark, the band stretches, the tension increasing with the extension. Eventually, the energy stored in the band becomes so large as to allow creation of a quark-antiquark pair, replacing one stretched band by two unstretched bands. Thus attempts to remove a quark from a proton result in the creation of a meson.

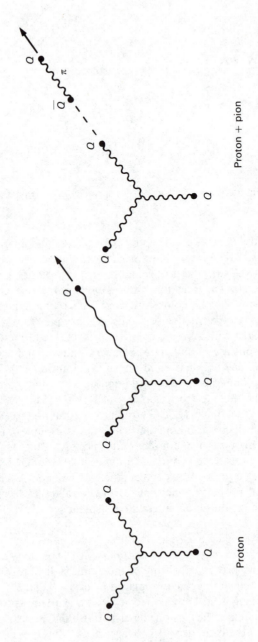

Fig. 1.11 Quarks in a proton, visualized as held together by elastic "gluon strings". When an attempt is made to knock out a quark, the string is stretched, and eventually the energy stored in it is sufficient to create a quark-antiquark pair (meson).

Models have been made of quarks interacting via the color lines of force (gluons) which are pulled together by the gluon-gluon interaction and thus do approximate to such a tube or band. Simple predictions follow, for example that the (mass)^2 of hadron states should be proportional to the spin J, as observed in practice.

1.12 HADRON-HADRON INTERACTIONS

As mentioned before, interactions between hadrons have to be considered as residual quark-quark interactions, just as interactions between nuclei can be considered as residual nucleon-nucleon interactions. Since there is no way of calculating the quark-quark interaction in the relatively "soft" hadron-hadron collisions, and we have in any case no formalism to handle the many-body problem, totally different approaches are needed. These in fact anteceded the quark model by many years.

One approach is basically an adoption of the methods of classical wave optics, and of the important concept of resonance in wave amplitudes. These ideas were taken over from similar methods applied in nuclear-structure physics. For example, diffraction-type maxima and minima are observed in the high-energy scattering angular distributions, reminiscent of the scattering of light by a totally or partially absorbing obstacle. Another approach is based on Yukawa's original concept of heavy quantum exchange mediating interactions between neutrons and protons, and is applicable particularly at high collision energies and for small momentum transfers between the hadrons involved. For example, in nucleon-nucleon collisions, it is found that part of the cross-section (that associated with small momentum transfers, or with partial waves of high angular momentum l and thus the longest-range component of the interaction) has the characteristics expected from single pion exchange (Fig. 1.12). The coupling constant f determined from such an analysis is given by $f^2/4\pi hc = 0.08$. While this description accounts well for high-l contributions, the exchange of many heavier mesons (ρ, ω, etc.) presumably dominates the partial waves of small l,

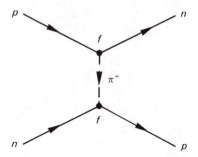

Fig. 1.12 Single-pion-exchange contribution to proton-neutron scattering.

TABLE 1.4 Fundamental interactions (M = nucleon mass)

Interaction	Field quantum	Spin-parity	Mass (mc^2) (GeV)	Range (m)	Source	Coupling	Dimensionless coupling constant	Typical cross-section (1 GeV)		Typical lifetime for decay (sec)
								m^2	μb	
Gravity	Graviton	2^+	0	∞	Mass	K (Newton)	$KM^2/\hbar c = 0.53 \times 10^{-38}$	–	–	–
Electro-magnetic	Photon	1^-	0	∞	Electric charge	–	$\alpha = e^2/4\pi\hbar c = \frac{1}{137}$	10^{-33}	10	10^{-20}
Weak	Intermediate bosons $W^{\pm}, Z^0$	$1^-, 1^+$	80–90	10^{-18}	"Weak charge"	G (Fermi)	$(Mc/\hbar)^2 G/\hbar c = 1.02 \times 10^{-5}$	10^{-44}	10^{-10}	10^{-8}
Strong	Gluon	1^-	0	$\leqslant 10^{-15}$	"Color charge"	–	$\alpha_s \sim 1$, large r; < 1, small r	10^{-30}	10^4	10^{-23}

and no simple description in terms of the properties of known particles seems possible. Considerable progress has been made, however, using such properties as analyticity (continuity) of the scattering amplitude, treating l as a continuous, complex variable — the so-called Regge-pole approach to single "particle" exchange (Chapter 4).

1.13   CONSERVATION RULES IN FUNDAMENTAL INTERACTIONS

A summary of the characteristics of the fundamental interactions described above is given in Table 1.4. Included is a list of typical collision cross-sections, and of typical lifetimes for decay via the various interactions.

TABLE 1.5   Conservation Rules

Conserved quantity	Interaction		
	Strong	Electromagnetic	Weak
Energy/momentum Charge Baryon number Lepton number	Yes	Yes	Yes
I (isospin)	Yes	No	No ($\Delta I = 1$ or $\frac{1}{2}$)
S (strangeness)	Yes	Yes	No ($\Delta S = 1, 0$)
C (charm)	Yes	Yes	No ($\Delta C = 1, 0$)
P (parity)	Yes	Yes	No
C (charge-conjugation parity)	Yes	Yes	No
CP (or T)	Yes	Yes	Yes[a]
CPT	Yes	Yes	Yes

[a] But 10^{-3} violation in K^0 decay.

Table 1.5 shows a list of some of the quantities conserved in each type of interaction. Some of the conservation rules are absolute, or very nearly so. Other properties, such as invariance under spatial inversion (parity conservation), are observed exactly in some interactions but not in others. Often the violation occurs in a regular way. For example, weak interactions may or may not conserve strangeness S. If it is violated, the rule is $\Delta S = 1$.

1.14   CROSS-SECTIONS AND DECAY RATES

Much of our information about properties of particles and their interactions comes from experimental data on collision cross-sections and rates for spontaneous decay. As an example we consider a reaction of the form

$$\underbrace{a + b}_{i} \rightarrow \underbrace{c + d}_{f} \qquad (1.24)$$

with two particles in the initial state i, and two in the final state f. If we regard b as the target and a as the projectile particle — usually in a well-collimated beam — then the *cross-section* for the above reaction is defined as the transition rate W per unit incident flux per target particle. If n_a is the density of particles in the incident beam, and v_i the relative velocity of a and b, then the flux of particles per unit time through unit area normal to the beam is

$$\phi = n_a v_i. \qquad (1.25)$$

If there are n_b particles in the target per unit area, each of effective cross-section σ, the probability that any incident particle will hit a target is σn_b, and the number of interactions per unit area per second will be $n_a n_b \sigma v_i$. Per target particle, the transition rate is therefore

$$W = \sigma \phi = \sigma n_a v_i. \qquad (1.26)$$

The value of W is given by the product of the square of a matrix element M_{if} and a density of final states, or phase-space factor, ρ_f. M_{if} contains various dynamical features of the interaction: coupling strength, energy dependence, angular distribution and so forth. The formula for W is*

$$W = \frac{2\pi}{h} |M_{if}|^2 \rho_f. \qquad (1.27)$$

This formula holds equally for a decay process, in which the initial state i consists of an unstable particle decaying to the final state f (i.e. $X \rightarrow c + d$). In first-order perturbation theory, where the interaction is assumed to be "weak", M_{if} can be interpreted as the overlap integral over volume, $\int \psi_f^* H' \psi_i \, d\tau$, where H' is the interaction potential, and ψ_i and ψ_f the initial- and final-state wavefunctions. This interpretation holds provided $H' \ll H_0$, the unperturbed energy operator (Hamiltonian). When the interaction is strong, however, M_{if} cannot be calculated explicitly, and (1.27) can then be regarded as a definition of M_{if}.

To calculate the phase space factor ρ_f we restrict the particles to some arbitrary volume, which cancels in the calculation and which we shall take as unity. Then

$$\rho_f = \frac{dn}{dE_0} = \frac{d\Omega}{h^3} p_f^2 \frac{dp_f}{dE_0} g_f, \qquad (1.28)$$

where E_0 is the total energy in the center-of-mass frame, p_f is the final-state

---

* See for example L. I. Schiff, *Quantum Mechanics*, McGraw-Hill, New York, 1955, p. 197.

momentum in this frame ($|\mathbf{p}_c| = |\mathbf{p}_d| = p_f$), and $d\Omega$ is the solid angle containing the final-state particles; g_f is a spin multiplicity factor. Equation (1.24) refers to one particle of each type in the normalization volume. With $n_a = 1$, and integrating over all angles of the products c and d, one obtains, using (1.26),

$$\sigma = \frac{W}{v_i} = |M_{if}|^2 \frac{p_f^2 \, dp_f}{v_i \, dE_0} g_f, \tag{1.29}$$

where numerical constants are absorbed in $|M_{if}|^2$. Conservation of energy gives

$$\sqrt{p_f^2 + m_c^2} + \sqrt{p_f^2 + m_d^2} = E_0$$

and thus

$$\frac{dp_f}{dE_0} = \frac{E_c E_d}{E_0 p_f} = \frac{1}{v_f}, \tag{1.30}$$

where E_c, E_d are total energies, and $v_f = v_c + v_d$ is the relative velocity of c and d. If s_c and s_d are the spins of particles c and d, the numbers of possible substates for them are $2s_c + 1$ and $2s_d + 1$, and thus $g_f = (2s_c + 1)(2s_d + 1)$. Equation (1.29) then becomes

$$\sigma(a + b \to c + d) = |M_{if}|^2 \frac{(2s_c + 1)(2s_d + 1)}{v_i v_f} p_f^2. \tag{1.31}$$

Implied in this expression, integrated over angles, is that $|M_{if}|^2$ has been averaged over all possible spin states of a and b, and summed over all orbital-angular-momentum states involved.

As indicated above, the precise form of M_{if} is usually not known. However, if we compare the forward and backward reactions $a + b \rightleftharpoons c + d$ at the same center-of-mass energy, this information is not necessary. We rely on the principle of detailed balance, which states that

$$|M_{if}|^2 = |M_{fi}|^2. \tag{1.32}$$

This relation follows by assuming invariance of the interaction under time reversal and space inversion, both of which hold good for the strong (hadronic) interactions (see Chapter 3). Time reversal interchanges final and initial states, but reverses all momenta and spins; space inversion then changes back the signs of momenta but leaves spins unchanged. Thus, writing $M_{if} = \langle f|T|i \rangle$, where T is a suitable transition matrix operator, we have

$$\langle f(p_c, p_d, s_c, s_d)|T|i(p_a, p_b, s_a, s_b) \rangle$$

$$\xrightarrow[\text{+ space inversion}]{\text{time reversal}} \langle i(p_a, p_b, -s_a, -s_b)|T|f(p_c, p_d, -s_c, -s_d) \rangle. \tag{1.33}$$

Summing over all $2s + 1$ spin projections, running from $-s$ to $+s$, we obtain $|M_{if}|^2 = |M_{fi}|^2$ as in (1.32).

Invariance under time reversal and space inversion does not hold good in the weak interactions. In these, however, first-order perturbation theory can be applied ($H' \ll H_0$), and H' is a Hermitian operator (that is, $\langle f|H'|i \rangle = \langle i|H'|f \rangle^*$), so that the detailed balance relation (1.32) is again valid. The relations (1.31) and (1.32) will be used in Chapter 3 in connection with determining the spin-parity of the pion.

Typical cross-sections and decay rates for the various interactions are given in Table 1.4. For the strong hadronic interactions, the cross-sections σ are clearly determined by the hadron size ($R \sim 10^{-15}$ m) and thus are of magnitude $R^2 \sim 30$ mb (3×10^{-30} m^2). The lifetime τ of hadronic states decaying via strong interactions is typically 10^{-23} sec, corresponding to transition rates $\lambda = \tau^{-1} \sim c/R$. Thus in order of magnitude,

$$\frac{\sigma}{\lambda} \sim \frac{V}{c}, \tag{1.34}$$

where V is the volume of a hadron. Of course the actual values of σ and λ depend also on kinematic (phase-space) and other factors. Leaving these aside, the order-of-magnitude value of the ratio (1.34) is seen to hold also for cross-sections and decays mediated by the other fundamental interactions.

1.15 UNITS IN HIGH-ENERGY PHYSICS

The fundamental units in physics are of length, mass, and time, the familiar system (MKS) expressing these in metres, kilograms, and seconds. Such units are however not very appropriate in particle physics, where lengths are typically 10^{-15} m and masses 10^{-27} kg.

Lengths in particle physics are usually quoted in terms of the *femtometer* or *fermi* (1 fm = 10^{-15} m), and cross-sections in terms of the *barn* (1 b = 10^{-28} m^2), millibarn (1 mb = 10^{-31} m^2), or microbarn (1 μb = 10^{-34} m^2). The unit of energy is based on the electron volt (1 eV = 1.6×10^{-19} joules) with the larger units MeV (= 10^6 eV), GeV (= 10^9 eV), and TeV (= 10^{12} eV). Masses are usually measured in MeV/c^2, meaning that if the mass is M, the rest energy is Mc^2 MeV. For example, the proton has a rest energy of 938.28 MeV or 0.938 GeV. Often masses (meaning the rest-energy equivalents) are loosely quoted in MeV or GeV.

In calculations, the quantities $\hbar = h/2\pi$ and c occur frequently, and it is often advantageous to use a system of units in which $\hbar = c = 1$. We do this by choosing some standard mass m_0 (e.g. the proton mass) as the unit:

$$m_0 = 1.$$

The natural unit of length is then the Compton wavelength of the standard particle:

$$\lambdabar_0 = \frac{\hbar}{m_0 c} = 1;$$

that of time is

$$t_0 = \frac{\lambdabar_0}{c} = \frac{h}{m_0 c^2} = 1,$$

and that of energy is

$$E_0 = m_0 c^2 = 1.$$

In these units, it is seen that $h = c = 1$. In converting back, at the end of the calculation, to the more usual units, it is useful to remember that $hc = 197$ MeV fm. Thus, a particle of mass energy $m_0 c^2 = 197$ MeV has a Compton wavelength of $h/m_0 c = hc/m_0 c^2 = 1$ fm.

Throughout this text we shall be dealing with the coupling of charges – strong, electric and weak – to mediating bosons. In *SI* units, electric charge, e, is measured in Coulombs and the fine structure constant is then given by

$$\alpha = e^2/(4\pi\varepsilon_0 hc) \simeq 1/137.$$

For the general coupling of charges to bosons, such units are not useful and we define e in Heaviside-Lorentz units ($\varepsilon_0 = \mu_0 = 1$) so that

$$\alpha = e^2/4\pi hc \simeq 1/137$$

as in (1.16). A similar definition is used to relate charges and coupling constants in the other interactions.

PROBLEMS

1.1 (a) Show that a negative muon captured in an S-state by a nucleus of charge Ze and mass A will spend a fraction $f \simeq 0.25A(Z/137)^3$ of its time in nuclear matter, and that in time t it will travel a total distance $fct(Z/137)$ in nuclear matter.

(b) The law of radioactive decay of free muons is $dN/dt = -\lambda_d N$, where $\lambda_d = 1/\tau$ is the decay constant and the lifetime $\tau = 2.16\ \mu s$. For a negative muon captured in an atom Z, the decay constant is $\lambda = \lambda_d + \lambda_c$, where λ_c is the probability of nuclear capture per unit time. For aluminum ($Z = 13$, $A = 37$) the mean lifetime of negative muons is $0.88\ \mu s$. Calculate λ_c, and using the expression for f in (a), compute the interaction mean free path Λ for a muon in nuclear matter.

(c) From the magnitude of Λ in (b) estimate the magnitude of the coupling constant in the reaction $\mu^- + p \rightarrow n + \nu$, assuming that a coupling constant of unity corresponds to a mean free path equal to the range of nuclear forces.

1.2 (a) Deduce an expression for the energy of a γ-ray from the decay of a neutral pion, $\pi^0 \rightarrow 2\gamma$, in terms of the mass m, energy E, and velocity βc of the pion, and of the angle of emission θ in the CMS.

(b) Show that if the pion has zero spin, the distribution in θ will be isotropic and that the energy distribution of the γ-rays will be flat, extending from $E(1 + \beta)/2$ to $E(1 - \beta)/2$.

(c) Find an expression for the disparity D (the ratio of energies) of the two γ-rays from π^0 decay. For the case of relativistic pions, show that $D > 3$ in half of the decays and $D > 7$ in one quarter of them.

1.3 A negative muon, when brought to rest in liquid hydrogen, can form a molecular ion H_2^+ by displacing an electron. Why? If the hydrogen contains even a tiny amount of deuterium, it is found that the negative muons eventually form molecular ions HD^+. Why? What is the typical internuclear distance in such an ion? If the two nuclei react to form ^3He, what may happen to the muon?

1.4 It has been postulated that the neutrinos v_e, v_μ, v_τ, etc. may consist of linear combinations of discrete mass eigenstates v_1, v_2, v_3,... Suppose that, in the decay $\pi \to \mu + v_\mu$, the state v_μ consists of a combination of states of mass m_1 and m_2. Thus, in principle, the energy or momentum of the muon from pion decay at rest would consist of two discrete values. Assuming that the muon momentum can be measured with unlimited precision, show that two such discrete values could be demonstrated only if $m_2^2 - m_1^2 > 1$ $(eV/c^2)^2$. Show that the irreducible error of measurement of the muon momentum places a less stringent limit on the minimum detectable value of $m_2^2 - m_1^2$.

1.5 The cross-section for the reaction $\pi^- + p \to \Lambda + K^0$ at 1-GeV/c incident momentum is approximately 1 mb $(10^{-27}$ cm$^2)$. Both Λ- and K^0-particles decay with a mean lifetime of about 10^{-10} sec. From this information, estimate the relative magnitude of the couplings responsible for the production and decay, respectively, of the Λ- and K^0-particles.

1.6 State which of the following reactions are allowed by the conservation laws and which are forbidden, and give the reasons in either case:

$$\pi^0 \to e^+ + e^-, \tag{i}$$

$$p \to n + e^+ + v_e, \tag{ii}$$

$$\mu^+ \to e^+ + e^- + e^+, \tag{iii}$$

$$K^+ + n \to \Sigma^+ + \pi^0. \tag{iv}$$

1.7 It was once suggested that the numerical values of the charge of the electron and proton might differ by a small amount $|\Delta e|$, so that expansion of the universe could be attributed to an electrostatic repulsion between hydrogen atoms in space. Estimate the minimum value of $|\Delta e/e|$ required for this hypothesis. [References: Theory, H. Bondi and R. A. Lyttleton, *Nature* **184**, 974 (1959); *Proc. Roy. Soc.* **A252**, 313 (1959). Experimental disproof, A. M. Hillas and T. E. Cranshaw, *Nature* **184**, 892 (1959).]

Chapter 2

Particle Detectors and Accelerators

2.1 THE INTERACTION OF CHARGED PARTICLES AND RADIATION WITH MATTER

2.1.1 Ionization Loss of Charged Particles

The detection of nuclear particles depends ultimately on the fact that, directly or indirectly, they transfer energy to the medium they are traversing via the process of ionization or excitation of the constituent atoms. This can be observed as charged ions, for example in a gas counter, or as a result of the scintillation light, Čerenkov radiation, etc., which is subsequently emitted.

The Bethe-Bloch formula for the mean rate of ionization loss of a charged particle is given by*

$$\frac{dE}{dx} = \frac{4\pi N_0 z^2 e^4}{mv^2} \frac{Z}{A} \left[\ln\left(\frac{2mv^2}{I(1-\beta^2)}\right) - \beta^2 \right], \tag{2.1}$$

where m is the electron mass, z and v are the charge (in units of e) and velocity of the particle, $\beta = v/c$, N_0 is Avogadro's number, Z and A are the atomic number and mass number of the atoms of the medium, and x is the path length in the medium measured in $g\,cm^{-2}$ or $kg\,m^{-2}$. The quantity I is an effective ionization potential, averaged over all electrons, with approximate magnitude $I = 10Z\,eV$. Equation (2.1) shows that dE/dx is independent of the mass M of the particle, varies as $1/v^2$ at nonrelativistic velocities, and, after passing through a minimum for $E \simeq 3Mc^2$, increases logarithmically with $\gamma = E/Mc^2 = (1 - \beta^2)^{-1/2}$. The dependence of dE/dx on the medium is very weak, since $Z/A \simeq 0.5$ in all but hydrogen and the heaviest elements. Numerically, $(dE/dx)_{min} \simeq 1–1.5$ MeV $cm^2\,g^{-1}$ (or $0.1–0.15$ MeV $m^2\,kg^{-1}$).

Figure 2.1 shows the observed relativistic rise in ionization loss as a function of $p/Mc = (\gamma^2 - 1)^{1/2}$ for relativistic particles in a gas (argon-methane mixture). For $\gamma \sim 10^3$, it reaches 1.5 times the minimum value. The relativistic rise is associated with the fact that the transverse electric field of the particle is

---

* For a semi-classical derivation of this formula, see for example B. Rossi, *High Energy Particles*, Prentice-Hall Inc, Englewood Cliffs, N. J., 1961, p. 17.

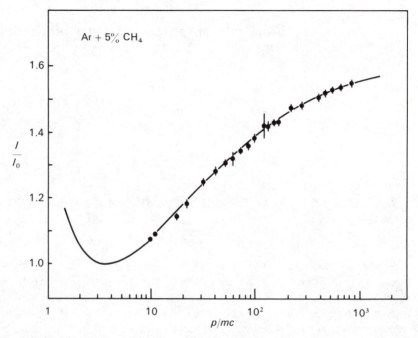

Fig. 2.1 Mean ionization energy loss of charged particles in argon-methane mixture, showing relativistic rise as a function of p/mc. Measurements by multiple ionization sampling (after Lehraus *et al*. 1978).

proportional to γ, so that more and more distant collisions became important as the energy increases. Eventually, when the impact parameter becomes comparable to interatomic distances, polarization effects in the medium (associated with the dielectric constant) halt any further increase. In solids, rather than gases, such effects become important at a much lower value of $\gamma \sim 10$, and this plateau value is only about 10% larger than $(dE/dx)_{\text{min}}$. Part of the energy loss of a relativistic particle may be reemitted from excited atoms in the form of coherent radiation at a particular angle. Such Čerenkov radiation is discussed in Section 2.3.6 below.

The bulk of the energy loss results in the formation of ion pairs (positive ions and electrons) in the medium. One can distinguish two stages to this process. In the first stage, the incident particle produces primary ionization in atomic collisions. The electrons knocked out in this process have a distribution in energy E' roughly of the form $dE'/(E')^2$: those of higher energy (called δ-rays) can themselves produce fresh ions in traversing the medium (secondary ionization). The resultant total number of ion pairs is 3–4 times the number of primary ionizations, and is proportional to the energy loss of the incident particle in the medium. Equation (2.1) gives the *average* value of the energy loss dE in a layer dx, but there will be fluctuations about the mean, dominated by the relatively small number of "close" primary collisions with large E'. This so-called Landau distribution about the mean value is therefore asymmetric, with a tail extending to values much greater than the average. Nevertheless, by sampling the number of ion pairs produced in many successive layers of gas and removing the "tail", the mean ionization loss can be measured within a few per cent. In this way γ can be estimated from the relativistic rise, and if the momentum is known, this provides a useful method for estimating the rest mass and thus differentiating between pions, kaons, and protons.

The total number of ions produced in a medium by a high-energy particle depends on dE/dx and the energy required to liberate an ion pair. In a gas, this varies from 40 eV in helium to 26 eV in argon. In semiconductors, on the other hand, it is only about 3 eV, so the number of ion pairs is much larger. If the charged particle comes to rest in the semiconductor, the energy deposited is measured by the total number of ion pairs, and such a detector therefore not only is linear but has extremely good energy resolution (typically 10^{-4}). Because of their small size, however, such solid-state counters have not found very general applications in high-energy physics.

2.1.2 Coulomb Scattering

In traversing a medium, a charged particle suffers electromagnetic interactions with both electrons and nuclei. As Eq. (2.1) indicates, dE/dx is inversely proportional to the target mass, so that in comparison with electrons, the energy lost in Coulomb collisions with nuclei is negligible. However, because of the larger target mass, transverse *scattering* of the particle is appreciable in the Coulomb field of the nucleus, and is described by the famous Rutherford formula for the differential cross-section at scattering angle θ:

$$\frac{d\sigma(\theta)}{d\Omega} = \frac{1}{4}\left(\frac{Zze^2}{pv}\right)^2 \frac{1}{\sin^4(\theta/2)}, \tag{2.2}$$

where p, v, z are the momentum, velocity, and charge of the incident particle, and Z is the charge on the nucleus, assumed to act like a point charge. For small scattering angles the cross-section is large, so that in any given layer of material the net scattering is the result of a large number of small deviations, which are independent of one another. The resultant distribution in the net angle of *multiple* scattering follows a roughly Gaussian distribution

$$P(\phi)\,d\phi = \frac{2\phi}{\langle\phi^2\rangle}\exp\left(\frac{-\phi^2}{\langle\phi^2\rangle}\right)d\phi. \tag{2.3}$$

The root-mean-square deflection in a layer t of the medium is given by

$$\phi_{\mathrm{rms}} = \langle\phi^2\rangle^{1/2} = \frac{zE_s}{pv}\sqrt{\frac{t}{X_0}}, \tag{2.4}$$

where

$$E_s = \sqrt{4\pi \times 137}\,mc^2 = 21\,\mathrm{MeV} \tag{2.5}$$

and

$$\frac{1}{X_0} = \frac{4Z(Z+1)r_e^2 N_0}{137A} \ln\left(\frac{183}{Z^{1/3}}\right). \tag{2.6}$$

In this formula, $r_e = e^2/mc^2$ is the classical electron radius. The quantity X_0, for reasons that appear below, is called the radiation length of the medium (see Table 2.1), and incorporates all the dependence of $\phi_{\rm rms}$ on the medium. Numerically therefore, a singly charged particle ($z = 1$) of momentum p and velocity v, with the product pv measured in MeV, suffers an rms deflection of $21/pv$ radians in traversing one radiation length. The rms angular deflection will be azimuthally symmetric about the trajectory: along one axis in the plane normal to the trajectory, it will be $1/\sqrt{2}$ times the value (2.4).

TABLE 2.1   Radiation lengths in various elements (after Bethe and Ashkin 1953)

Element	Z	E_c, MeV	X_0, g/cm²
Hydrogen	1	340	58
Helium	2	220	85
Carbon	6	103	42.5
Aluminum	11	47	23.9
Iron	26	24	13.8
Lead	82	6.9	5.8

Coulomb scattering is important in practice because it frequently limits the precision with which the direction of a particle can be determined. As an example, we consider the determination of the momentum of a high-energy charged particle by its deflection in the field **B** of a solid iron magnet. If there is no scattering, and the momentum does not change appreciably in traversing the magnet, the radius of curvature ρ is given by $pc = Be\rho$, so that in traversing a distance s, the deflection will be

$$\phi_{\rm mag} = \frac{s}{\rho} = \frac{Bes}{pc} = \frac{300Bs}{pc},$$

where B is in tesla (1 tesla $= 10\,{\rm kG}$), s is in metres, and pc is in MeV. The rms Coulomb scattering in the plane of the trajectory will be

$$\phi_{\rm scat} = \frac{21}{\sqrt{2}} \frac{1}{p\beta c} \sqrt{\frac{s}{X_0}},$$

so that

$$\frac{\phi_{\rm scat}}{\phi_{\rm mag}} = \frac{0.05}{B\beta\sqrt{X_0 s}},$$

independent of the particle momentum for a relativistic particle ($\beta = 1$). For example, in iron $X_0 = 0.02$ m, $B \simeq 1.5$ tesla, and $\phi_{scat}/\phi_{mag} = 0.25$ for $s = 1$ m, falling to 0.10 for $s = 6$ m.

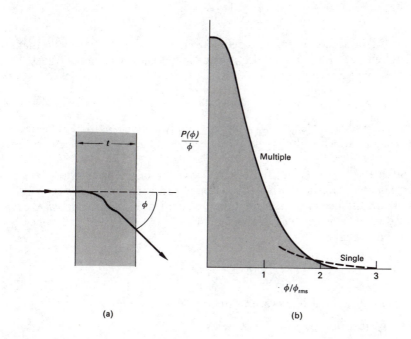

(a)   (b)

Fig. 2.2 (a) Multiple scattering of a charged particle in traversing a layer of material of thickness t. (b) If the number of particles scattered through angle $\phi \rightarrow \phi + d\phi$ is $P(\phi)\, d\phi$, the distribution of $P(\phi)/\phi$ is approximately Gaussian. For very large deflections, however, the main contribution comes from single scattering (shown dashed).

The multiple-scattering distribution in fact is only approximately Gaussian. As indicated in Fig. 2.2, it has a long tail due to occasional single large deflections, for which the distribution varies as ϕ^{-3} (as can be deduced from (2.2)).

2.1.3 Radiation Loss of Electrons

Electrons lose energy in traversing a medium in two ways: the ionization energy loss (2.1) and the process of radiation loss or *bremsstrahlung*. The radiative collisions of electrons occur principally with the atomic nuclei of the medium. The nuclear electric field decelerates the electron, and the energy change appears in the form of a photon; hence the term bremsstrahlung = braking radiation. The photon spectrum has the approximate form dE'/E', where E' is the photon energy. Integrated over the spectrum, the total radiation loss of an electron in traversing a thickness dx of medium is

$$\left(\frac{dE}{dx}\right)_{\text{rad}} = -\frac{E}{X_0}, \tag{2.7}$$

where X_0, the radiation length, is defined in (2.6). From (2.7) it follows that the average energy of a beam of electrons of initial energy E_0, after traversing a thickness x of medium, will be

$$\langle E \rangle = E_0 \exp(-x/X_0). \tag{2.8}$$

Thus, the radiation length X_0 may be simply defined as that thickness of the medium which reduces the mean energy of a beam of electrons by a factor e.

Since the rate of ionization energy loss for fast electrons, $(dE/dx)_{\text{ion}}$ is approximately constant, while the average radiation loss $(dE/dx)_{\text{rad}} \propto E$, it follows that at high energies, radiation loss dominates. The *critical energy* E_c is defined as that at which the two are equal. From (2.1), (2.6), and (2.7) it is easy to show that, roughly,

$$E_c \simeq \frac{600}{Z} \text{ MeV}. \tag{2.9}$$

Values of X_0 and E_c in various materials are given in Table 2.1.

2.1.4   Absorption of γ-rays in Matter

There are three types of process responsible for attenuation of γ-rays in matter: photoelectric absorption, Compton scattering, and pair production. The photoelectric cross-section varies with photon energy E as $1/E^3$, and the Compton cross-section as $1/E$, so that for $E > 10$ MeV, the process of pair production, with a cross-section essentially independent of energy, is dominant (see Fig. 2.3).

The process of conversion of a high-energy photon to an electron-positron pair (in the field of a nucleus to conserve momentum) is closely related to that of electron bremsstrahlung. The attenuation of a beam of high-energy photons of intensity I_0 by pair production in a thickness x of absorber is described by

$$I = I_0 \exp\left(-\frac{7x}{9X_0}\right), \tag{2.10}$$

so that the intensity is reduced by a factor e in a distance $9X_0/7$, sometimes called the *conversion* length. It is to be emphasized that, although the threshold for pair production is $E_{\text{th}} = 2m_e c^2 \simeq 1$ MeV, the asymptotic value (2.10) for the absorption coefficient is not attained until one reaches photon energies of almost 1 GeV.

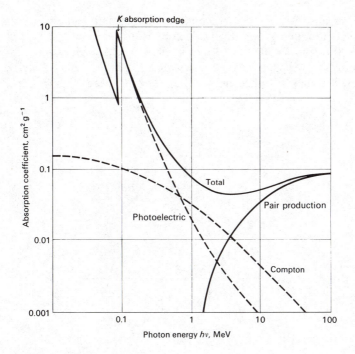

Fig. 2.3 The absorption coefficient per $g\,cm^{-2}$ of lead for γ-rays as a function of energy.

2.2 ACCELERATORS

2.2.1 Linear and Cyclic Accelerators: Phase and Transverse Oscillations

We make here only brief comments to illustrate the basic principles, possibilities, and limitations of electron and proton accelerators, and how they affect the execution of high-energy physics experiments. For details of this subject, the reader should consult the bibliography at the end of this chapter.

Two main types of device are in use, both forming components in virtually all accelerator facilities in use today. In the *linear* accelerator, a particle beam, for example of protons from an ion source, passes down a straight vacuum tube, through a chain of radio-frequency cavities. These provide oscillating electric fields, the relative phases of which are arranged so that a particle is accelerated continuously – it "rides the crest of the wave". Linear accelerators are also used as injectors to the higher-energy, cyclic stages of acceleration. In the *cyclic* accelerator, or *synchrotron*, the beam is constrained in a circular or near-circular path of fixed radius, by means of a ring of electromagnets, and acceleration is accomplished as it repeatedly (typically 10^5 times) traverses one or more RF cavities placed in the ring. If the magnetic field and RF frequency are suitably increased during the acceleration cycle, the beam will stay in an orbit of fixed radius. Thus it can be contained in a small vacuum tube, and the magnetic field required can be provided by relatively small magnets enclosing the tube. Since the distance traveled by an individual particle during acceleration is up to 10^6 km, the orbit must be very stable. The important considerations are the phase stability of the beam and transverse focusing.

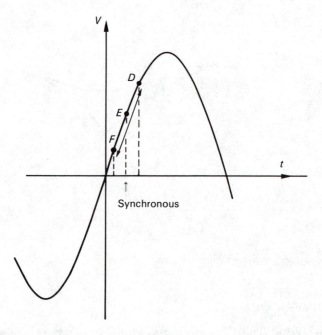

Fig. 2.4 Particles arriving early (D) or late (F) receive more or less energy in RF acceleration than a synchronous particle (E). This decreases (increases) the rotation frequency, and the effect is that the particle performs oscillations (shown by arrow) with respect to the synchronous particle.

Particles in a synchrotron are accelerated in discrete bunches of finite length, each bunch being synchronized with the RF field. Fig. 2.4 depicts three particles D, E, F as they encounter the field. Suppose particle E is in synchronous orbit, i.e., the electric impulse it receives increases its momentum p just enough to keep the circulating frequency

$$\omega_0 = eBc/p \qquad (2.11)$$

in step with the RF as the magnetic field B is increased during one rotation. Particle D arrives "early", and attains a slightly higher-momentum, larger orbit and thus a lower frequency ω than E; so it starts to fall back in phase, towards that of the synchronous particle E. A "late" particle F receives a smaller momentum from the RF field, thus has a slightly smaller orbit and higher frequency than E, and hence will "catch up". The effect is that particles remain in the bunch but execute *synchrotron oscillations* about the equilibrium position.

Particles also undergo transverse or *betatron oscillations* both in the radial direction r and in a direction z perpendicular to the orbit plane. The condition for stable radial oscillations (so that the beam does not "blow up" and hit the vacuum tube) is that the z-component of the magnetic field should vary as r^{-n} where $0 < n < 1$. The frequencies of horizontal and vertical oscillations are

$$Q_H = \omega_0(1 - n)^{1/2} \tag{2.12}$$

and

$$Q_V = \omega_0 n^{1/2}. \tag{2.13}$$

On the other hand, the *amplitude* of the vertical betatron oscillations is proportional to $n^{-1/2}$; hence the field gradient n should be as large as possible, in order to keep the particles inside the smallest possible vacuum tube. These two conflicting requirements were resolved with the invention of the strong-focusing principle by Courant, Livingston, and Snyder in 1952. The field gradient n is large, but has opposite sign for alternate magnets around the ring. Such gradients are achieved with quadrupole magnets (Fig. 2.5), so that one magnet is vertically focusing (F) and horizontally defocusing (D), while the reverse is true for its neighbors. The result is a net focusing in both horizontal and vertical planes. This is achieved if the particle passes close to the axis of a D-magnet, where the field is weak, and far from the axis of an F-magnet, where the field is strong.

To summarize, particles oscillate sinusoidally in phase and in radial and vertical position about the synchronous, equilibrium orbit. As the energy

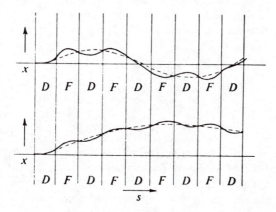

Fig. 2.5 Net focusing effect of a system of alternate focusing and defocusing quadrupole magnets.

increases during the accelerating cycle, particles bunch together in phase and the amplitude of transverse oscillations decreases. To give an example, the CERN PS (25-GeV proton synchrotron) has ring radius 100 m, and the protons are injected into the machine from a linear accelerator at 50-MeV energy. Essentially the whole of the cross-section of the vacuum tube (15 cm × 7 cm) is filled at this stage. The mean magnetic field varies from 0.015 T at injection to 1.4 T at the end of the cycle, which lasts about 1 second. The RF bunches of protons are ∼ 10 ns (= 10 feet) long and 100 ns apart, with 20 bunches distributed around the ring. The RF "kick" per revolution is 220 keV, and ∼ 100,000 revolutions complete the cycle. By that time, the beam has been compacted to 2-mm lateral dimension. In the CERN PS design, the ring magnets both bend and focus the beam. In the larger accelerators, it is usual to separate these functions by using physically distinct dipole (bending) and quadrupole (focusing) magnets.

The full-energy protons are either extracted and injected into another accelerator or storage ring, or ejected onto an external target by a special kicker magnet. Two methods of ejection can be used (separately or together). In fast spill, all the protons in the machine are ejected as they hit the kicker magnet (i.e. in 20 × 100 ns = 2 μs). In slow spill, the protons are peeled off with the aid of a thin foil and emerge over a period of about 1 second.

2.2.2   Beam Intensity, Defocusing, and Resonances

The objective in building an accelerator is to reach not only high energy but the highest possible intensity. The latter is ultimately limited by defocusing processes which blow up the beam, originating from repulsive forces of the space charge in the bunch, as well as image charges on the vacuum tube. Apart from such limitations, the accelerator can only be successfully operated if betatron resonance effects are avoided. Let us denote the lateral (horizontal or vertical) displacement of a particle by y, the arc distance around the ring by s; then the equation of motion has the form

$$\frac{d^2y}{ds^2} = -K(s)y \qquad (2.14)$$

with solution

$$y = \sqrt{\varepsilon}\,\sqrt{\beta(s)}\,\sin\phi(s). \qquad (2.15)$$

Here the azimuthal phase $\phi(s) = Qs/2\pi R$, with R the orbit radius and Q the frequency of betatron oscillations (depending on the magnetic field gradient n) with values Q_H and Q_V in the horizontal and vertical planes. The quantity ε is the emittance of the beam, that is, its extent in phase space, which fills up as one tries to cram more particles into the beam. ε decreases as acceleration proceeds.

Values of Q_H, Q_V which are integers or simple fractions must be avoided; otherwise effects due to any magnet imperfections will build up turn after turn

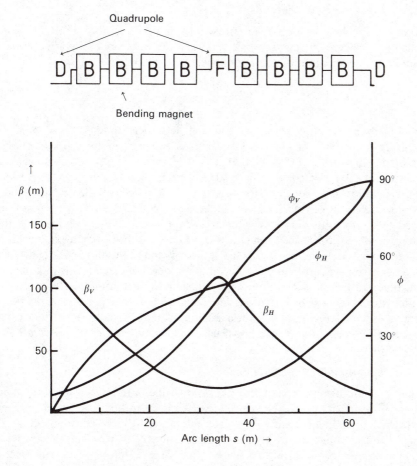

Fig. 2.6 Variation of the values, in horizontal and vertical planes, of the β-function (beam envelope) and azimuthal phase ϕ, in a typical *FDFD*...quadrupole lattice.

and such resonances will result in loss of the beam. The accelerator can only operate in regions of the Q_H, Q_V plot not crossed by such resonance lines. The effects of space charge are to introduce a spread – the so-called tune shift – in the values of Q_H or Q_V, so that they eventually cross resonance lines and the beam is lost. The function $\beta(s)$ describes the envelope of the beam. Typical values of β_V and β_H, together with ϕ_V and ϕ_H, are given in Fig. 2.6 for an *FDFD*... quadrupole lattice. In addition to lateral instabilities, longitudinal beam instabilities also occur and limit the permissible beam intensity.

In proton synchrotrons, achievable beam intensities are of the order of 10^{12}–10^{13} particles per second on to the target.

2.2.3 Electron Synchrotrons

Electron synchrotrons have an important limitation, absent in a proton machine. Under the circular acceleration, an electron emits synchrotron radiation, the energy radiated per particle per turn being

$$\Delta E = \frac{4\pi}{3} \frac{e^2 \beta^2 \gamma^4}{\rho}, \tag{2.16}$$

where ρ is the bending radius, β is the particle velocity, and $\gamma = (1 - \beta^2)^{-1/2}$. Thus, for relativistic protons and electrons of the same momentum the energy loss is in the ratio $(m/M)^4$, so that it is 10^{13} times smaller for protons than electrons. For an electron of energy 10 GeV circulating in a ring of radius 1 km, this energy loss is 1 MeV per turn – rising to 16 MeV per turn at 20 GeV. Thus, even with very large rings and low guide fields, synchrotron radiation and the need to compensate this loss with large amounts of RF power become the dominant factor for an electron machine. Indeed, for this very reason, one large (2-mile long) electron linac of 20 GeV has been built at Stanford. The emittance in an electron synchrotron does not fall with increasing energy, as in a proton machine; rather it is determined by the lateral scattering of the electrons following energy loss by radiation, which increases with energy. Thus the beam size tends to be larger than in proton machines, and e^+e^- colliders include extra focusing magnets (low β insertions) near intersection regions to decrease the beam size.

2.2.4 Colliding-Beam Machines

The vast bulk of present experimental knowledge in the high-energy field has been obtained with proton and electron accelerators in which the beam has been extracted and directed on to an external target – the so-called fixed-target experiments. In particular, high-energy proton synchrotrons can thus provide intense secondary beams of hadrons (π, K, p, $\bar{p}$) and leptons (μ, ν), and several beam lines from one or more targets can be used simultaneously for a range of experiments.

During the last two decades, colliding-beam machines have become important. In these accelerators, two counterrotating beams of particles collide in several intersection regions around the ring. Their great advantage is in terms of the large center-of-mass energy available for the creation of new particles. A fixed-target machine provides particles of energy E which collide with a nucleon of mass M in a target, and the square of the CMS energy W is (see Appendix A)

$$s = W^2 = 2ME + M^2. \tag{2.17}$$

Thus, for $E \gg M$, the kinetic energy available in the CMS for new particle creation rises only as $E^{1/2}$. The remaining energy is not wasted – it is converted into kinetic energy of the secondary particles in the laboratory system, and this allows the production of high-energy secondary beams.

If two relativistic particles of energy E_1 and E_2 circulate in opposite directions in a storage ring, then in a head-on collision the value of W is given by

$$s = W^2 = 4E_1 E_2. \tag{2.18}$$

If $E_1 = E_2$, the CMS of the collision is at rest in the laboratory. Virtually all the energy is available for new-particle creation, and W rises as E instead of as $E^{1/2}$. Indeed, the value of W is the same as that in a fixed-target machine of energy $E = 2E_1 E_2/M$. As an example, the CERN ISR (Fig. 2.7) consists of two rings of magnets close together, containing proton beams of up to 30 GeV which intersect in eight points. Thus $W = 60$ GeV, and a fixed-target proton synchrotron of energy $E = 2000$ GeV would be required to provide the same value of W.

Colliding-beam machines also possess severe disadvantages. The colliding particles must be stable, limiting one to collisions of protons (or heavier nuclei), antiprotons, electrons, and positrons. All colliders so far built are of the pp, e^+e^-, or $p\bar{p}$ variety, although ep is also planned. Secondly, the collision rate in

Fig. 2.7 The CERN ISR consists of two oppositely circulating proton beams of energy up to 30 GeV, confined in two magnet rings of 940-m circumference. The beams intersect at eight crossing points at a 15° angle. At the intersection region shown, an experiment is being set up to search for charmed particle production. (Photograph courtesy CERN.)

the intersection region is low. The reaction rate is given by

$$R = \sigma L, \tag{2.19}$$

where σ is the cross-section for the beam-beam interaction and L is the luminosity (in units $\text{cm}^{-2}\,\text{sec}^{-1}$). For two oppositely directed beams of relativistic particles the formula for L is

$$L = fn\frac{N_1 N_2}{A}, \tag{2.20}$$

where N_1 and N_2 are the numbers of particles in each bunch, n is the number of bunches in either beam, and A is the cross-sectional area of the beams, assuming them to overlap completely. f is the revolution frequency. Obviously, L is largest if the beams have small cross-sectional area A. The luminosity is however limited by the beam-beam interaction. The maximum L-values are $\sim 10^{31}\ \text{cm}^{-2}\,\text{sec}^{-1}$ for e^+e^- colliders and $\sim 10^{32}\ \text{cm}^{-2}\,\text{sec}^{-1}$ for pp machines. These values may be compared with that of a fixed-target machine. A beam of 2×10^{12} protons sec^{-1} from a proton synchrotron, in traversing a liquid-hydrogen target 1 m long, provides a luminosity $L \sim 10^{37}\ \text{cm}^{-2}\,\text{sec}^{-1}$.

Figure 2.8 shows the general layout of the secondary beam lines around the 500-GeV proton accelerator at Fermilab, near Chicago. The 2.5-km-long beam to the neutrino area provides beams of muons as well as neutrinos. Figure 2.9 shows the layout of the accelerator complex at CERN, Geneva. The 26-GeV PS

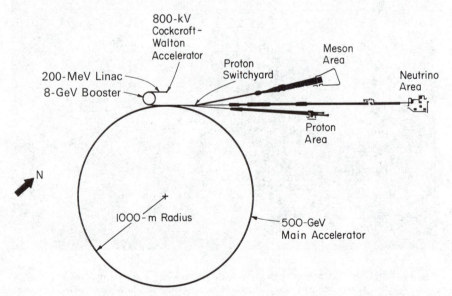

Fig. 2.8 Layout of the 500-GeV proton synchrotron and beamlines at Fermilab, near Chicago. This machine is being upgraded to 1000 GeV by use of high-field superconducting magnets.

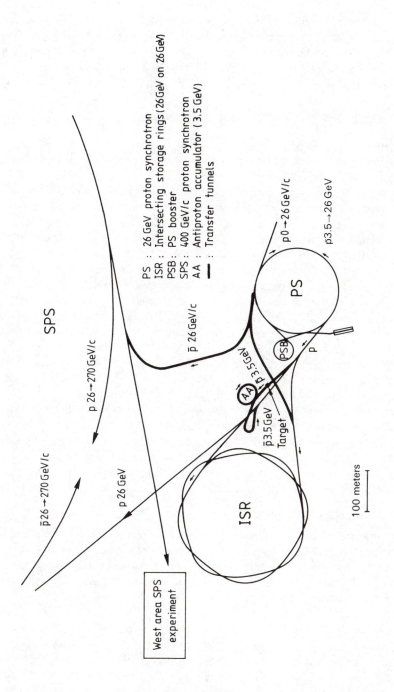

SPS

p 26→270 GeV/c

p̄26→270 GeV/c

p 26 GeV

West area SPS experiment

p 26→270 GeV/c

p̄ 26 GeV/c

PS : 26 GeV proton synchrotron
ISR : Intersecting storage rings (26 GeV on 26 GeV)
PSB: PS booster
SPS : 400 GeV/c proton synchrotron
AA : Antiproton accumulator (3.5 GeV)
━━ : Transfer tunnels

p0→26 GeV/c

p̄3.5→26 GeV

PS

AA 3.5 GeV

p̄ 3.5 GeV

PSB

p

p̄3.5 GeV
Target

ISR

100 meters

Fig. 2.9 Layout of the accelerator complex at CERN, near Geneva.

accelerator is used to fill the ISR with oppositely circulating proton beams, as an injector to the 400-GeV SPS, and also to make secondary antiprotons, which are transferred, stacked, and "cooled" in a 3.5-GeV antiproton accumulator, before being accelerated in the PS and injected into the SPS, used as a 270-GeV + 270-GeV $p\bar{p}$ collider.

2.3 DETECTORS OF SINGLE CHARGED PARTICLES

The detectors employed in experiments in high-energy physics are required to record the position, arrival time, and identity of charged particles. Precise evaluation of position coordinates is required to determine the particle trajectory and, in particular, its momentum (from the deflection in a magnetic field); precise timing is often required in order to associate one particle with another from the same interaction, frequently in situations where the total interaction rate per unit time may be very high. The identity of a particle may be established from simultaneous measurement of velocity (by time-of-flight or Čerenkov radiation) and momentum, and hence the rest mass; from the observation of decay modes, if the particle is unstable; and from its observed interaction with matter via strong, electromagnetic, or weak forces. Neutral particles are detected through their decay (e.g. $K^0 \rightarrow \pi^+\pi^-$) and/or interaction with matter (e.g. $\pi^0 \rightarrow 2\gamma$, $\gamma \rightarrow e^+e^-$), leading to secondary charged particles.

No single detector is in general able to meet all these requirements, and a combination of detectors of different types is required. We first discuss the principal types of detector in current use in the field, and then give some examples of how they may be combined in an integrated system.

Over the last two decades, very few radically new concepts in basic detection methods have been developed. Rather, progress has been in the exploitation and adaptation of well-established methods, a revolution in electronics and computer technology to select, record and analyse huge amounts of data at high speed, and the development of hybrid systems involving many different types of detector, frequently on a massive scale.

2.3.1 Proportional Counters

The proportional counter is one of the oldest devices for the recording of ionization. Single counters consist of a gas-filled cylindric metal or glass tube of radius r_2 maintained at negative potential, with a fine central anode wire of radius r_1 at positive potential. The electric field in the gas for a potential difference V_0 is then

$$E(r) = \frac{V_0}{r \ln(r_2/r_1)}. \tag{2.21}$$

An electron liberated by ionization at radius r_a will drift towards the anode, gaining an energy $T = e \int_{r_b}^{r_a} E(r)\, dr$ when it reaches radius r_b. If T exceeds the ionization energy of the gas, then fresh ions are liberated and a chain of such

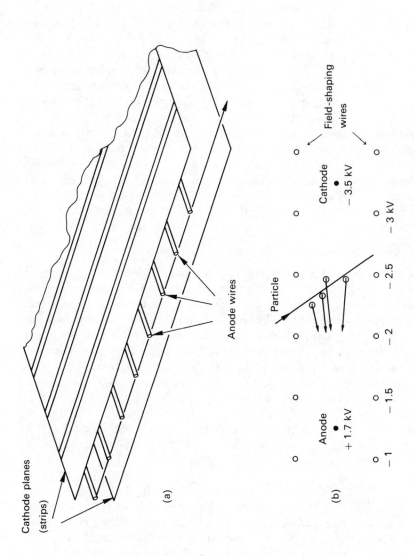

Cathode planes (strips)

Anode wires

(a)

Field-shaping wires

Cathode
− 3.5 kV

Particle

Anode
+ 1.7 kV

− 1   − 1.5   − 2   − 2.5   − 3 kV

(b)

Fig. 2.10 (a) Schematic layout of multiwire proportional chamber. (b) Typical arrangement of electrodes for a drift-chamber cell.

processes leads to an avalanche of electrons and positive ions. The gas amplification factor, equal to the total number of secondary electrons reaching the anode per initial ion pair, is typically $\sim 10^5$, independent of the number of primary ions; hence the name "proportional counter".

The most significant advance in this field was the introduction of the multiwire proportional counter (MWPC) by Charpak (1968, 1970). This device consists of many parallel anode wires stretched in a plane between two cathode planes (Fig. 2.10). The different anode wires act as independent detectors. A typical structure has wires of 20 μm diameter with 2-mm spacing, between cathode planes 12 mm apart, operating at a potential difference of 5 kV, and containing an argon-isobutane gas mixture. In general, several electrons from the primary ionization will drift towards an anode wire and create separate avalanches yielding negative pulses with very fast rise times (~ 0.1 ns). The positive ions have much lower mobility and induce pulses on both the cathode and neighboring anode wires of duration ~ 30 ns. The effective spatial resolution from the anode pulses is of order 0.7 mm. If the cathode is in the form of strips, the center of gravity of the cathode pulses may be used to obtain accurate spatial position of the avalanche (to within 0.05 mm).

2.3.2 Drift Chambers

As indicated above, the MWPC typically has spatial resolution of 1 mm or less, and a time resolution of 30 ns. However, to achieve this resolution over large areas, an enormous number of wires (together with amplifiers) are required. A great reduction in cost can be achieved by drifting the electrons from the primary ionization over (typically) 10 cm in a low-field region (1 kV/cm) before reaching the high-field amplification region near the anode wire; the collection time of the avalanche then gives a measure of the position of the ionization column. Figure 2.10 shows a typical arrangement of anode, cathode, and field wires required to give a uniform drift field. Space resolutions of order 0.1 mm are attained, with drift velocities of order 40 μm/ns in argon-isobutane mixtures, almost independent of the drift field. Since, over a distance of 10 cm, the typical drift time is then 2 μs, such chambers are only useful with fairly low beam intensities or interaction rates, for example at e^+e^- colliders.

2.3.3 Scintillation Counters

The scintillation counter has been a universal detector in a very wide range of high-energy physics experiments for over 30 years. The excitation of the atoms of certain media by ionizing particles results in luminescence (scintillation), which can be recorded by a photomultiplier. The scintillators in most common use are inorganic single crystals and organic liquids and plastics, although the phenomenon occurs also in liquids and gases. The decay times of the fastest (organic) scintillators are of order 1 ns.

Inorganic crystal scintillators, such as sodium iodide, are doped with activator centers (e.g. thallium). Ionizing particles traversing the crystal produce free electrons and holes, which move around until captured by an activator center. This is transformed into an excited state and decays with emission of light, over a broad spectrum in the visible region and with a decay time of order 250 ns.

In organic materials (either solid or liquid), on the other hand, the mechanism is excitation of molecular levels which decay with emission of light in the UV region. The conversion to light in the blue region is achieved via fluorescent excitation of dye molecules known as wavelength shifters, incorporated into the primary scintillator medium. Table 2.2 gives a list of a few organic and inorganic scintillators in common use.

TABLE 2.2 Characteristics of typical scintillators

	Pulse height (relative to anthracene)	Decay time, ns	λ_{max} Å	Density, g/cm^3
Polystyrene + p-terphenyl	0.28	3	3550	0.9
+ tetraphenyl-butadiene	0.38	4.6	4800	
Sodium iodide (+ thallium)	2.1	250	4100	3.7
Anthracene	1.0	32	4100	3.7
Toluene	0.7	< 3	4300	0.9

The light from the scintillating medium is recorded by a photomultiplier tube or tubes (Fig. 2.11). These consist of a photocathode coated with alkali metals, where electrons are liberated by the photoelectric effect. The electrons travel to a chain of secondary-emission electrodes (dynodes) at successively larger potentials. Since about 4 secondary electrons are emitted per incident electron, amplification factors of 10^8 are achieved with 14 dynodes. The transit time from the cathode to the output dynode is typically 50 ns, with a jitter of order 1 ns. The jitter is determined mostly by the variation in transit time to the first dynode from different points on the cathode. Photocathode quantum efficiencies are typically $\leqslant 25\%$, peaking at $\lambda = 400$ nm.

The light from the scintillator slab travels down it by internal reflection, and the traditional way of directing this onto the photocathode is via multiple reflections down a suitably shaped plastic light guide. For very large area scintillators, the light guides consist of bent plastic rods or strips and the device can become very bulky. An alternative method is to place shifter bars along the edge of the scintillator slab. Blue light from the scintillator enters the bar, consisting of acrylic material doped with molecules (e.g. BBQ) which absorb the blue light and reemit isotropically in the green. Part of this light travels down the

Scintillator

Light guide

Mu-metal screen

Steel outer casing

Complete scintillation counter

Photomultiplier

Associated electronics

Fig. 2.11 Plastic scintillation counter, light pipe, and photomultiplier. (Photograph courtesy Rutherford Laboratory.)

bar by internal reflection and is recorded by a photomultiplier glued to the end. The output is much less than that using light guides, but there is a large saving in space, number of photomultipliers, and construction of complex light-guide structures. The output pulse from the photomultiplier is fed into suitable amplifiers, discriminators, and scalers, the essential function of which is to store the number of photomultiplier pulses of a particular magnitude.

The performance of a scintillation counter can be illustrated with a few numbers. In traversing 1 cm of plastic scintillator, a minimum-ionizing particle will lose about 1.5 MeV in ionization, liberating on average 10^4 photons of mean quantum energy $hv = 3$ eV (thus converting 2% of the energy loss to fluorescence). Assuming 10% of the light is collected, and that the photo-multiplier cathode efficiency (number of photoelectrons per incident photon) is also 10%, approximately 2000 photoelectrons will be liberated. Clearly there-fore, even with very thin scintillators, the efficiency for recording particles is essentially 100%. For inorganic crystals (NaI/Tl) the light output is very closely proportional to the ionization energy loss. Organic scintillators are not nearly so linear; for example, the pulse height per unit energy loss of a 5-MeV α-particle is only 10% of that for a singly charged particle at minimum ionization.

The response time of a scintillator, as measured by the width of the output pulse, is determined by a number of factors. One must consider the time required for the light to travel through the material by various routes, and the fact that the light emitted from the fluorescing material decays exponentially with a characteristic decay constant. This is 3 to 30 ns for organic materials and some 250 ns for sodium iodide. There are also fluctuations in transit time through the photomultiplier, typically a few nanoseconds. Under favorable conditions, therefore, one can obtain final electrical pulses of width down to about 10 ns (for plastics). The great advantages of the scintillation counter are that it is robust, simple, and efficient, giving large, sharp output pulses. Its spatial resolution is poor, since the pulse is not related in an obvious way to the location of the trajectory of the charged particle through the counter. Thus, if spatial information is required, it is necessary to use a large hodoscope array of very small counters. Because all photomultipliers generate random noise, it is usual, when counting particles by this method, to place two or more scintillators in coincidence. The accidental coincidence rate is then small because of the good resolution time.

2.3.4 Bubble Chambers

The bubble chamber has been an indispensible tool of high-energy physics for 30 years, particularly for studying complex interactions involving many secondary particles. Conceived by Glaser in 1952, it relies for its operation on the fact that, in a superheated liquid, boiling will start with formation of gas bubbles at nucleation centers in the liquid, and particularly along trails of ions left by the passage of a charged particle. The liquid filling is maintained under an overpressure (typically 5–20 atmospheres), and the superheating achieved by

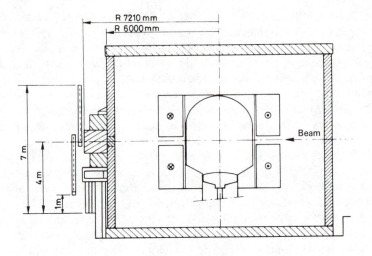

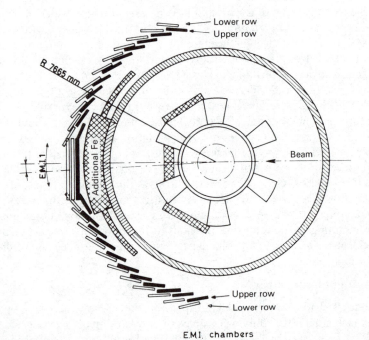

E.M.I. chambers

Fig. 2.12 Elevation and plan views of the 3.7-m-diameter bubble chamber (BEBC) at CERN. The chamber is filled with liquid hydrogen, deuterium, or neon-hydrogen mixture, and is equipped for neutrino experiments with an external muon identifier. This consists of 150 m² of multiwire proportional chambers placed outside the magnet yoke.

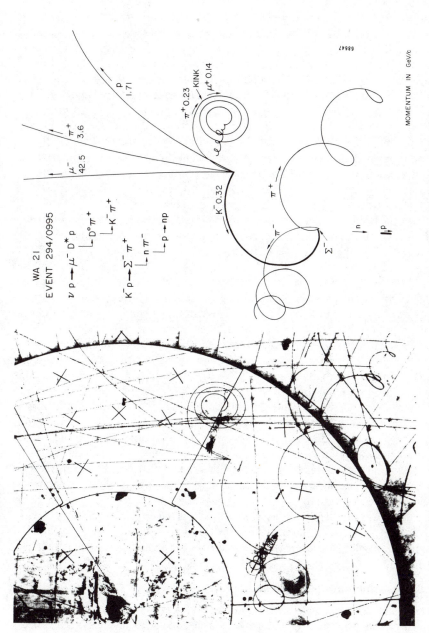

WA 21
EVENT 294/0995

$\nu p \rightarrow \mu^- D^* p$
$\qquad \hookrightarrow D^\circ \pi^+$
$\qquad\quad \hookrightarrow K^- \pi^+$

$K^- p \rightarrow \Sigma^- \pi^+$
$\qquad \hookrightarrow n \pi^-$
$\qquad\quad \hookrightarrow p \rightarrow np$

μ^- 42.5

π^+ 3,6

p 1,71

π^+ 0.23   KINK

μ^+ 0.14

K^- 0.32

π^-

π^+

Σ^-

$\downarrow \nu$

$\parallel p$

68647

MOMENTUM IN GeV/c

Fig. 2.13 Example of charmed-particle production and decay in the hydrogen bubble chamber BEBC exposed to a neutrino beam at the CERN SPS. (Courtesy CERN)

sudden expansion of a piston, bellows, or diaphragm placed at the rear of the chamber. After expansion, bubbles along the tracks are allowed to grow over a period of order 10 ms, and photographed by an array of stereo cameras using flash illumination. The bubbles then collapse under a recompression stroke. Since the cycle time is of order 1 second, the bubble chamber is well matched to pulsed, cyclic accelerators, with repetition rates of the same order.

The most usual liquid fillings for bubble chambers are hydrogen, deuterium, and heavy liquids such as neon-hydrogen mixture, propane (C_3H_8), and Freon (CF_3Br). The entire chamber is immersed in a strong magnetic field (2–3.5 tesla) provided by an electromagnet with conventional or superconducting coils, to permit momentum measurement from track curvature. The bubble images are recorded on photographic film from several cameras in stereo. Subsequent measurement of images on film are digitized, and a geometry program is used to reconstruct tracks and event vertices in three dimensions.

Over the last decade, medium-sized bubble chambers have been phased out, and the few remaining chambers fall into two categories. Giant chambers, like BEBC (Big European Bubble Chamber), 3.7 m in diameter (see Fig. 2.12) contain many cubic metres of liquid, forming a large and homogeneous detector, principally for the study of neutrino interactions, with low event rates. The great detail which is recorded in a complex production and decay pattern is well illustrated in the example of Fig. 2.13. The second category of chamber is the small, rapid-cycling chamber serving as the interaction vertex detector in a hybrid system, which incorporates large electronic detectors and spectrometers for recording the high-energy forward secondaries, downstream of the chamber. Small chambers, with volumes up to a few litres, are capable of extremely good optical resolution ($\sim 6\ \mu$m if holographic readout is employed). This type of chamber is "triggerable" in the sense that flash illumination and film advance are operated only for events in those expansions which the surrounding electronic detectors indicate to be of interest. The evolution of bubble chambers as part of an integrated system is part of the recent trend in detector technology.

The main disadvantages of a bubble chamber are, firstly, that it has a low repetition rate, and analysis of film is a lengthy process (typical experiments rarely involve more than 10^5 useful analysed events); secondly, that many present and essentially all new accelerators are of the colliding-beam type, with an effectively d.c. interaction rate, and the very low duty cycle (10^{-2}) of bubble chambers, as well as the interaction geometry of the beams, will exclude their use in this field.

2.3.5 Streamer and Flash Chambers

Proportional chambers are operated with anode-cathode potentials of order 5 kV, where the efficiency for recording charged particles reaches a plateau of practically 100%. Further increase in voltage eventually leads to electric breakdown of the gas. This takes place when the space charge inside the avalanche is strong enough to shield the external field; recombination of ions

Fig. 2.14 Interaction of a 200-GeV proton in a xenon gas target placed in a streamer chamber. The proton enters the target at right. The streamers are normal to the plane of the picture, as is the applied magnetic field. The gas filling is 90% Ne, 10% He. (Courtesy V. Eckhardt, MPI, Munich.)

then occurs, resulting in photon emission and the birth of secondary ionization and new avalanches outside the initial one. This process propagates until an ion column links anode and cathode and a spark discharge occurs. The counter is then said to operate in the Geiger region.

If however a short (10-ns) high-voltage pulse (10–50 kV cm^{-1}) is applied between transparent parallel-plate electrodes, then only short (2–3-mm) streamer discharges develop from the ion trail of a crossing particle, and a direct track image (rather like that in a bubble chamber) can be obtained and photographed through the electrodes (Fig. 2.14). Such a *streamer chamber* possesses good multitrack efficiency and space resolution; an advantage over the bubble chamber is that the high potential is triggered (by outside scintillators) and that it has a very fast response, limited only by the speed of the film transport system.

The *flash chamber* is a simple and cheap detector, consisting of a large number of tubes or plastic channels filled with neon-helium mixture placed between planar electrodes to which a triggered high-voltage pulse is applied. A glow discharge is generated in the cells where ionization has been produced by a crossing charged particle. The discharge is recorded photographically or by electronic readout. Spatial resolution is determined by the tube diameter – typically several millimetres – but the method allows large-volume calorimeter-type detectors to be constructed at low cost.

2.3.6 Čerenkov Counters

When high-energy charged particles traverse dielectric media, part of the light emitted by excited atoms appears in the form of a coherent wavefront at fixed angle with respect to the trajectory – a phenomenon known as the Čerenkov

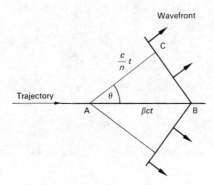

Fig. 2.15

effect, after its discoverer. Such radiation is produced whenever the velocity βc of the particle exceeds c/n, where n is the refractive index of the medium. From the Huyghens construction of Fig. 2.15 one sees that the wavefront forms the surface of a cone about the trajectory as axis, such that

$$\cos\theta = \frac{ct/n}{\beta ct} = \frac{1}{\beta n}, \qquad \beta > \frac{1}{n}. \tag{2.22}$$

Čerenkov radiation appears as a continuous spectrum. In a dispersive medium, both n and θ will be functions of the frequency v. The total energy content of the radiation, per unit track length, is

$$\frac{dE}{dx} = \frac{4\pi^2 z^2 e^2}{c^2} \int \left(1 - \frac{1}{\beta^2 n^2}\right) v \, dv. \tag{2.23}$$

The number of photons at a particular frequency or wavelength is proportional to dv or to $d\lambda/\lambda^2$. Thus, blue light predominates. Over a small frequency range, we can neglect the dependence of n on v and (2.23) becomes

$$\frac{dE}{dx} = \frac{z^2}{2} \left(\frac{e^2}{hc}\right)^2 \left(\frac{mc^2}{e^2}\right) \left[\frac{(hv_1)^2 - (hv_2)^2}{mc^2}\right] \left(1 - \frac{1}{\beta^2 n^2}\right)_{av}. \tag{2.24}$$

For a particle of $z = 1$ and $\beta \simeq 1$ in water ($n = 1.33$) this expression gives $dE/dx = 400\,\text{eV cm}^{-1}$ for visible light ($\lambda = 400\text{--}700$ nm) and thus 200 photons cm^{-1}. Note that this is small compared with the total energy loss of about $2\,\text{MeV cm}^{-1}$.

The usefulness of the Čerenkov effect lies in the fact that measurement of the angle in (2.22) provides a direct measurement of the velocity βc. Table 2.3 gives a list of some radiating media, demonstrating that most of the range of γ-values from 1.2 to 100 can be covered by means of solids, liquids, gases, or aerogels.

Threshold Čerenkov counters can be used to discriminate between two relativistic particles of the same momentum p and different masses m_1 and m_2, if the heavier, slower particle (m_2) is just below threshold. In this case, $\beta_2^2 = 1/n^2$ and it is straightforward to show that the production rate of photons from the particle m_1 is given by the previous equation with

$$\sin^2 \theta_1 = 1 - \frac{1}{\beta_1^2 n^2} \simeq \frac{m_2^2 - m_1^2}{p^2}. \tag{2.25}$$

TABLE 2.3 Čerenkov radiators

Medium	$n - 1$	γ (threshold)
Helium (NTP)	3.3×10^{-5}	123
CO_2 (NTP)	4.3×10^{-4}	34
Pentane (NTP)	1.7×10^{-3}	17.2
Aerogel	$0.075 \to 0.025$	$2.7 \to 4.5$
H_2O	0.33	1.52
Glass	$0.75 \to 0.46$	$1.22 \to 1.37$

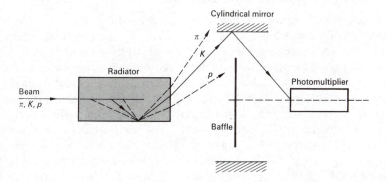

Fig. 2.16 Early design of differential Čerenkov counter. The arrangement is intended to select light from one of three components of the beam (K-mesons, in the case shown).

Thus the length of radiator to produce a given number of photoelectrons (assuming one with the right refractive index can be found) increases as the square of the momentum. For high-momentum beams, large, pressurized gas radiators of several metres length may be required.

In the differential Čerenkov counter, the angle of Čerenkov emission is measured in order to identify particles. The cone of light from the radiating particle is focused by a lens or spherical mirror into a ring image, and an adjustable diaphragm at the focus transmits the light to a phototube. Differential counters with velocity resolution $\Delta\beta/\beta \sim 10^{-7}$ have been built, and separation of charged pions, kaons, and protons up to several hundred GeV/c is achievable. As an alternative to changing the radius of the diaphragm, this may be fixed and a velocity scan carried out by varying the gas pressure. By integrating the photomultiplier output over time, such a counter may be used to measure secondary particle yields in high-intensity beams.

2.4 SHOWER DETECTORS AND CALORIMETERS

The energy and position coordinates of secondaries from high-energy interactions can also, under suitable conditions, be measured by total-absorption methods. In the absorption process, the incident particle interacts in a large detector mass, generating secondary particles which in turn generate tertiary particles, and so on, so that all (or most) of the incident energy appears as ionization or excitation in the medium — hence the term "calorimeter". Such devices are essential in recording the energy of neutral hadrons; and since the fractional energy resolution varies as $E^{-1/2}$, calorimeters provide, even for charged hadrons, a precision at high energies (10–100 GeV) comparable with or better than what can be achieved by magnetic deflection. Just as important is the fact that total-absorption calorimeters provide fast (100-ns) "total energy" signals useful for quick decisions on event selection.

2.4.1 Electromagnetic Shower Detectors

For electrons and photons of high energy, a dramatic result of the combined phenomena of bremsstrahlung and pair production is the occurrence of cascade showers. A parent electron will radiate photons, which convert to pairs, which radiate and produce fresh pairs in turn, the number of particles increasing exponentially with depth in the medium. The development of such a shower can be discussed according to the following very simplified model. Starting off with a primary electron of energy E_0, suppose that, in traversing one radiation length, it radiates half its energy, $E_0/2$, as one photon. Assume that, in the next radiation length, the photon converts to a pair, the electron and positron each receiving half the energy (that is, $E_0/4$), and that the original electron radiates a further photon carrying half the remaining energy, $E_0/4$. Thus, after two radiation lengths, we shall have a photon of energy $E_0/4$ and two electrons and one positron, each of $E_0/4$. By proceeding in this way, it is easily seen that, after t

radiation lengths, there will be $N = 2^t$ particles, with photons, electrons, and positrons approximately equal in number. We have here neglected ionization loss and the dependence of radiation and pair-production cross sections on energy. The energy per particle at depth t will then be $E(t) = E_0/2^t$. This process continues until $E(t) = E_c$, when we suppose that ionization loss suddenly becomes important and no further radiation is possible. The shower will thus reach a maximum and then cease abruptly. The maximum will occur at

$$t = t_{max} = \frac{\ln(E_0/E_c)}{\ln 2},$$ (2.26)

the number of particles at the maximum being

$$N_{max} = \exp[t_{max} \ln 2] = \frac{E_0}{E_c}.$$ (2.27)

The number of particles of energy exceeding E will be

$$N(> E) = \int_0^{t(E)} N \, dt = \int_0^{t(E)} e^{t \ln 2} \, dt$$

$$\simeq \frac{e^{t(E)\ln 2}}{\ln 2} = \frac{E_0/E}{\ln 2},$$

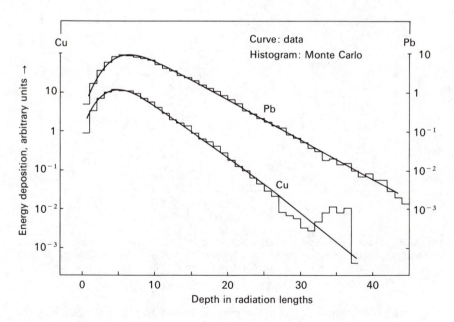

Fig. 2.17 Longitudinal distribution of energy deposition in a 6-GeV electron shower (after Bathow *et al.* 1970).

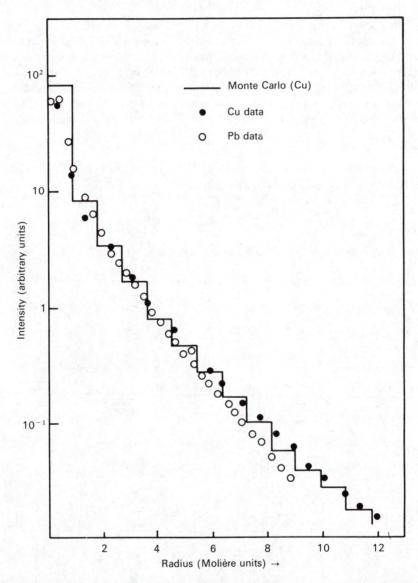

Fig. 2.18 Lateral distribution of energy deposition in 6-GeV electron shower. The points indicate data, and the histogram the Monte Carlo predictions. The horizontal scale is in Molière units, $R_m = 21(X_0/E_c)$, with X_0 the radiation length and E_c the critical energy in MeV (after Bathow *et al.* 1970).

where $t(E)$ is the depth at which the particle energy has fallen to E. Thus, the differential energy spectrum of particles $dN/dE \propto 1/E^2$. The total integral track length of *charged* particles (in radiation lengths) in the whole shower will be

$$L = \frac{2}{3} \int_0^{t_{max}} N \, dt = \frac{2}{3 \ln 2} \frac{E_0}{E_c} \simeq \frac{E_0}{E_c}. \tag{2.28}$$

The last result also follows from the definition of E_c and conservation of energy; nearly all the energy of the shower must eventually appear in the form of ionization loss of charged particles in the medium.

In practice, the development of a shower consists of an initial exponential rise, a broad maximum, and a gradual decline. Nevertheless, the above equations indicate correctly the main qualitative features, which are:

(a) a maximum at a depth increasing logarithmically with primary energy E_0;
(b) the number of shower particles at the maximum being proportional to E_0; and
(c) a total track-length integral being proportional to E_0.

The observed longitudinal development of a 6-GeV electron-initiated shower in different absorbers is given in Fig. 2.17, together with the prediction of a Monte Carlo program incorporating the known energy-dependent cross-sections for radiation and ionization loss by electrons and for absorption of photons by pair production and other processes.

Because of Coulomb scattering, a shower spreads out laterally. The radial spread is determined by the radiation length in the medium and the angular deflection per radiation length at the critical energy. In all materials, this spread is of order one Moliere unit $R_m = 21(X_0/E_c)$, with E_c in MeV (see Fig. 2.18).

Electromagnetic shower detectors are built from high-Z materials of small X_0, so as to contain the shower in a small volume. In lead-loaded glass (55% PbO, 45% SiO_2) detectors, the Čerenkov light from relativistic electrons is used to measure the shower energy. The resolution is typically $\Delta E/E = 0.05/\sqrt{E \text{ (GeV)}}$, and is determined by the fluctuation $\sqrt{N}$ in the number N of particles at maximum, where $N \propto E$. Calorimeters built from alternate sheets of lead and plastic scintillator are also used; the resolution depends on the sampling frequency but is comparable with that from lead glass.

2.4.2 Hadron-Shower Calorimeters

A hadron shower results when an incident hadron undergoes an inelastic nuclear collision with production of secondary hadrons, which again interact inelastically to produce a further hadron generation, and so on. The scale for longitudinal development is set by the nuclear absorption length λ, varying from 80 g cm^{-2} (C) through 130 g cm^{-2} (Fe) to 210 g cm^{-2} (Pb). This scale is big compared with the radiation length X_0 in heavy elements, so that, in comparison with electromagnetic shower detectors, hadron calorimeters are large. For an iron-scintillator sandwich, for example, the longitudinal and transverse dimensions are of order 2 m and 0.5 m respectively.

In an electromagnetic cascade, the bulk of the incident energy appears eventually in the form of ionization. However, in a hadron cascade roughly 30%

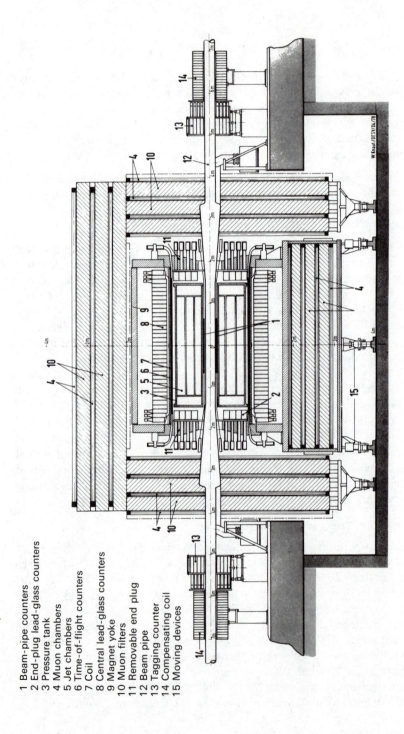

1 Beam-pipe counters
2 End-plug lead-glass counters
3 Pressure tank
4 Muon chambers
5 Jet chambers
6 Time-of-flight counters
7 Coil
8 Central lead-glass counters
9 Magnet yoke
10 Muon filters
11 Removable end plug
12 Beam pipe
13 Tagging counter
14 Compensating coil
15 Moving devices

Fig. 2.19 The spectrometer detector JADE used at the PETRA e^+e^- storage ring at DESY, Hamburg. A vertical section through the beam pipe is shown. (Courtesy DESY)

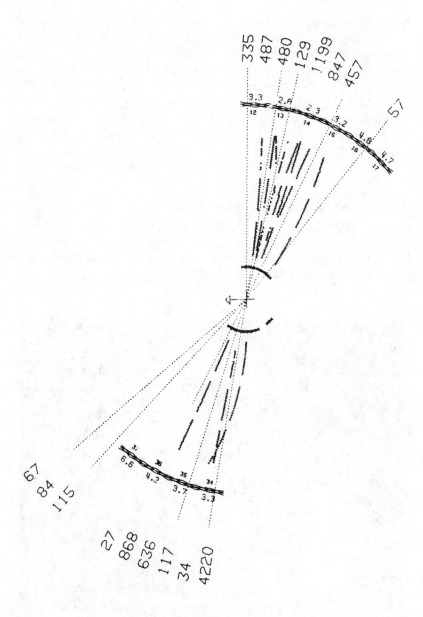

Fig. 2.20 Example of hadron production in e^+e^- annihilation in PETRA at 30-GeV CMS energy. The computer reconstruction shows trajectories of charged particles as lines of crosses where they traverse drift chambers, while γ-rays from neutral-pion production and decay, recorded in lead-glass shower counters, are indicated by dotted lines. Projection in plane normal to beam pipe. The concentration of hadrons into two oppositely directed jets is well displayed. Event from JADE detector of Fig. 2.19.

of the incident hadron energy is lost by the breakup of nuclei, nuclear excitation, and evaporation neutrons (and protons), and does not give an observable signal. One successful method of compensating for this is the use of ^{238}U as the cascade medium, the extra energy released by fast neutron fission of ^{238}U making up for the "invisible" energy losses from nuclear breakup. Typically the energy resolution of hadron calorimeters is $\Delta E/E \simeq 0.5/\sqrt{E}$ with E in GeV.

Various types of hadron calorimeter are in current use. The iron-plus-scintillator sandwich has already been mentioned, but proportional tubes, flash tubes, and drift chambers have also been used for sampling. Calorimeters containing liquid argon as the cascade and recording medium are instrumented with multiwire proportional chambers.

2.4.3 Examples of Large Hybrid Detectors

Typical experiments in high-energy physics involve the simultaneous detection, measurement, and identification of many particles, both charged and neutral, from each interaction which occurs. They therefore usually incorporate several types of detection technique in a single detector array.

Fig. 2.21 Magnetized calorimeter employed by CDHS collaboration in study of neutrino interactions at the CERN SPS. It is instrumented with magnetized iron toroids, scintillation counters, and drift chambers, for the identification and momentum measurement of secondary muons, and to measure the nuclear cascade energy of secondary hadrons. Total mass is 1400 tons. (Photograph courtesy CERN.)

As one example, Fig. 2.19 shows the large spectrometer detector employed by the JADE collaboration at the e^+e^- storage ring PETRA at DESY, Hamburg. A vertical section containing the beam axis is shown. Charged particles from the beam intersection region are recorded by beam pipe counters, by an array of cylindrical drift chambers ("jet chambers"), and by time-of-flight counters. The drift chambers provide ionization information as well as track positions. All these are inside a solenoid magnet providing a field of 0.5 T parallel to the beam axis, extending over 3.5 m length by 2 m diameter. Outside the coil, lead-glass shower counters record electrons and photons, and the outermost layers of drift chambers ("muon chambers") identify and record muons by their penetration through iron and concrete absorber. Tagging counters placed upstream and downstream around the beam pipe record particles at small angles (see Fig. 2.20 for example of event).

Figure 2.21 shows the neutrino detector used by the CDHS collaboration at the CERN SPS. It is essentially a magnetized calorimeter of mass 1400 tons, and consists of a succession of iron plates 3.5 m in diameter and 15 cm thick, separated by scintillation counters and drift chambers. The latter record the trajectories of charged secondaries of neutrino reactions, while the integrated scintillator pulse height indicates the hadron (shower) energy. High-energy muon secondaries are identified by their penetration, and their momentum is determined from their deflection in the magnetized iron.

2.5 EXAMPLES OF THE APPLICATION OF DETECTION TECHNIQUES TO EXPERIMENTS

In Chapter 1, some of the early discoveries in the field of particle physics were outlined. Progress, as in any field of science, has depended on the close interplay between the discovery of experimental phenomena, usually resulting from the application of new techniques, and the introduction of fresh ideas and principles which they stimulate. In the early days, as we have seen, the introduction of the counter-controlled cloud chamber led to the discovery of muons. The observation of the vital link of the $\pi \rightarrow \mu$ decay chain, proving the essential correctness of Yukawa's ideas, had to await the introduction of a different technique – in this case, special photographic emulsions, with a stopping power and spatial resolution superior to that of the cloud chamber. In more recent times, the invention of the hydrogen bubble chamber, ideally matched to pulsed cyclic accelerators, and with a momentum resolution for charged particles of order 1%, has led to the discovery of a host of new baryon and boson resonant states.

Discussion of the application of the detection methods outlined above to particular experiments appears whenever appropriate throughout the text of this book. At this point, however, it is perhaps worth singling out a couple of experiments, because they illustrate two quite different aspects of the experimental method.

The first experiment led to the discovery of the antiproton. At the time there was no evidence for its existence; two cosmic-ray events in nuclear emulsion had previously been ascribed to antiprotons, but it is now virtually certain that they were wrongly interpreted. If such a particle did exist, however, it was expected to have well-defined properties of charge and mass, and to undergo annihilation in ordinary matter. The problem was then to observe a very specific signal due to antiprotons, against a very large background due to more prolifically produced particles.

The other experiment we describe is the determination of the neutral-pion lifetime. In this case, numerous attempts had been made to determine this quantity directly, based on small numbers of events, using the nuclear emulsion technique stretched to the very limit of its capability. Successive estimates had given smaller and smaller values, ranging from 10^{-14} sec in 1950 to $\simeq 2 \times 10^{-16}$ sec in 1961. There was no unambiguous theoretical prediction of the lifetime, although 10^{-16} sec was thought to be an upper limit. The experiment described, using a counter technique and exploiting the very large intensities available at an accelerator, under carefully controlled conditions, was able to detect an extremely small effect (1.5% in counting rate) and measure it with precision.

2.5.1   Discovery of the Antiproton

The experiment was performed at the Berkeley Bevatron in 1955 by Chamberlain, Segre, Wiegand, and Ypsilantis. The Bevatron had been expressly designed to accelerate protons of momentum up to 6.3 GeV/c, approximately the threshold for the production of a nucleon-antinucleon pair in a proton-proton collision. The value of the total energy for an incident proton in collision with a stationary proton is $E/Mc^2 = 7$, or $p_{\text{threshold}} = 6.5$ GeV/c. In the actual experiment, the circulating protons collided with a copper target. In the copper nucleus, a nucleon has Fermi momentum $\mathbf{p}_f$, so that if the latter is in the opposite sense to that of the incident proton, the available center-of-mass energy of the collision is increased, and the incident threshold momentum is reduced by a factor of approximately $1 - p_f/Mc$, i.e. to about 4.8 GeV/c, if we take $p_f/Mc \simeq \frac{1}{4}$ as a typical value. (See Problem 2.6, and Appendix A.)

Figure 2.22 shows the experimental setup employed. Secondary negative particles of 1.2-GeV/c momentum, from an internal copper target T, are bent out in the fringe field of the Bevatron magnet ring, further deflected by a bending magnet, and focused by means of a quadrupole magnet onto a scintillation counter $S1$. The negative beam is further deflected by a second bending magnet and refocused by another quadrupole onto scintillator $S2$. $C1$ is a threshold Čerenkov counter, recording particles of $\beta > 0.79$, and $C2$ is a differential counter, of the type shown in Fig. 2.16, recording particles of $0.78 > \beta > 0.75$. When account is taken of energy loss in the counters and other material traversed, it turns out that negative pions of the beam momentum have $\beta = 0.99$, and particles of protonic mass have $\beta = 0.76$. Thus, the coincidence $S1 + S2 + C1 + S3 - C2$ (meaning a signal from the first four, not accompanied by a signal from $C2$) corresponds to the passage of a negative pion, while $S1 + S2 + C2 + S3 - C1$ corresponds to an antiproton. In this experiment, the ratio of negative pions to antiprotons in the beam was of order 100,000, so that there are severe background problems. Thus, an independent check of the events was made by measuring the time of flight between counters $S1$ and $S2$, separated by some 40 ft, which was expected to be 40 ns for a negative pion ($\beta = 0.99$) and 51 ns for an antiproton ($\beta = 0.76$). The time of flight could be determined by displaying the $S1 - S2$ pulses on an oscilloscope. Only those antiproton candidates (from the Čerenkov counter signals) with correct flight times were accepted as bona fide antiprotons.

Several further checks were then made. First, the beam was tuned to slightly different momenta, and the antiproton yield plotted as a function of momentum. Figure 2.23 shows the result. The abscissae plotted are particle masses, as

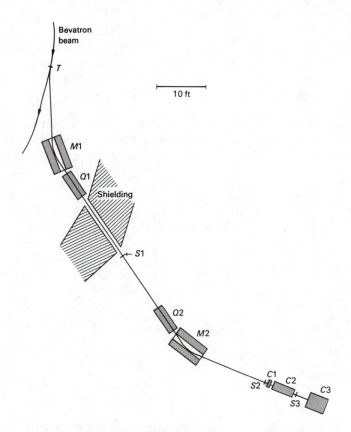

Fig. 2.22 Experimental setup used in the discovery of the antiproton. Q, quadrupole magnet; M, bending magnet; $S1, S2, S3$, scintillators; $C1, C2$, Čerenkov counters; $C3$, total-absorption Čerenkov counter (after Chamberlain *et al.* 1955).

deduced from momentum and velocity. Because the mass resolution of the apparatus is finite, one obtains a mass spectrum of finite width. By changing the position of the target T and reversing the current in the bending magnets, so that π^+ and positive protons are transported by the beam, a similar mass spectrum was deduced for protons. The agreement between the distributions for proton and antiproton events confirms that the negative heavy particles have the same mass as the proton within about 1%.

Secondly, a search was made by Brabant and others for large pulses from the final 14-in. lead-glass Čerenkov counter $C3$, contemporaneous with antiprotons from the previous selection system. In order to exclude Čerenkov light from the enormous flux of fast negative pions, only Čerenkov radiation emitted in the backward direction was recorded, by placing the photomultipliers at the front of the radiator. Pulses corresponding to energy release > 1 GeV were indeed observed, thus proving that some of the antiprotons annihilated in flight in the lead-glass counter, giving charged and neutral pions. Neutral pions from annihilation would decay to γ-rays, generating a cascade shower in the lead glass, and hence a large Čerenkov pulse. Since the energy released in neutral pions alone exceeded the kinetic energy of the negative protons, this proved that the latter were annihilating with nucleons, as expected. (Such annihilations were shortly afterwards observed in stacks of nuclear emulsion exposed to an antiproton beam.)

Finally, the yield of antiprotons relative to pions in the beam was measured as a function of the momentum of the circulating proton beam in the Bevatron. The observed excitation curve (Fig. 2.23) corresponds closely to what would be

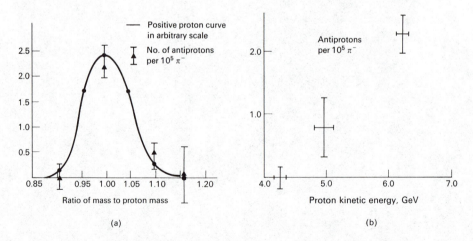

Fig. 2.23 (a) Mass spectrum of particles of near-protonic mass, obtained by varying the momentum of the beam in the antiproton experiment: ▲, negative particles; ●, positive particles. (b) Rate of production of antiprotons relative to negative pions, as a function of the kinetic energy of the incident proton beam on the target.

expected for the production of a nucleon-antinucleon pair in a nucleon-nucleon collision, taking account of Fermi motion of the target nucleon.

2.5.2 Lifetime of the Neutral Pion

The experiment was carried out at the CERN proton synchrotron in 1963 by Von Dardel, Dekkers, Mermod, Van Putten, Vivargent, Weber, and Winter. The neutral pion decays into γ-rays, $\pi^0 \to 2\gamma$, with a lifetime now known to be $\simeq 10^{-16}$ sec. To see what this implies for the experimenter, we note that the mean distance traveled in the laboratory system, for a neutral pion of momentum p, is $\lambda = \gamma\beta c\tau = c\tau p/m_\pi c$. Thus, even for very energetic pions, with $p = 5$ GeV/c, for example, $\lambda \sim 10^{-4}$ cm or 1 μ only.

The principle of the experiment is first to select neutral pions of a nearly unique momentum. Suppose for the moment one is able to do this. In Fig. 2.24, t represents the thickness of a thin metal foil bombarded by 18-GeV protons; neutral pions are produced uniformly through the foil, so that the rate of production in dx is $K\,dx$. The probability that a pion survives to the layer dy and there decays is $\exp(-y/\lambda)\,dy/\lambda$, and the probability that either photon then converts inside the foil is $(t - y - x)/X$, where X is the γ-ray conversion length ($X \gg t$). These considerations lead to a pair-production rate, as a function of thickness t, given by

$$R(t) = Kt\left\{B + \frac{1}{X}\left[\frac{t}{2} - \lambda + \frac{\lambda^2}{t}(1 - e^{-t/\lambda})\right]\right\}, \qquad (2.29)$$

where the first term arises from the internal-conversion contribution, i.e. from pairs produced in the Dalitz decay mode $\pi^0 \to e^+e^-\gamma$. If we neglect a small correction term due to energy degradation in the foil, Eq. (2.29) gives the t-dependence of the number of positrons emerging from the foil, assuming they originate from π^0-decay. We note that, for large t, the effect of a finite lifetime λ is to reduce the average path length for conversion, $t/2$, by the quantity λ. In the experiment, foils of platinum were used, for which $X \sim 1$ cm, $\lambda/X \sim 10^{-4}$, whilst the Dalitz-pair branching ratio $B \sim 10^{-2}$. Thus $\lambda/X \ll B$. In principle, it might

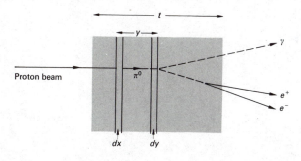

Fig. 2.24

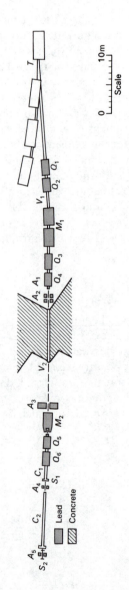

Fig. 2.25 Experimental arrangement for measurement of the π^0 lifetime. T, target; Q_1 to Q_6, quadrupole focusing magnets; $M_{1,2}$, bending magnets; A_1 to A_5, scintillators; $C_{1,2}$, hydrogen Čerenkov counters; $S_{1,2}$, collimators (after Von Dardel *et al.* 1963).

be possible to measure λ by using thick targets of different Z- and hence X-values, but this would require absolute measurements of the constant term in the curly brackets $(B - \lambda/X)$ to a precision better than 1% in different materials. On the other hand, if one expands Eq. (2.29) for small t/λ, one obtains

$$R(t) = Kt\left\{B + \frac{t^2}{6\lambda X} + \cdots\right\},\qquad (2.30)$$

and one is faced with the somewhat easier problem of detecting a quadratic term in t using very thin targets ($t \sim \lambda$) of a given material, but of a different and accurately measurable thickness. This was the method employed by Von Dardel *et al.*

The experimental setup is shown in Fig. 2.25. The target system T consisted of four strips of platinum, thicknesses 3, 4, 18, and 58 μ, mounted on a rotatable head, so that different foils could be flipped in turn into identical radial and azimuthal positions relative to the circulating proton beam. The bending magnets, focusing magnets, and collimators serve to define a beam of positive particles of 5-GeV/c momentum at 6° angle, and they are recorded by scintillators $S1$ and $S2$ in coincidence. Positrons among the heavy particles (π^+, protons) were detected by the Čerenkov counters $C1$ and $C2$. These were

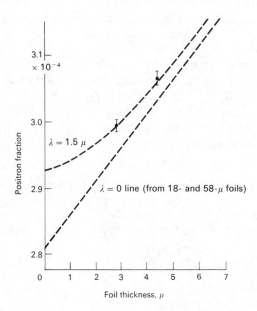

Fig. 2.26   Result of Von Dardel *et al.* on π^0 lifetime. The straight line is the expected variation in positron fraction with foil thickness, assuming the decay length $\lambda = 0$, using the data from two thick foils. For the foils of 3 and 4.5 μ, one obtains a significant (1.5%) deviation from the line, corresponding to a finite decay length $\lambda = 1.5 \pm 0.25\ \mu$.

threshold counters, 10 m long, filled with hydrogen gas, and with a 100% efficiency for positrons and an acceptance for heavier particles of less than 5×10^{-6}. The coincidence ratio $(S_1 + S_2 + C_1 + C_2)/(S_1 + S_2)$ measures the fraction of positrons, i.e. the quantity within the curly brackets in Eq. (2.29). Positrons of 5-GeV/c momentum must come from neutral pions of higher momentum. On the other hand, the flux of pions of momentum p falls off with increasing p in a known way (roughly, exponentially) and if one folds into this spectrum the kinematics of the decay $\pi^0 \to 2\gamma$ and the conversion $\gamma \to e^+e^-$, it is found that the effective π^0-spectrum (contributing 5-GeV/c positrons) is narrow and centered around 7.1 GeV/c. It is, near enough, a unique energy, so that a given value of τ, the proper lifetime, corresponds to a unique decay path λ.

Special precautions were necessary to ensure that the measured coincidence rates were not influenced by counting losses dependent on the length or intensity of the proton burst on the target, and that the accidental coincidence rates were subtracted in a rate-independent manner. These are technical details which we do not discuss. The results on the positron fraction of the beam, for different foil thicknesses, were then fitted with the parameters of formula (2.29). Figure 2.26 shows the result, and proves that the neutral pion has a finite lifetime, with $\lambda = 1.5 \pm 0.25 \ \mu$, and thence

$$\tau_{\pi^0} = (1.05 \pm 0.18) \times 10^{-16} \text{ sec.} \tag{2.31}$$

In this experiment, the value of B deduced from the fit was reasonably consistent with the Dalitz decay rate found from independent measurements. Furthermore, from the rates it could be shown that the effect measured was due principally to neutral pions and not to other processes (for example, $\eta \to 2\gamma$).

We should emphasize that it is possible to determine the lifetime *indirectly* by other methods. In 1951, Primakoff had shown that the cross-section for photoproduction of a neutral pion in the Coulomb field of a heavy nucleus would be inversely proportional to the lifetime. Experiments based on this method give an average value

$$\tau_{\pi^0} = (0.83 \pm 0.06) \times 10^{-16} \text{ sec.} \tag{2.32}$$

The Coulomb cross-section has to be extracted from the total cross-section, which includes nuclear effects, and corrections must be made for pion absorption in the nucleus. It is believed that such corrections can be made accurately and have a small effect on τ. Obviously, the values (2.32) and (2.31) are consistent.

PROBLEMS

2.1 The average number $\bar{n}$ of ionizing collisions suffered by a fast particle of charge ze in traversing an interval dx (g cm^{-2}) of a medium, and resulting in energy transfers $E' \to E' + dE'$, is

$$\bar{n} = f(E') \, dE' \, dx = \frac{2\pi z^2 e^4 N_0 Z}{mv^2 A} \frac{dE'}{(E')^2} \left(1 - \frac{v^2}{c^2} \frac{E'}{E'_{max}}\right) dx,$$

where the symbols are as in Eq. (2.1) and the maximum transferable energy is $E'_{max} = 2mv^2/(1 - \beta^2)$, with $\beta = v/c$. For individual particles, the distribution in number of collisions n follows the Poisson law, so that $\langle (n - \bar{n})^2 \rangle = \bar{n}$. If we multiply the above equation by $(E')^2$ and integrate, we obtain the mean squared deviation in energy loss, $\varepsilon^2 = \langle (\Delta E - \overline{\Delta E})^2 \rangle$, about the mean $\overline{\Delta E}$. Show that

$$\varepsilon^2 = 0.6 \frac{Z}{A} (mc^2)^2 \gamma \left(1 - \frac{\beta^2}{2} \right) \Delta x.$$

Calculate the fractional rms deviation $\varepsilon/\overline{\Delta E}$ in energy loss for protons of kinetic energy 500 MeV traversing (a) 0.1, (b) 1.0, and (c) 10 g cm^{-2} of plastic scintillator ($Z/A = \frac{1}{2}$). Take dE/dx as 3 MeV g^{-1} cm^2.

2.2 A narrow pencil beam of singly charged particles of momentum p, traveling along the x-axis, traverses a slab of material s radiation lengths in thickness. If ionization loss in the slab may be neglected, calculate the rms lateral spread of the beam in the y-direction, as it emerges from the slab. [*Hint*: consider an element of slab of thickness dx at depth x, and find the contribution $(dy)^2$ which this element makes to the mean squared lateral deflection; then integrate over the slab thickness.] Use the formula you derive to compute the rms lateral spread of a beam of 10-GeV/c muons in traversing a 100-m pipe filled with (a) air, (b) helium, at NTP.

2.3 Extensive air showers in cosmic rays contain a "soft" component of electrons and photons, and a "hard" component of muons. Suppose the central core of a shower, at sea level, contains a narrow, vertical, parallel beam of muons of energy 1000 GeV, which penetrate underground. Assume the ionization loss in rock is constant at 2 MeV g^{-1} cm^2. Find the depth in rock at which the muons come to rest, assuming the rock density to be 3.0. Using the formula of the preceding problem, estimate their radial spread in meters, taking account of the change in energy of the muons as they traverse the rock. (Radiation length in rock = 25 g cm^{-2}.)

2.4 Show that in the head-on collision of a beam of relativistic particles of energy E_1 with one of energy E_2, the square of the energy in the center-of-momentum frame is $4E_1E_2$, and that for a crossing angle θ between the beams, this is reduced by a factor $(1 + \cos\theta)/2$. Show that the available kinetic energy in the head-on collision of two 25-GeV protons is equal to that in the collision of a 1300-GeV proton with a stationary nucleon.

2.5 A high-energy electron collides with an atomic electron. What is the threshold energy for production of an e^+e^- pair?

2.6 A proton of momentum $\mathbf{p}$, large compared with its rest mass M, collides with a proton inside a target nucleus, with Fermi momentum $\mathbf{p}_f$. Find the available kinetic energy in the collision, as compared with that for a free-nucleon target, when $\mathbf{p}$ and $\mathbf{p}_f$ are (a) parallel, (b) antiparallel, (c) orthogonal.

2.7 It is sometimes possible to differentiate between the tracks due to relativistic pions, protons, and kaons in a bubble chamber by virtue of the high-energy δ-rays which are produced. For a pion of momentum 5 GeV/c, what is the minimum energy of a δ-ray which must be observed to prove it is not produced by a kaon or proton? What is the probability of observing such a knock-on electron in 1 m of liquid hydrogen (density 0.06)? Refer to Problem 2.1 for formulae.

2.8 An experiment on proton decay is to be carried out using a large cubical tank of water as the proton source, and the possible decay mode $p \rightarrow e^+ + \pi^0$ is to be detected by the Čerenkov light emitted when the electromagnetic showers from the decay products traverse the water. How big should the water tank be in order to contain such showers? Estimate the total track-length integral (TLI) of the showers in a decay event and hence the total number of photons emitted in the visible region ($\lambda = 400–700$ nm). The light is to be detected by an array of photomultipliers placed at the water surfaces. If the optical transmission of the water is 20% and the photocathode efficiency is 15%, what fraction of the surface must be covered by photocathode to give an energy resolution of 10%?

BIBLIOGRAPHY

Bethe, A. A., and J. Ashkin, "Passage of radiations through matter", in Segre (ed.), *Experimental Nuclear Physics*, John Wiley, New York, 1953, Vol. 1, p. 166.

Blewett, M. H., "Characteristics of typical accelerators," *Ann. Rev. Nucl. Science* **17**, 427 (1967).

Bradner, H., "Bubble chambers", *Ann. Rev. Nucl. Science* **10**, 109 (1960).

Charpak, G., "Evolution of automatic spark chambers", *Ann. Rev. Nucl. Science* **20**, 195 (1970).

Courant, E. D., "Accelerators for high intensities and high energies", *Ann. Rev. Nucl. Science* **18**, 435 (1968).

Fabjan, C. W., and H. G. Fischer, "Particle detectors", *Rep. Prog. Physics* **43**, 1003 (1980).

Glaser, D., "The bubble chamber", in S. Fluegge (ed.), Encyclopaedia of Physics, Springer (Berlin), 1955, Vol. 45.

Hutchinson, G., "Čerenkov detectors", *Prog. Nucl. Phys.* **8**, 195 (1960).

Kleinknecht, K., "Particle detectors", Proc. Advanced Study Inst. on Techniques and Concepts in High Energy Physics, St Croix, USVI (ed. T. Ferbel), 1980.

Livingstone, M. S., and J. P. Blewett, *Particle Accelerators*, McGraw-Hill, New York, 1962.

McMillan, E. M., "Particle accelerators", in Segre (ed.), *Experimental Nuclear Physics*, John Wiley, New York, 1959, Vol. 3, p. 639.

Panofsky, W. K. H., "High energy physics horizons", *Physics Today*, **26** (June 1973).

Pellegrini, C., "Colliding beam accelerators", *Ann. Rev. Nucl. Science* **22**, 1 (1972).

Price, W. J., *Nuclear Radiation Detection*, McGraw-Hill, New York, 1964.

Rossi, B., *High Energy Particles*, Prentice-Hall, New York, 1952.

Segre, E., *Nuclei and Particles*, Benjamin, New York, 1977, Chapters 2–4.

Sternheimer, R. M., "Interaction of radiation with matter", in Yuan and Wu (eds.), *Methods of Experimental Physics*, Academic Press, 1961, Vol. 5A, p. 1. (This volume also contains articles on the different types of particle detector.)

Chapter 3

Invariance Principles and Conservation Laws

One of the most important concepts in physics is that of symmetry or invariance of the equations describing a system under an operation – which might be, for example, a translation or rotation in space. Intimately connected with such invariance properties are conservation laws – in the above cases, conservation of linear and angular momentum. A particular type of interaction is generally observed to obey many different conservation laws, so that the mathematical description of the interaction has to fulfil several invariance requirements, which severely restricts the possible forms of the interaction, and furthermore allows one for example to obtain relations between cross-sections for different processes mediated by that interaction. Thus, the conservation of isospin in strong interactions is equivalent to invariance under rotations in "isospin space", and leads to relations between cross-sections for the various possible charge states in pion-nucleon scattering.

In this chapter, we discuss the more important examples of conservation rules and invariance in particle physics. Invariance under space-time transformations in relativity is discussed in Appendix A on relativistic mechanics.

3.1 INVARIANCE IN CLASSICAL MECHANICS

The conservation laws of classical mechanics provide the most familiar examples of invariance. The most general equations of nonrelativistic motion for classical systems are those of Lagrange and Hamilton. They describe each of the n particles of an isolated system by six generalized coordinates, which may be chosen as the Cartesian position coordinates x, y, z, together with the momentum components p_x, p_y, p_z. These coordinates are usually denoted q_i for position and p_i for momentum, where $i = 1, 2, \ldots, 3n$. The Hamiltonian function H represents (under certain conditions) the total energy (kinetic plus potential) of the whole system, and the Hamilton equations of motion have the form

$$\dot{p}_i = \frac{dp_i}{dt} = -\frac{\partial H}{\partial q_i},$$

(3.1a)

$$\dot{q}_i = \frac{\partial H}{\partial p_i}. \tag{3.1b}$$

As a simple example, consider a single particle in a harmonic-oscillator potential in one dimension (so that $i = 1$), for which the total energy is

$$H = \frac{p^2}{2m} + Kx^2,$$

where K is the spring constant. For an isolated system, H cannot change with time, so $\partial H/\partial t = 0$, and thus $\dot{p}p/m + 2K\dot{x}x = 0$, or since $p = m\dot{x}$, $\dot{p} = -2Kx$. Hence, as in (3.1),

$$-\frac{\partial H}{\partial x} = -2Kx = \dot{p}$$

and

$$\frac{\partial H}{\partial p} = \frac{p}{m} = \dot{x}.$$

Suppose now that in the n-particle system described by (3.1), a translation δq is applied to all space coordinates:

$$q_i \rightarrow q_i + \delta q,$$

so that the change in total energy is

$$\delta H = \delta q \sum \frac{\partial H}{\partial q_i} = -\delta q \sum \dot{p}_i. \tag{3.2}$$

If the system is isolated, the momentum is a conserved quantity, so $\sum \dot{p}_i = 0$ and $\delta H = 0$. Thus conservation of momentum may equally be described as invariance of the Hamiltonian under a space translation. This is a commonsense result. If there are no external forces on the system, the energy cannot be changed by moving the whole system bodily in space. (Equally, if a system is subjected to an external field or force, its total momentum is not constant with time, and the total energy of the system will be changed under a space translation, as work will be done by the external field.) Had we taken the generalized coordinates as those of energy and time, we could similarly have shown that conservation of energy corresponds to invariance under time translations; or taking angular momentum and the angular coordinate about an axis, conservation of angular momentum would correspond to invariance of the system under rotations about that axis.

3.2 INVARIANCE IN QUANTUM MECHANICS

In quantum mechanics a system of particles is represented by a wavefunction or state vector ψ. The Schrödinger and Heisenberg equations of motion describe the temporal development of such a system. The result of a physical measurement on the system corresponds to the expectation value of some operator Q acting on the wavefunction and has the value $q = \int \psi^* Q \psi \, d\tau$. The development of q with time can be attributed either to the time dependence of ψ (Schrödinger representation) or equivalently to that of the operator Q (Heisenberg representation). For the former we write $\psi = \psi_s(t)$ in obvious notation. The Schrödinger equation of motion for ψ is

$$i\hbar \frac{\partial}{\partial t} \psi_s(t) = H\psi_s(t) \tag{3.3}$$

where H is the total energy, or Hamiltonian, operator: it gives the energy eigenvalues E of a stationary state, $H\psi = E\psi$. The time dependence $\psi_s(t)$ is clearly

$$\psi_s(t) = T(t, t_0)\psi_s(t_0) \tag{3.4}$$

with

$$T(t, t_0) = \exp[-i(t - t_0)H/\hbar] \tag{3.5}$$

and where it has been assumed that H does not depend on t explicitly. The operator T preserves the norm of the wavefunction and is unitary, with $T^{-1} \equiv T^* = \exp[i(t - t_0)H/\hbar]$ and $T^{-1}T = 1$. Thus the complex conjugate wavefunction satisfies

$$\psi_s(t)^* = \psi_s(t_0)^* T^{-1}(t, t_0). \tag{3.6}$$

The Heisenberg description attributes the time dependence to the operator Q rather than to ψ. Now the physically observable expectation value must be the same in either picture, so that

$$q = \int \psi_s(t_0)^* Q \psi_s(t_0) \, d\tau \equiv \int \psi_s(t)^* Q_0 \psi_s(t) \, d\tau,$$

$$\underset{\text{Heisenberg}}{} \qquad\qquad \underset{\text{Schrödinger}}{}$$

or from (3.4) and (3.6)

$$\psi_s(t_0)^* Q \psi_s(t_0) = \psi_s(t_0)^* T^{-1} Q_0 T \psi_s(t_0),$$

or

$$Q = T^{-1}Q_0 T, \tag{3.7}$$

giving an expression for the time dependence of the operator Q. So

$$ih\frac{dQ}{dt} = ih\frac{dT^{-1}}{dt}Q_0T + ihT^{-1}Q_0\frac{dT}{dt}$$

$$= -HT^{-1}Q_0T + T^{-1}Q_0TH$$

$$= -HQ + QH = [Q, H], \tag{3.8}$$

using (3.5) and (3.7). Here, $[Q, H]$ is the commutator of Q and H. If Q depends explicitly on the time, $\partial Q/\partial t \neq 0$, then (3.8) generalizes to

$$ih\frac{dQ}{dt} = ih\frac{\partial Q}{\partial t} + [Q, H], \tag{3.9}$$

which is the Heisenberg equation of motion for the operator Q. When $\partial Q/\partial t = 0$, we have $dQ/dt = 0$ if $[Q, H] = 0$. Thus an operator, not depending explicitly on the time, will be a constant of the motion if it commutes with the Hamiltonian operator. Generally, *conserved quantum numbers are associated with operators commuting with the Hamiltonian.*

3.3 TRANSLATIONS AND ROTATIONS

The transformations produced by quantum-mechanical operators are broadly of two types: continuous transformations (e.g. translations in space and time) and discrete transformations (e.g. spatial inversion through the origin). As an example of a continuous transformation we first discuss space translations, since these were treated in the classical case. The effect of an infinitesimal translation δr in space on a wavefunction ψ will be

$$\psi' = \psi(r + \delta r) = \psi(r) + \delta r\frac{\partial\psi(r)}{\partial r} = \left(1 + \delta r\frac{\partial}{\partial r}\right)\psi = D\psi,$$

where

$$D = \left(1 + \delta r\frac{\partial}{\partial r}\right) \tag{3.10}$$

is an infinitesimal space translation operator. Since the momentum operator is $p = -ih\,\partial/\partial r$, we can write this as

$$D = (1 - \delta r\,p/ih). \tag{3.11}$$

A finite translation Δr can be obtained by making n steps in succession ($\Delta r = n\,\delta r$), giving

$$D = \lim_{n\to\infty}\left(1 - \delta r\frac{p}{ih}\right)^n = \exp\left(\frac{ip\,\Delta r}{h}\right). \tag{3.12}$$

Thus D is a unitary operator, with $D^*D = D^{-1}D = 1$. The momentum operator p is called the generator of the operator D of space translations. If the

Hamiltonian H is independent of such space translations,

$$[D, H] = 0,$$

$$HD\psi = DH\psi, \tag{3.13}$$

$$D^{-1}HD\psi = D^{-1}DH\psi = H\psi.$$

From the form (3.11) it is clear that if D commutes with H, so also does the generator p:

$$[p, H] = 0. \tag{3.14}$$

Thus, if the Hamiltonian is invariant under space translations (3.13), then the momentum operator p (which generates these translations) commutes with the Hamiltonian, and the expectation value of p (the momentum of the system) is conserved. So the following statements are all equivalent:

(a) Momentum is conserved in an isolated system.
(b) The Hamiltonian is invariant under space translations.
(c) The momentum operator commutes with the Hamiltonian.

In complete analogy with the operator D of space translations (3.10), the generator of infinitesimal rotations about some axis may be written

$$R = 1 - \delta\phi \frac{\partial}{\partial\phi}. \tag{3.15}$$

The operator of the z-component of angular momentum is (see Appendix C)

$$J_z = -i\hbar\left(x\frac{\partial}{\partial y} - y\frac{\partial}{\partial x}\right) = i\hbar\frac{\partial}{\partial\phi},$$

where ϕ measures the azimuthal angle about the z-axis. So

$$R = 1 + \frac{i}{\hbar}J_z\,\delta\phi.$$

A finite rotation $\Delta\phi$ is obtained by repeating the infinitesimal rotation n times: $\Delta\phi = n\,\delta\phi$, where $n \to \infty$ as $\delta\phi \to 0$. Then

$$R = \lim_{n\to\infty}\left(1 + \frac{i}{\hbar}J_z\,\delta\phi\right)^n = \exp\left(\frac{i}{\hbar}J_z\,\Delta\phi\right). \tag{3.16}$$

Conservation of angular momentum about an axis corresponds to invariance of the Hamiltonian under rotations about that axis, and is also expressed by the commutation relation $[J_z, H] = 0$.

3.4 PARITY

The operation of spatial inversion of coordinates $(x, y, z \rightarrow -x, -y, -z)$ is an example of a discrete transformation. This transformation is produced by the parity operator P, where

$$P\psi(\mathbf{r}) \rightarrow \psi(-\mathbf{r}).$$

Repetition of this operation clearly implies $P^2 = 1$, so that P is a unitary operator. The eigenvalue of the operator (if there is one) will be ± 1, and this is also called the parity P of the system. A wavefunction may or may not have a well-defined parity, which can be even $(P = +1)$ or odd $(P = -1)$. For example, for

$$\psi = \cos x, \quad P\psi \rightarrow \cos(-x) = \cos x = +\psi: \qquad \text{even} \quad (P = +1),$$
$$\psi = \sin x, \quad P\psi \rightarrow \sin(-x) = -\sin x = -\psi: \quad \text{odd} \quad (P = -1),$$

while for

$$\psi = \cos x + \sin x, \quad P\psi \rightarrow \cos x - \sin x \neq \pm \psi,$$

so that the last function has no definite parity eigenvalue. As usual, the parity of a system will be a conserved quantum number if $[H, P] = 0$. For example, any spherically symmetric potential has the property that $H(-\mathbf{r}) = H(\mathbf{r}) = H(r)$, so that $[P, H] = 0$: the bound states of the system have definite parity. A familiar example is provided by the hydrogen-atom wavefunctions (neglecting spin effects), where the angular solutions are spherical harmonics (see Appendix, Table II):

$$\psi(r, \theta, \phi) = \chi(r) Y_l^m(\theta, \phi)$$

$$= \chi(r) \sqrt{\frac{(2l + 1)(l - m)!}{4\pi(l + m)!}} P_l^m(\cos \theta) e^{im\phi}. \qquad (3.17)$$

The spatial inversion $\mathbf{r} \rightarrow -\mathbf{r}$ is equivalent to

$$\theta \rightarrow \pi - \theta,$$
$$\phi \rightarrow \pi + \phi,$$

as in Fig. 3.1, with the result

$$e^{im\phi} \rightarrow e^{im(\pi + \phi)} = (-1)^m e^{im\phi},$$
$$P_l^m(\cos \theta) \rightarrow P_l^m(\cos(\pi - \theta)) = (-1)^{l+m} P_l^m(\cos \theta),$$

or

$$Y_l^m(\theta, \phi) \rightarrow Y_l^m(\pi - \theta, \pi + \phi) = (-1)^l Y_l^m(\theta, \phi). \qquad (3.18)$$

Thus the spherical harmonic functions have parity $(-1)^l$. So $s, d, g, \ldots$ atomic states have even parity, while $p, f, h, \ldots$ have odd parity. Electric dipole transitions between states are characterized by the selection rule $\Delta l = \pm 1$, so that, as a result of the transition, the parity of the atomic state must change. The parity of the electromagnetic $(E1)$ radiation (photons) emitted in this case must be -1, so that the parity of the whole system (atom + photon) is conserved.

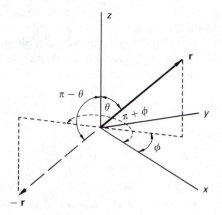

Fig. 3.1

Parity is a multiplicative quantum number, so that the parity of a composite system $\psi = \phi_a \phi_b \cdots$ is equal to the product of the parities of the parts.

In strong as well as electromagnetic interactions, parity is conserved. This is true, for example, in the reaction $p + p \rightarrow \pi^+ + p + n$ in which a single boson (pion) is created. In such a case, it is necessary to assign an *intrinsic parity* to the pion in order to ensure the same parity in initial and final states, in just the same way that one has assigned a charge to the pion in order to ensure charge conservation in the same reaction. The measurement of the intrinsic spin and parity of the pion is now described.

3.5 SPIN AND PARITY OF THE PION

3.5.1 Pion Spin

Before discussing the determination of the intrinsic parity of pions, we must discuss the experiments measuring the pion spin. For positive pions, the spin was determined by applying detailed balance to the reversible reaction

$$p + p \rightleftharpoons \pi^+ + d. \tag{3.19}$$

Applying Eq. (1.31) to the forward reaction, we obtain

$$\sigma_{pp \rightarrow \pi^+ d} = |M_{if}|^2 \frac{(2s_\pi + 1)(2s_d + 1)}{v_i v_f} p_\pi^2 \tag{3.20}$$

where p_π stands for the absolute value of the momentum $\mathbf{p}_\pi = -\mathbf{p}_d$ in the center-of-momentum system (CMS) and $s_d = 1$. For the backward reaction

$$\sigma_{\pi^+ d \to pp} = \tfrac{1}{2}|M_{fi}|^2 \frac{(2s_p + 1)^2}{v_f v_i} p_p^2. \tag{3.21}$$

At a given energy in the CMS, detailed balance ensures that the factors $|M_{if}|^2/v_i v_f$ for the forward and backward reactions will be the same. The factor $\tfrac{1}{2}$ in (3.21) arises from the fact that the integration of the phase-space factor (1.28) over 2π (rather than 4π) solid angle gives all the states when a pair of identical protons is involved. Thus the cross-section ratio

$$\frac{\sigma(pp \to \pi^+ d)}{\sigma(\pi^+ d \to pp)} = 2\frac{(2s_\pi + 1)(2s_d + 1)}{(2s_p + 1)^2}\frac{p_\pi^2}{p_p^2}. \tag{3.22}$$

The differential cross section $d\sigma/d\Omega$ for the reaction $p + p \to \pi^+ + d$ was first measured by Cartwright *et al.* (1953), using 340-MeV incident protons, corresponding to a pion CMS kinetic energy $T_\pi = 21.4$ MeV. The total cross section for the inverse reaction $\pi^+ + d \to p + p$ had previously been measured by Clark *et al.* (1951, 1952), with $T_\pi = 23$ MeV, giving $\sigma = 4.5 \pm 0.8$ mb, and by Durbin *et al.* (1951), with $T_\pi = 25$ MeV, giving $\sigma = 3.1 \pm 0.3$ mb. Integration of the Cartwright measurements predicted for this reaction $\sigma = 3.0 \pm 1.0$ mb for $s_\pi = 0$, and $\sigma = 1.0 \pm 0.3$ mb for $s_\pi = 1$. These data clearly showed that $s_\pi = 0$, and later experiments confirmed this.

For neutral pions, the existence of the decay

$$\pi^0 \to 2\gamma$$

proves that s_π must be integral (since $s_\gamma = 1$) and that $s_\pi \neq 1$. A photon has only two possible spin states, either parallel or antiparallel to the direction of motion (see Section 3.6). Taking the common line of flight of the photons in the π^0 rest frame as the quantization axis, the z-component of total photon spin can have the value $s_z = 0$ or 2. Suppose $s_\pi = 1$, then only $s_z = 0$ is possible. In this case the two-photon amplitude must behave under spatial rotations like the polynomial $P_1^m(\cos\theta)$ with $m = 0$, where θ is the angle relative to the z-axis. Under a $180°$ rotation about an axis normal to z, $\theta \to \pi - \theta$; since $P_1^0 \propto \cos\theta$, it therefore changes sign. For $s_z = 0$, the situation corresponds to two right circularly polarized (or two left circularly polarized) photons, traveling in opposite directions. The above rotation is therefore equivalent to interchange of the two photons, for which the wavefunction must be symmetric. Hence, the neutral-pion spin cannot be unity, and $s_\pi = 0$, or $\geqslant 2$. In high-energy nucleon-nucleon collisions, it is observed that neutral, positive and negative pions are produced in equal numbers, indicating that the neutral pion has spin zero, with the same spin multiplicity as the charged pions.

3.5.2 Pion Parity

The intrinsic parity of the charged pion has been determined from observations of absorption of slow negative pions in deuterium, leading to the following reactions:

$$\pi^- + d \rightarrow n + n \tag{3.23a}$$

$$\rightarrow 2n + \gamma. \tag{3.23b}$$

The existence of the reaction (3.23a) proves that the pion has odd parity. This follows from the fact that capture of the pion by the deuteron takes place from an atomic S-state, as is proved by studies of mesonic X-rays as well as by direct calculations. Since $s_d = 1$ and $s_\pi = 0$, the total angular momentum is then $J = 1$ on either side of the reaction. Here, $\mathbf{J} = \mathbf{L} + \mathbf{S}$, where L is the orbital angular momentum of the two final-state neutrons, and S their total spin. As indicated in (3.37) below, the triplet spin state of $S = 1$ is symmetric, while the singlet state $S = 0$ is antisymmetric. Hence the overall symmetry under interchange of the two identical neutrons will be $(-1)^{L+S+1}$, and this must be negative, from (1.12), so that $L + S$ is even. The overall angular momentum $J = 1$ requires either $L = 0$, $S = 1$, or $L = 1$, $S = 0$ or 1, or $L = 2$, $S = 1$. Of these, only $L = S = 1$ has $L + S$ even. So the only possibility is the $^3P_1$ state of two neutrons, with parity $(-1)^L = -1$, which requires negative parity also for the initial state.

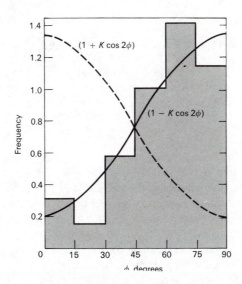

Fig. 3.2 Plot of weighted frequency distribution of angle ϕ between planes of polarization of the pairs in the "double Dalitz decay," $\pi^0 \rightarrow (e^+ + e^-) + (e^+ + e^-)$. For a scalar π^0, this should have the form $1 + K\cos 2\phi$, and for a pseudoscalar π^0, $1 - K\cos 2\phi$. (After Plano et al., 1959.)

Neutrons and protons are, by convention, assigned the same intrinsic parity + 1. (Since baryon number is conserved, the actual assignment is immaterial; the nucleon parities cancel in any reaction.) Thus the deuteron parity is even, and negative parity must be assigned to the pion.

The parity of the neutral pion has been established from observations of the γ-ray polarization in the decay $\pi^0 \to 2\gamma$. Using arguments identical to those in Section 3.11 for the 2γ-decay of positronium, odd parity for the pion would predict plane polarization vectors (**E** vectors) of the two photons to be preferentially orthogonal. Observations were actually made on the "double Dalitz decay"

$$\pi^0 \to (e^+ + e^-) + (e^+ + e^-),$$

in which each photon internally converts to a pair. The branching ratio, as compared with $\pi^0 \to 2\gamma$, is $\alpha^2 \sim 10^{-4}$. Since the plane of each electron-positron pair lies predominantly in the plane of the **E**-vector, measurement of the angular distribution between the plane of the pairs allows one to demonstrate the odd parity of the neutral pion (see Fig. 3.2).

In summary, the pion has zero spin and odd intrinsic parity, that is, $J^P = 0^-$. It is represented by a wavefunction which has the space transformation properties under inversions and rotations of a pseudoscalar. Such mesons are therefore called *pseudoscalar* mesons. By similar reasoning, particles of $J^P = 0^+$ are referred to as *scalar*, $J^P = 1^-$ as *vector*, and $J^P = 1^+$ as *axial-vector*.

3.6   PARITY OF PARTICLES AND ANTIPARTICLES

While the intrinsic parity of the proton is a matter of convention, the relative parity of proton and antiproton (or of any fermion and antifermion) is not. For example, one can produce a proton-antiproton pair in the reaction

$$p + p \to p + p + (p + \bar{p}),$$

so that, just as in the case of production of a single pion, the intrinsic parity of a nucleon-antinucleon pair is a measurable quantity. The Dirac theory of relativistic fermions (see Appendix F) predicts *opposite parity for fermion and antifermion*. This prediction has been verified experimentally, for example by observations on positronium (Section 3.12). For bosons, on the contrary, particles and antiparticles have the same intrinsic parity.

While the intrinsic parity assignment for the pion arises because pions can be created or destroyed singly, strange mesons must be created in association, for example in a reaction such as $p + p \to K^+ + \Lambda + p$. Thus, only the parity of a ΛK pair (relative to the nucleon) can be measured, and it is found to be odd. By convention, the Λ-hyperon is assigned the same (even) parity as the nucleon, so that of the kaon is odd.

3.7 TESTS OF PARITY CONSERVATION

While the purely strong and electromagnetic interactions are observed to be parity-conserving, the weak interactions are not. In such interactions, the matrix elements contain superpositions of amplitudes of even and odd parities. Thus nuclear β-decay is described by the so-called $V - A$ theory, in which the odd and even parity amplitudes have approximately the same magnitude. This is called the principal of maximal parity violation. A fuller discussion is given in Chapter 6.

In experimental studies of both strong and electromagnetic interactions, tiny degrees of parity violation are in fact observed. These arise, not from the breakdown of parity conservation in these interactions as such, but because the Hamiltonian describing the interaction inevitably contains contributions as well from the weak interactions between the particles concerned:

$$H = H_{\text{strong}} + H_{\text{electromagnetic}} + H_{\text{weak}}. \tag{3.24}$$

In nuclear transitions, the degree of parity violation will obviously be of the order of the ratio of weak to strong couplings, that is, $\sim 10^{-7}$ typically. As one example, we quote the observation of the circular polarization of γ-rays emitted in the reaction

$$n + p \rightarrow d + \gamma. \tag{3.25}$$

The observation of a net circular polarization (Lobashov et al. 1972) of value $P = (1.3 \pm 0.5) \times 10^{-6}$ clearly implies noninvariance under space inversion. As a second example, we cite the α-decay of the excited state of ^{16}O at 8.87 MeV:

$$\begin{array}{ccc} ^{16}O^* & \rightarrow & ^{12}C + \alpha, \\ J^P = 2^- & & J^P = 2^+ \end{array} \tag{3.26}$$

where the initial state is known to have odd parity, and the final state, even parity. The extremely narrow partial width for this decay, $\Gamma_\alpha = (1.0 \pm 0.3) \times 10^{-10}$ eV (Neubeck et al. (1974), is consistent with the magnitude expected from the parity-violating (weak-interaction) contribution, and may be contrasted with the width for γ-decay, $^{16}O^* \rightarrow {}^{16}O + \gamma$, of 3×10^{-3} eV. For more examples of parity violation in nuclear reactions, see the review by Tadic (1980).

Very small parity-violating effects have also been observed in atomic transitions. For example, a small (10^{-7} radian) rotation is observed to the plane of polarization of light traversing, and inducing optical transitions in, bismuth vapour. The parity violation arises through the interference between the weak-neutral-current and the purely electromagnetic contributions to the transition amplitude, and its magnitude is roughly in accord with that expected from the neutral-current couplings in the Weinberg-Salam-Glashow theory. For a recent review, see Fortson and Wilets (1981). Neutral-current effects are discussed in detail in Chapter 8.

3.8 CHARGE CONSERVATION, GAUGE INVARIANCE, AND PHOTONS

An important class of continuous transformations in particle physics are the gauge transformations, which are connected with conservation of electric charge. Electric charge is known to be very accurately conserved. For example, in the decay of the neutron, $n \to p + e^- + \bar{\nu}$, charge is conserved to an accuracy of better than 1 part in 10^{22} (see Problem 1.7).

The concept of gauge invariance and charge conservation may be introduced using an argument of Wigner (1949). In electrostatics, the potential ϕ of a system is arbitrary. The equations are always concerned with *changes* of potential, and independent of the absolute value of ϕ at any point in space. Suppose now that charge is not conserved, that it can be created or destroyed by some magic process. To create a charge Q, say, will require work W, which can be recovered when the charge is destroyed. Let the charge be created at a point where the potential on the chosen scale is ϕ. The work done will be W, independent of ϕ, since by hypothesis no physical process can depend on the absolute potential scale. If the charge is now moved to a point where the potential is ϕ', the energy change will be $Q(\phi - \phi')$. Thus, when the charge is destroyed, we recover the original system but have gained a net energy $W - W + Q(\phi - \phi')$. So, conservation of energy implies that we cannot create or destroy charge if the scale of electrostatic potential is arbitrary. In other words, conservation of electric charge allows us the freedom to choose potential scales at will.

In electromagnetism, the fields **B**, **E** can be expressed as derivates of vector and scalar potentials **A**, ϕ where

$$\mathbf{B} = \mathbf{V} \times \mathbf{A}, \tag{3.27}$$

$$\mathbf{E} = -\mathbf{V}\phi - \frac{1}{c}\frac{\partial \mathbf{A}}{\partial t}. \tag{3.28}$$

B and **E** satisfy Maxwell's equations, which incorporate the conservation of electric charge via a relation between charge and current densities: $\partial \rho / \partial t = -\mathbf{V} \cdot \mathbf{j}$. The values of **B** and **E** are invariant under a gauge transformation of the form

$$\mathbf{A} \to \mathbf{A}' = \mathbf{A} + \mathbf{V}\alpha, \tag{3.29}$$

$$\phi \to \phi' = \phi - \frac{1}{c}\frac{\partial \alpha}{\partial t}, \tag{3.30}$$

where α is *any* scalar function of space and time. This freedom to choose the value of α, called gauge invariance, means that **A**, ϕ can be defined in various ways. For example, the Coulomb gauge is chosen to fulfil

$$\mathbf{V} \cdot \mathbf{A} = 0. \tag{3.31}$$

Using (3.28) and the Maxwell equation $\mathbf{V} \cdot \mathbf{E} = 4\pi\rho$, one obtains Poisson's equation

$$\nabla^2\phi = -4\pi\rho,$$

so that in this gauge ϕ is determined just from the static charge distribution ρ.

In free space, the vector potential **A** obeys the wave equation

$$\nabla^2\mathbf{A} - \frac{1}{c^2}\frac{\partial^2\mathbf{A}}{\partial t^2} = 0, \tag{3.32}$$

corresponding to the propagation of free, massless photons (see Eq. (1.2)). This has a plane-wave solution

$$\mathbf{A} = \mathbf{e}A_0 \exp i(\mathbf{k}\cdot\mathbf{r} - \omega t),$$

where **k** is the propagation vector and **e** is a unit vector (polarization vector) giving the direction of the **E**-field. The x-component of **A** is

$$A_x = e_x A_0 \exp i(\mathbf{k}\cdot\mathbf{r} - \omega t)$$
$$= e_x A_0 \exp i(k_x x + k_y y + k_z z - \omega t),$$

and

$$\frac{\partial A_x}{\partial x} = ie_x k_x A_0 \exp i(\mathbf{k}\cdot\mathbf{r} - \omega t),$$

with similar expressions for the y- and z-components. Using (3.31),

$$\mathbf{\nabla}\cdot\mathbf{A} = \frac{\partial A_x}{\partial x} + \frac{\partial A_y}{\partial y} + \frac{\partial A_z}{\partial z} = 0,$$

we therefore obtain the relation

$$\mathbf{e}\cdot\mathbf{k} = 0. \tag{3.33}$$

Thus, the field **E** (and **B**) associated with a plane electromagnetic wave in free space is transverse to the propagation vector (Fig. 3.3). If one takes **k** along the z-axis, then e_z and A_z are zero, and one is left with two linearly independent components

$$A_x = e_x A_0 \exp i(kz - \omega t + \delta),$$
$$A_y = e_y A_0 \exp i(kz - \omega t),$$

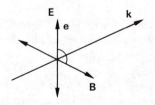

Fig. 3.3

where $e^2 = e_x^2 + e_y^2 = 1$, and where the amplitudes and the relative phase δ between the components are arbitrary. For example, if $\delta = 0$ one obtains plane polarization, and for $\delta = \frac{1}{2}\pi$ and $e_x = e_y$, circular polarization. The latter can be interpreted in terms of a rotating **e**-vector. For right and left circularly polarized waves, the combinations are

$$e_R = \frac{1}{\sqrt{2}}(e_x + ie_y),$$

$$e_L = \frac{1}{\sqrt{2}}(e_x - ie_y). \tag{3.34}$$

Note that $|e_R|^2 = |e_L|^2$ for all values of e_x and e_y, as required by parity conservation. The polarization vectors of the classical field can be associated with the spin states of the free photons of the field. For an infinite plane wave, propagating in the z-direction, it is obvious that since the momentum components $p_x = p_y = 0$, the z-component of orbital angular momentum $L_z = xp_y - yp_x = 0$. Thus the total angular momentum about the z-axis, J_z, must refer to the spin of the associated photons.

Consider now a rotation through angle θ about the z-axis, which can be obtained via the rotation operator (3.16), $R = \exp(iJ_z\theta)$ – where we have dropped the factor $\hbar$ for brevity. The transformed values of the components e_x, e_y, and e_z are

$$e'_x = e_x \cos\theta - e_y \sin\theta,$$

$$e'_y = e_x \sin\theta + e_y \cos\theta,$$

$$e'_z = e_z.$$

The first two equations can be rearranged as

$$e'_R = \frac{1}{\sqrt{2}}(e'_x + ie'_y) = \frac{1}{\sqrt{2}}(e_x + ie_y)e^{i\theta},$$

$$e'_L = \frac{1}{\sqrt{2}}(e'_x - ie'_y) = \frac{1}{\sqrt{2}}(e_x - ie_y)e^{-i\theta}.$$

Thus the states e_R, e_L, and e_z are eigenstates of the rotation operator $\exp(iJ_z\theta)$ with $J_z = +1, -1$, and 0 respectively, and correspond to the $2J + 1 = 3$ possible substates of a spin-1 photon. Because of the transversality condition (3.33), however, we have $e_z = 0$, and the $J_z = 0$ substate of a free photon does not exist. In general, Lorentz invariance allows only two substates for a massless particle of spin J: $J_z = \pm J$.

Real photons travel at the speed of light and are therefore strictly massless (the experimental limit on the photon mass is $< 10^{-47}$ g (Goldhaber and Nieto 1971)). They are called *transverse* photons, meaning that the associated **E** and

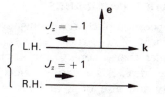

Transverse photon ($m = 0$)

Longitudinal photon ($m \neq 0$)

Fig. 3.4

B fields are perpendicular to the propagation vector **k**. However, electromagnetic disturbances can also travel at $v < c$ (for example, in a waveguide), and both longitudinal and transverse field components are then possible. In such a situation, the photons cannot be exactly massless. They have a spin component $J_z = 0$ as well as ± 1 (Fig. 3.4). A photon with $J_z = 0$ is called *longitudinal* (or scalar). An example of a longitudinal photon is the virtual photon mediating the static interaction between two charges, where the **e** vector lies along the line joining the charges.

The spin polarization of particles is often referred to by the term *helicity*. A particle with spin vector **σ**, momentum **p**, and energy E has helicity

$$H = \frac{\boldsymbol{\sigma} \cdot \mathbf{p}}{E|\sigma|}. \tag{3.35}$$

Thus right-handed, left-handed, and scalar photons have $H = +1, -1, 0$ respectively. Since, under space inversion, $\boldsymbol{\sigma} \cdot \mathbf{p}$ changes sign (see Table 3.2), the net helicity of photons associated with the parity-conserving electromagnetic interactions must be zero: so right-handed and left-handed photons always occur with equal amplitudes, and the term "transverse" refers to both helicity states taken together.

To summarize this section: gauge invariance is a property of electromagnetic fields associated with conservation of electric charge. It leads to the transversality condition for plane electromagnetic waves in free space. As a result, free photons—the spin-1 carriers of a vector field—can exist only in two substates and must be massless.

More complicated gauge transformations, of a type introduced by Yang and Mills (1954), are considered when both charged and neutral fields (bosons) are involved. This is the case for the unified gauge theories of electromagnetic and weak interactions, discussed in Chapter 8.

3.9   CHARGE-CONJUGATION INVARIANCE

As the name implies, the operation of charge conjugation reverses the sign of charge and magnetic moment of a particle (leaving all other coordinates unchanged). Symmetry under charge conjugation in classical physics is evidenced by the invariance of Maxwell's equations under change in sign of the charge and current density and also of **E** and **H**. In relativistic quantum mechanics the term "charge conjugation" also implies the interchange of particle and antiparticle. For baryons and leptons, a reversal of charge entails a change in sign of the baryon number or lepton number. As examples, we show in Table 3.1 the effects of charge conjugation on electron and proton. Note that μ positive (negative) means that the magnetic moment is parallel (antiparallel) to the spin vector.

TABLE 3.1   Charge conjugation

	Proton	Antiproton
Q	$+e$	$-e$
B	$+1$	-1
μ	$+2.79(e\hbar/2Mc)$	$-2.79(e\hbar/2Mc)$
σ	$\frac{1}{2}\hbar$	$\frac{1}{2}\hbar$

	Electron	Positron
Q	$-e$	$+e$
L_e	$+1$	-1
μ	$-e\hbar/2mc$	$+e\hbar/2mc$
σ	$\frac{1}{2}\hbar$	$\frac{1}{2}\hbar$

Experimental evidence for invariance of the strong and electromagnetic interactions under the C-operation is discussed in Section 3.13. On the contrary, weak interactions violate charge-conjugation invariance just as they violate invariance under the parity operation. These effects are discussed in detail in Chapter 6, but it is perhaps worthwhile mentioning them briefly at this point. The noninvariance of weak interactions under the parity and charge-conjugation operations is exemplified by the longitudinal polarization of neutrinos (ν) and antineutrinos ($\bar{\nu}$) emitted in β-decay, in company with positrons and electrons respectively. Neutrinos have spin $\frac{1}{2}$ and zero (or almost zero) mass, so that, as pointed out in Section 3.8, the possible spin eigenstates are of $J_z = \pm\frac{1}{2}$, where z denotes the direction of the momentum vector **p**. Experimentally, it is found that for neutrinos $J_z = -\frac{1}{2}$ only, and for antineutrinos $J_z = +\frac{1}{2}$ only. In other words, neutrinos are "left-handed" and

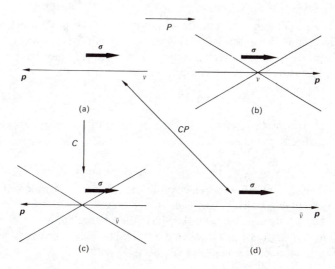

Fig. 3.5 Results of the C- and P-operations on neutrino states. Only states (a) and (d) are observed in nature.

antineutrinos "right-handed". This is shown in Figure 3.5, where **p** is along the negative z-axis in (a). Under the parity operation, which in this case corresponds to the inversion $z \to -z$, the polar vector **p** changes sign, whereas, as Table 3.2 indicates, the axial vector **σ** remains unchanged. This situation is shown in diagram (b). It corresponds to a right-handed neutrino, which does not exist in nature. This tells us that the weak interaction is not invariant under space inversions. One can contrast this with photons emitted in electromagnetic processes. Both right-handed and left-handed photons exist, but they occur with equal probability, so that the interaction is invariant under a spatial inversion.

Similarly, if we apply the charge-conjugation operation to the v-state (a), we obtain a left-handed $\bar{v}$, as in (c). This also does not exist. However, if, in addition, we make the spatial inversion of this state, we end up with a right-handed antineutrino (d), which *is* observed. Thus the weak interactions are not invariant under C or P separately, but do exhibit CP-invariance. Although we come to this conclusion by considering neutrino states, it may be remarked that it is a general property of all weak interactions, whether they involve neutrinos or not.*

3.10 EIGENSTATES OF THE CHARGE-CONJUGATION OPERATOR

Consider the operation of charge conjugation performed on a charged-pion wave function, which we write as $|\pi^{\pm}\rangle$:

$$C|\pi^{+}\rangle \to |\pi^{-}\rangle \neq \pm |\pi^{+}\rangle.$$

---

* The CP-invariance of weak interactions is not in fact exact. The small departures from CP-invariance are discussed in Chapter 6.

In this operation, an arbitrary phase may enter; this is not important for the present discussion. We note that $|\pi^+\rangle$ and $|\pi^-\rangle$ are *not* C-eigenstates. However, for a neutral system, the charge-conjugation operator may have a definite eigenvalue. Thus, for the neutral pion,

$$C|\pi^0\rangle = \eta|\pi^0\rangle,$$

since the π^0 transforms into itself; η is a constant. Clearly, repeating the operation gives us $\eta^2 = 1$, so

$$C|\pi^0\rangle = \pm 1|\pi^0\rangle.$$

To find the sign, we note that electromagnetic fields are produced by moving charges (currents) which change sign under charge conjugation. As a consequence, the photon has $C = -1$. Since the charge-conjugation quantum number is multiplicative, this means that a system of n photons has C-eigenvalue $(-1)^n$. The neutral pion undergoes the decay

$$\pi^0 \to 2\gamma,$$

and thus has even C-parity:

$$C|\pi^0\rangle = +|\pi^0\rangle.$$

The decay

$$\pi^0 \to 3\gamma$$

will therefore be forbidden if the electromagnetic interactions are invariant under C. Experimentally, the branching ratio

$$\frac{\pi^0 \to 3\gamma}{\pi^0 \to 2\gamma} < 5 \times 10^{-6}.$$

3.11 POSITRONIUM DECAY

We now consider the restrictions imposed by C-invariance on the states of *positronium*, which undergoes annihilation in the modes

$$e^+e^- \to 2\gamma, 3\gamma.$$

The bound state of electron and positron possesses energy levels similar to the hydrogen atom (but with about half the spacing, because of the factor 2 in the reduced mass). We write down the total wave function of the positronium state as the product of three wave functions depending on the spin, space, and charge coordinates:

$$\psi(\text{total}) = \Phi(\text{space})\,\alpha(\text{spin})\,\chi(\text{charge}) \qquad (3.36)$$

as in Eq. (1.14), and consider how these functions behave under particle interchange. If electron and positron had been identical fermions, rather than fermion and antifermion, we should have been able to conclude that ψ(total) was antisymmetric, as in Eq. (1.12).

The spin functions for a combination of two spin-$\frac{1}{2}$ particles are written as follows. Denoting these as $\psi_1(s, s_z)$ and $\psi_2(s, s_z)$ for particles 1 and 2, and that of the combination by $\alpha(S, S_z)$ (where s, S refer to spins and s_z, S_z to third components), the four possible combinations are

$$\alpha(1, 1) = \psi_1(\tfrac{1}{2}, \tfrac{1}{2})\psi_2(\tfrac{1}{2}, \tfrac{1}{2}), \tag{3.37a}$$

$$\alpha(1, 0) = \frac{1}{\sqrt{2}}[\psi_1(\tfrac{1}{2}, \tfrac{1}{2})\psi_2(\tfrac{1}{2}, -\tfrac{1}{2}) + \psi_2(\tfrac{1}{2}, \tfrac{1}{2})\psi_1(\tfrac{1}{2}, -\tfrac{1}{2})], \tag{3.37b}$$

$$\alpha(1, -1) = \psi_1(\tfrac{1}{2}, -\tfrac{1}{2})\psi_2(\tfrac{1}{2}, -\tfrac{1}{2}), \tag{3.37c}$$

$$\alpha(0, 0) = \frac{1}{\sqrt{2}}[\psi_1(\tfrac{1}{2}, \tfrac{1}{2})\psi_2(\tfrac{1}{2}, -\tfrac{1}{2}) - \psi_2(\tfrac{1}{2}, \tfrac{1}{2})\psi_1(\tfrac{1}{2}, -\tfrac{1}{2})], \tag{3.37d}$$

where the first three form a spin triplet of $S = 1$ and $S_z = 1, 0, -1$, and the last is a singlet of $S = S_z = 0$. The triplet states have the property that they are *symmetric* under particle label interchange $1 \leftrightarrow 2$ (α does not change sign), while the singlet state is *antisymmetric* (α changes sign). Thus, the symmetry of the spin function α under particle interchange is $(-1)^{S+1}$, where S is the total spin.

The space wavefunction Φ is expressed as a spherical harmonic $Y_l^m(\theta, \phi)$ as in Eq. (3.17). Particle interchange is equivalent to space inversion, introducing a factor $(-1)^l$, where l is the orbital angular momentum of the system.

Finally, let the charge wavefunction χ acquire a factor C under interchange. The product of the factors applying to the separate spin, space, and charge functions must then be that of the total wavefunction ψ, which we denote by K, so that

$$K = C(-1)^{S+1}(-1)^l. \tag{3.38}$$

To find C and K, we must appeal to additional information. As indicated before, two separate decay modes, into two and three γ-rays respectively, are observed for positronium annihilation from the ground state ($l = 0$). These modes obviously must correspond to the two possible spin states, the singlet ($J = 0$) and triplet ($J = 1$). By appealing either to the Bose symmetry of the two-photon system or to the masslessness of the photon (Fig. 3.4), we know that the 2γ decay must have $J = 0$, so the 3γ decay has to be assigned $J = 1$. Thus, from the formula $C = (-1)^n$ for a system of n photons, we find $C = +1$ for the $J = 0$ state and $C = -1$ for the $J = 1$ state. Inserting the appropriate factors in (3.38), we then obtain

Decay	$S = J$	l	C	K	Lifetime τ (sec)
Singlet ($^1S_0$) 2γ	0	0	$+1$	-1	1.25×10^{-10}
Triplet ($^3S_1$) 3γ	1	0	-1	-1	1.5×10^{-7}

We note that the factor acquired by the total wavefunction ψ is $K = -1$ under interchange, i.e., if we take account of the charge-conjugation operation, the total wavefunction is antisymmetric, just as for two identical fermions.

The annihilation rate to two γ-rays has been calculated to be

$$\frac{1}{\tau(2\gamma)} = 4\pi r_e^2 c |\psi(0)|^2, \tag{3.39}$$

where $r_e = e^2/4\pi mc^2$ is the classical electron radius, $\psi(0)$ is the amplitude of the electron-positron radial wavefunction at the origin. From the solution of the Schrödinger equation for the ground state of the hydrogen atom, we know that

$$|\psi(0)|^2 = \frac{1}{\pi a^3}, \tag{3.40}$$

where a is the Bohr radius. States of angular momentum l contain a factor r^l in the radial wave function – hence, for all except the ground state, they vanish at the origin. Remembering the factor 2 for the reduced mass effect, the Bohr radius in positronium will have a value

$$a = \frac{2r_e}{\alpha^2}. \tag{3.41}$$

From (3.39), (3.40), and (3.41), and using the values $r_e = 2.8 \times 10^{-13}$ cm, $\alpha^{-1} = 4\pi\hbar c/e^2 = 137$, one obtains for the mean lifetime,

$$\tau(2\gamma) = \frac{2r_e}{c\alpha^6} = 1.25 \times 10^{-10} \text{ sec.} \tag{3.42}$$

For the 3γ-decay, the annihilation rate is slower by a factor of order α. The calculated value is

$$\tau(3\gamma) = \frac{9\pi}{4(\pi^2 - 9)} \frac{\tau(2\gamma)}{\alpha} = 1.4 \times 10^{-7} \text{ sec.} \tag{3.43}$$

Both the 2γ and 3γ periods were first detected in the work of Deutsch (1953), who measured the annihilation rates of positrons stopping in gases. The long period due to the 3γ-mode was found to be $(1.45 \pm 0.15) \times 10^{-7}$ sec, in agreement with (3.43).

3.12   PHOTON POLARIZATION IN POSITRONIUM DECAY

To complete the discussion of the quantum numbers of positronium, we discuss the polarization of the two γ-rays emitted in the decay of the singlet state ($J = 0$). The electron and positron are fermion and antifermion, and as such must have opposite intrinsic parities (see Appendix F). The relative parity of e^+ and e^- will not affect the foregoing analysis, since under particle interchange, the total parity stays the same. However, it does imply that the two γ-rays emitted must have odd parity.

Let $\mathbf{k}$, $-\mathbf{k}$ be the momentum vectors of the two photons, and $\mathbf{e}_1$ and $\mathbf{e}_2$ their polarization vectors (**E**-vectors). The initial state has $J = 0$, and the simplest linear combinations one can form to include both **E**-vectors and satisfy requirements of exchange symmetry for identical bosons are

$$\psi(2\gamma) = a(\mathbf{e}_1 \cdot \mathbf{e}_2), \tag{3.44a}$$

and

$$\psi(2\gamma) = b(\mathbf{e}_1 \times \mathbf{e}_2) \cdot \mathbf{k}, \tag{3.44b}$$

where a and b are constants. The first product is a scalar and is therefore even under space inversions. The second is a pseudoscalar (the product of a polar vector with an axial vector) and therefore has the odd parity required. If $\psi(2\gamma)$ is to be finite, $\mathbf{e}_1$ and $\mathbf{e}_2$ cannot be parallel. Thus the planes of polarization of the two photons should be preferentially at right angles, the probability of observing an angle ϕ between the planes being from (3.44b), $|\psi|^2 \propto \sin^2 \phi$. Experimentally, this can be investigated by observing the angular distribution of the Compton scattering of the γ-rays, which depends strongly on polarization, being more probable in a plane normal to the electric vector. This is clear if one considers that the incident photon sets up oscillations of the target electron in the direction of the **E**-vector. In the subsequent dipole radiation of the scattered photon, the radiated intensity is greatest normal to **E**.

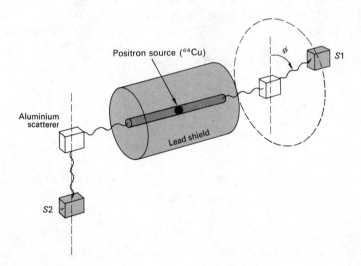

Fig. 3.6 Sketch of the method used by Wu and Shaknov (1950) to measure the relative orientation of the polarization vectors of the two photons emitted in decay of $^1S_0$-positronium. $S1$ and $S2$ are anthracene counters, recording the γ-rays after Compton scattering by aluminum cubes. The results proved that fermion and antifermion have opposite intrinsic parity, as predicted by the Dirac theory.

The experimental setup used by Wu and Shaknov (1950) is shown in Fig. 3.6. The coincidence rate of γ-rays scattered by aluminum blocks was measured in the anthracene counters $S1$ and $S2$ as a function of their relative azimuthal angle ϕ. The expected ϕ-dependence depends on the (polar) angle of scattering, θ, of the γ-rays by the cubes, and is maximum for $\theta \sim 81°$ (Pryce and Ward 1947, Snyder *et al.* 1948). The observed ratio was

$$\frac{\text{rate}(\phi = 90°)}{\text{rate}(\phi = 0°)} = 2.04 \pm 0.08, \tag{3.45}$$

compared with a theoretically expected value of 2.00. Such experiments therefore bear out the prediction of preferentially orthogonal polarizations of the γ-rays, and thus demonstrate the correctness of the assumption that fermions and antifermions have opposite intrinsic parities. Similar arguments have been made to demonstrate that the neutral pion has odd intrinsic parity (Fig. 3.2).

The argument used above represents the properties of each photon by a momentum vector **k** and a plane-polarization vector **e** normal to **k**, giving the direction of the associated electric (or magnetic) field. An alternative and equivalent discussion is possible in terms of the spin vectors of the photons. As in Fig. 3.4, a photon may be either right-handed (R) or left-handed (L) according as the spin points in the direction of motion or against it. This corresponds to the classical description of circularly polarized light with rotating **e**-vectors. We write RR to denote the amplitude for two right-handed photons, LL for two left-handed, and RL for a right-left combination. RL and LR obviously correspond to $J_z = 2$, while LL or RR have $J_z = 0$, where J_z is the total photon spin projection along the propagation vector **k**. To find eigenstates of the parity operator P, we try the linear combinations

$$|\alpha\rangle = RR + LL,$$

where

$$P|\alpha\rangle = LL + RR = + |\alpha\rangle,$$

and

$$|\beta\rangle = RR - LL,$$

where

$$P|\beta\rangle = LL - RR = - |\beta\rangle.$$

Thus the state $|\beta\rangle = RR - LL$ has odd parity, and can also be written $|\beta\rangle = a_+ a_-$, where $a_+ = R + L$ and $a_- = R - L$ are the decompositions, along the directions $+ \mathbf{k}$ and $- \mathbf{k}$ respectively, of each photon into a superposition of left- and right-handed states. Each combination of circular polarizations is of course equivalent to a plane polarization, as in (3.34). Since $R + L \propto e_x$ and $R - L \propto e_y$, these plane-polarization vectors must again be at right angles.

3.13   EXPERIMENTAL TESTS OF C-INVARIANCE

Experimental tests of charge conjugation invariance compare reactions in which particles are replaced by their antiparticles. For example, in strong interactions comparisons have been made of the rates and spectra of positive and negative mesons in the reactions

$$p + \bar{p} \rightarrow \pi^+ + \pi^- + \cdots$$
$$\rightarrow K^+ + K^- + \cdots,$$

and any possible violation of C-invariance has been found to be $\ll 1\%$.

The search for possible C-violation in electromagnetic interactions has an interesting history. In 1964, a small (0.1%) violation of CP invariance was observed in the decay of neutral kaons (see Section 6.13.3). The origin was unknown, but clearly such an effect might imply C-violation in electromagnetic (or strong) interactions, since in these reactions parity is exactly conserved. An intensive analysis was therefore made of the decay products of the η-meson (mass 550 MeV/c^2), which decays through electromagnetic transitions. Some possible decay modes are

$$\eta \rightarrow \gamma\gamma \qquad\qquad (3.46a)$$
$$\rightarrow \pi^+ \pi^- \pi^0 \qquad\qquad (3.46b)$$
$$\rightarrow \pi^+ \pi^- \gamma \qquad\qquad (3.46c)$$
$$\rightarrow \pi^0 e^+ e^-. \qquad\qquad (3.46d)$$

The decay $\eta \rightarrow \gamma\gamma$, with a branching ratio of 38%, establishes the C-parity of the η to be $+ 1$. Hence, the decay mode $\eta \rightarrow \pi^0 e^+ e^-$ should be forbidden, if we interpret it as $\eta \rightarrow \pi^0\gamma$ with internal conversion of the γ-ray to an $e^+ e^-$ pair, since $C_\gamma = - 1$ and $C_{\pi^0} = + 1$. The existence of this decay mode would therefore be definite evidence for C-violation. Present limits for the branching ratio are $< 5 \times 10^{-4}$.

Unfortunately, the nonexistence of the decay (3.46d) does not unambiguously prove the absence of C-violation, since the interpretation of the limit is model-dependent. A better test is the comparison of the π^+ and π^- spectra in the decays $\eta \rightarrow \pi^+\pi^-\gamma$ and $\eta \rightarrow \pi^+\pi^-\pi^0$. Indeed, early experiments showed an apparently significant asymmetry, but subsequent and more careful work verified the equivalence of π^+ and π^- spectra down to the 0.5% level.

In summary, during the later 1960s a great deal of effort went into investigation of possible C-violation in strong and electromagnetic interactions, but no such violations have been detected.

3.14   TIME-REVERSAL INVARIANCE

Both invariance and noninvariance under time reversal are familiar in classical physics. For example, Newton's law $F = m\,d^2x/dt^2$ is invariant under change of

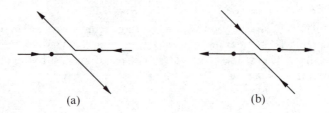

(a)   (b)

Fig. 3.7 (a) Collision between two molecules. (b) Time-reversed collision.

sign of the time coordinate: a film of the trajectory of a projectile in the earth's gravitational field looks equally realistic whether we run it backward or forward (if we neglect air resistance). The situation is different for the laws of heat conduction or diffusion, which depend on the first derivative of the time coordinate. In this case, the elementary collisions between molecules do obey the principle of microscopic reversibility. For each collision, there exists a time-reversed collision as in Fig. 3.7. For a gas in equilibrium, the two types of collision occur with equal probability and the entropy is constant. A film of an order-disorder transition, such as of gas containing many molecules expanding through a nozzle to a region of low pressure, does specify an arrow of time and looks unreal if run in the reverse direction. This is simply a consequence however of the initial conditions and has nothing to do with time-reversal invariance in elementary collisions.

The transformations of common quantities in classical physics under space inversion P and time reversal T are given in Table 3.2. We note that an "elementary" particle of spin $\boldsymbol{\sigma}$ is not expected to possess a static electric dipole moment, if the interaction of the particle with the electromagnetic field is to be invariant under the T- or P-operations. At first sight this may seem strange, in

TABLE 3.2

Quantity	T	P	
$\mathbf{r}$	$\mathbf{r}$	$-\mathbf{r}$	
$\mathbf{p}$	$-\mathbf{p}$	$-\mathbf{p}$	Polar vector
$\boldsymbol{\sigma}$ (spin)	$-\boldsymbol{\sigma}$	$\boldsymbol{\sigma}$	Axial vector $(\mathbf{r} \times \mathbf{p})$
$\mathbf{E}$ (electric field)	$\mathbf{E}$	$-\mathbf{E}$	$(\mathbf{E} = -\partial V/\partial \mathbf{r})$
$\mathbf{B}$ (magnetic field)	$-\mathbf{B}$	$\mathbf{B}$	(As for $\boldsymbol{\sigma}$; e.g. consider a ring current)
$\boldsymbol{\sigma} \cdot \mathbf{B}$	$\boldsymbol{\sigma} \cdot \mathbf{B}$	$\boldsymbol{\sigma} \cdot \mathbf{B}$	Magnetic dipole moment
$\boldsymbol{\sigma} \cdot \mathbf{E}$	$-\boldsymbol{\sigma} \cdot \mathbf{E}$	$-\boldsymbol{\sigma} \cdot \mathbf{E}$	Electric dipole moment
$\boldsymbol{\sigma} \cdot \mathbf{p}$	$\boldsymbol{\sigma} \cdot \mathbf{p}$	$-\boldsymbol{\sigma} \cdot \mathbf{p}$	Longitudinal polarization
$\boldsymbol{\sigma} \cdot (\mathbf{p}_1 \times \mathbf{p}_2)$	$-\boldsymbol{\sigma} \cdot (\mathbf{p}_1 \times \mathbf{p}_2)$	$\boldsymbol{\sigma} \cdot (\mathbf{p}_1 \times \mathbf{p}_2)$	Transverse polarization

view of the fact that molecules and even atoms may possess large electric dipole moments, while the interactions involved are surely invariant under the T and P operations. In the case of molecular or atomic systems, electric dipole moments are always associated with the existence of two degenerate, or nearly degenerate, energy eigenstates. As an example, the first excited state of the hydrogen atom ($n = 2$) contains $s_{1/2}$ and $p_{1/2}$ levels which are degenerate in energy (apart from a small effect associated with the Lamb shift). These states have opposite parity, but, under an applied electric field E, a Stark-effect mixing occurs. The new energy eigenstates are linear combinations of the s ($l = 0$) and p ($l = 1$) states and clearly do not have a definite parity. They are associated with an asymmetric (i.e. pear-shaped) electron charge distribution, and hence an electric dipole moment μ_e and a corresponding energy perturbation $\pm \mu_e E$. (No states are *exactly* degenerate: what we mean here is that the original energy separation between the levels is small compared with $\mu_e E$).

In the case of a neutron, for example, an electric dipole moment could only arise, in the absence of P, T violation, if there existed a "ghost" neutron state of opposite parity degenerate with the first. A superposition of these states could result in a "handedness" and allow an asymmetric charge distribution and thus an electric dipole moment (see Fig. 3.9). There is absolutely no evidence for two types of neutron state; if such existed, the whole of nuclear structure physics

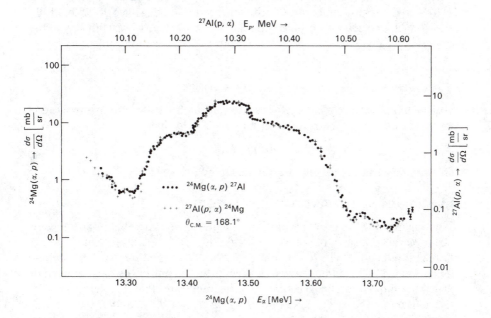

Fig. 3.8 The differential cross sections for the reaction ^{24}Mg$(\alpha, p)^{27}$Al and its inverse, as measured by Von Witsch *et al.* (1968).

would be quite different. We consider therefore that for the neutron we are dealing with one state of fixed parity. The spin vector $\boldsymbol{\sigma}$ prescribes the only possible direction in space of any electric (or magnetic) dipole moment (i.e. $\boldsymbol{\mu}_e$ or $\boldsymbol{\mu}_m$ must be parallel or antiparallel to $\boldsymbol{\sigma}$). The measurement of any neutron electric dipole moment is therefore an important test of T-invariance, and is discussed in Section 3.16 below.

Experimentally, time reversal has been verified in strong interactions by application of the principle of detailed balance, as explained in Section 1.14. Figure 3.8 shows the results of a test of detailed balance in the reaction

$$p + {}^{27}\text{Al} \rightleftarrows \alpha + {}^{24}\text{Mg},$$

from which it was concluded that the T-violating amplitude was $< 0.3\%$ of the T-conserving amplitude.

Another consequence of T-invariance is the so-called polarization-asymmetry equality in pp elastic scattering. In scattering through an angle θ, from an unpolarized target, an initially unpolarized proton beam will acquire a polarization

$$P(\theta) = \frac{N_+ - N_-}{N_+ + N_-},$$

where N_+ and N_- represent the number of protons with spin up and spin down relative to the scattering plane. On the other hand, if one starts off with a fully (transversely) polarized proton beam, one obtains a right-left asymmetry in the scattering at angle θ in the plane normal to the polarization direction:

$$A(\theta) = \frac{N_R - N_L}{N_R + N_L},$$

where R and L refer to right and left scattering angles. If the strong interactions are invariant under time reversal, it may be shown that

$$A(\theta) = P(\theta).$$

This equality has been verified to an accuracy of $\simeq 1\%$.

3.15 *CP*-VIOLATION AND THE *CPT* THEOREM

Experimental tests of T-invariance have acquired considerable importance in recent years, because of the observed breakdown of CP-invariance in K^0-decay in weak interactions. The reason for this is that the T- and CP-transformations are connected by the famous *CPT theorem*, which is one of the most important principles of quantum field theory. This theorem states that *all* interactions are invariant under the succession of the three operations C, P, and T taken in any order. The proof of this theorem is based on very general assumptions and is cherished by theorists in the sense that it is difficult to formulate field theories which are not automatically CPT-invariant. However, the CPT theorem is not on quite such experimentally solid ground as conservation of energy, for example, and bearing in mind the history of long-respected conservation laws which have fallen by the wayside, we must make experimental checks.

Some consequences of the CPT theorem which may be verified experimentally relate to the properties of particles and antiparticles, which should have the same mass and lifetime, and magnetic moments equal in magnitude but opposite in sign. These results would follow from charge-conjugation invariance alone, if it held universally. However, weak interactions are not C-invariant, so that the prediction rests on the more general theorem.

The experimental consequences of the CPT theorem seem to be well verified. Current results are as shown in Table 3.3. The best experimental limit comes from the K^0-$\bar{K}^0$ comparison.

Another consequence of the CPT theorem is that particles would be expected to obey the "normal" spin-statistics relationship, i.e. integral spin and half-integral spin particles should follow Bose and Fermi statistics respectively.

Until 1964, it was believed that all types of interaction were invariant under the combined operation CP. Weak interactions were known to violate C- and P-invariance separately, but to respect the CP-symmetry. In that year, however, it was discovered by Christenson $et\ al.$ (1964) that the long-lived neutral K-particle, which normally decays by a weak interaction into three pions of CP eigenvalue -1, could occasionally (with probability 2×10^{-3}) decay into two pions, of $CP = +1$. These results are discussed in detail in Section 6.13.3.

The origin of CP-violation is not at present established, although it is a feature of theories of fundamental interactions incorporating at least six quark flavors. If one believes in the "big bang" theory of the universe with no initial asymmetry, the results of CP-violation (or T-violation) are all around us: the universe has evolved with time in such a way as to result in a huge preponderance of matter over antimatter. But the discussion of these questions has to be

TABLE 3.3 Tests of CPT theorem

		Limit on fractional difference
Lifetime	$\tau_{\pi^+} - \tau_{\pi^-}$	$< 10^{-3}$
	$\tau_{\mu^+} - \tau_{\mu^-}$	$< 2 \times 10^{-3}$
	$\tau_{K^+} - \tau_{K^-}$	$< 10^{-3}$
Magnetic moment	$\lvert\mu_{\mu^+}\rvert - \lvert\mu_{\mu^-}\rvert$	$< 3 \times 10^{-9}$
	$\lvert\mu_{e^+}\rvert - \lvert\mu_{e^-}\rvert$	$< 10^{-5}$
Mass	$M_{\pi^+} - M_{\pi^-}$	$< 10^{-3}$
	$M_{\bar{p}} - M_p$	$< 8 \times 10^{-3}$
	$M_{K^+} - M_{K^-}$	$< 10^{-3}$
	$M_{K^0} - M_{\bar{K}^0}$	$< 10^{-14}$

deferred until Chapter 8. As indicated above, the magnitude of the violation of CP-symmetry is very small and so far has only been (and possibly will only be) observed in the laboratory using the very precise "interferometer" provided by the $K^0 \bar{K}^0$ system. Nevertheless, attempts have been made to find evidence of T-violation in other processes, and we next discuss an experiment on the neutron electric dipole moment.

3.16 ELECTRIC DIPOLE MOMENT OF THE NEUTRON

We previously noted that the existence of an electric dipole moment implies violation of both T- and P-invariance. Since a dipole moment can be measured with great precision, this provides a sensitive test of T-invariance. Before describing the experiment on the electric dipole moment (EDM) of the neutron, let us try to make a guess at the magnitude of the effect, just from dimensional arguments. One can write

EDM = charge (e) × a length (l) × T-violation parameter (f).

The neutron is uncharged, so that the dipole moment could result from an asymmetry between positive and negative charge clouds, of net value zero, relative to the spin direction σ (Fig. 3.9). Since P-invariance is also violated, we must somehow bring in the weak interactions; the natural length involving the weak interaction is $l = GM$, where M is some chosen mass – the obvious one being the nucleon mass M – and $G = 10^{-5}/M^2$ is the weak-interaction coupling constant. We have used units $\hbar = c = 1$ here. Then

$$\text{EDM} = 10^{-5}\frac{ef}{M} \sim 10^{-19}f \ e\,\text{cm}, \qquad (3.47)$$

where for $1/M$ we put the proton Compton wavelength $\hbar/Mc = 2 \times 10^{-14}\,\text{cm}$. What do we take for f? From the parity-conserving electromagnetic interactions, the CPT theorem tells us that if T-invariance is violated, so is

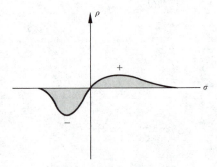

Fig. 3.9 An asymmetric distribution of positive and negative charge density ρ in the neutron would give rise to an electric dipole moment.

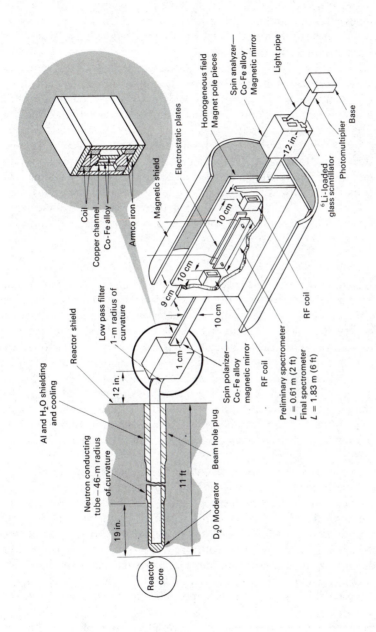

Fig. 3.10 Sketch of the apparatus used by Dress *et al.* (1968) to measure the electric dipole moment of the neutron.

C-invariance, and hence the η-decay results (Section 3.13) suggest $f < 10^{-2}$. Or we may take the *T*-violation parameter from K^0 decay, giving $f < 10^{-3}$. Either way, one can expect a maximum value of the EDM of order 10^{-22} *e* cm. Note that the effective length of the dipole is very small compared with the "size" of an elementary particle, of about 10^{-13} cm.

In the experiment by Dress *et al.* (1968), a reactor was used as a source of (predominantly) thermal neutrons. In order to make the experiment more sensitive, these neutrons were "cooled" by passage through a narrow, curved tube of highly polished nickel of 1-m radius of curvature (see Fig. 3.10). The critical angle for total internal reflection of neutrons by the tube is inversely proportional to velocity, so that for a beam of finite divergence, only low-velocity neutrons are transmitted with high intensity. The beam emerging from the tube then falls on a polarizing magnet, consisting of a polished, magnetized mirror of cobalt-iron alloy, the direction of the field **B** being normal to the surface. Total internal reflection of neutrons will occur for an angle of incidence (relative to the surface) less than the critical angle θ_c, where

$$\sin^2 \theta_c = 1 - n^2 = \frac{\lambda^2 N a}{\pi} \pm \frac{\mu B}{T}. \tag{3.48}$$

Here λ, T, and μ are the wavelength, kinetic energy, and magnetic moment of the neutrons, n is the refractive index of the mirror, N is the number of scattering nuclei per unit volume, and a is the coherent nuclear scattering length. Because of the second term in (3.48), θ_c depends on the sign of μ and therefore the neutron spin direction. So the reflection angle can be chosen to provide a transversely spin-polarized beam. In a typical case, for neutrons of velocity $v = 100$ m sec^{-1} (temperature 1°K), $\theta = 2°$ gives a beam with 70% polarization.

After traversing a spectrometer, the neutrons are reflected from the analyzing magnet (similar to the polarizer), and recorded in the detector, consisting of a ^{6}Li-loaded glass scintillator, sensitive to neutrons. The transmitted intensity I is a maximum for neutrons which do not suffer depolarization in the spectrometer.

The spectrometer consists firstly of a uniform magnetic field H ($\simeq 10$ G) which causes the neutrons to precess with the Larmor frequency $v_L = \mu H/h$, where μ is the neutron magnetic moment ($v_L \simeq 25$ kHz). Secondly, a RF field of frequency v is applied by means of two coils, so that at resonance, when $v = v_L$, spin-flip transitions occur, the neutron beam is partly depolarized, and the transmitted intensity I changes. Two coils are used to give an interferometer effect, producing several maxima and minima in the resonance curve (Fig. 3.11), and providing a rapid change in counting rate with RF frequency. Finally, a reversible electric field E of 100 kV/cm is applied in the direction of the steady magnetic field H.

The experiment consists essentially of sitting in a region of the resonance curve where dI/dv is large, and observing the change in I when the electrostatic field E is reversed. If the neutron possesses an electric dipole moment in the

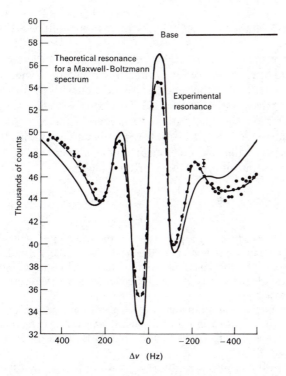

Fig. 3.11 Resonance curve obtained with the apparatus of Fig. 3.10 as the RF frequency is varied about the resonance value. At the steepest part of the curve, the counting rate varies by 1% for a change of frequency $\Delta v = 1$ Hz. The full-line curve is that calculated assuming a simple Maxwellian distribution in the velocity of the neutron beam.

direction of the spin, the field E will produce an additional small precession and consequently a change in I when the frequency v is held constant. dI/dv is proportional to the time spent by the neutron between the RF coils and is thus greatest for large coil separations and low velocities – hence the advantage of using "cold" neutrons. The final result of the most recent version of the experiment sets a limit on the electric dipole moment of

$$\text{EDM of neutron} < 2 \times 10^{-24} \; e \, \text{cm.} \qquad (3.49)$$

It may be remarked that one limitation of the experiment is the difficulty of obtaining **E** and **H** exactly parallel. If there is a small component $E_\perp$ perpendicular to **H**, this will induce an extra magnetic field $\Delta H = (v/c)E_\perp$ in the direction of **H** and hence a spurious effect.

From (3.48) we note that for ultracold neutrons ($T \simeq 0.002°$K, $v \simeq 6$ m sec^{-1}), θ_c can exceed $\pi/2$, so that neutrons can be reflected at normal incidence. Such neutrons can be captured and stored in magnetic "bottles", allowing longer observation times and more precise measurement of the EDM.

An improvement on the limit (3.49) by 2 orders of magnitude is expected using such techniques.

Theoretical estimates of the EDM vary over many orders of magnitude. Since there is no direct evidence for C-violation in electromagnetism, it is usual to assume that any EDM of the neutron is associated with CP violation, such as is observed in K^0 decay (see Section 6.13). This leads to predictions in the region of 10^{-28} e cm for the so-called "standard model" with six quark flavors, but values as high as 10^{-24} e cm are possible with other assumptions. It is clear that future experiments on the neutron electron dipole moment will be of fundamental significance for our understanding of CP-violation.

PROBLEMS

3.1 Λ-hyperons are produced by a pion beam in the reaction $\pi^- + p \rightarrow K^0 + \Lambda$, and observed via their decay $\Lambda \rightarrow p + \pi^-$. Let J denote the spin of the Λ, z the beam direction, and θ the angle of a decay product, relative to z, measured in the Λ rest frame.

(a) In the case where the Λ is produced exactly along the z direction, what are the possible values of J_z?

(b) For this configuration, and assuming $J_\Lambda = \frac{1}{2}$, calculate the angular distribution of the decay pion if it is emitted in (1) an s-state, (2) a p-state. If the target proton is unpolarized, show that the angular distribution must be isotropic for s- and p-waves and for any admixture of them.

(c) Show that, for unpolarized protons, the decay angular distributions for the forward-produced Λs as a function of Λ spin will be as follows:

$$J_\Lambda = \tfrac{1}{2} \quad \text{(isotropic)},$$
$$J_\Lambda = \tfrac{3}{2} \quad (3\cos^2\theta + 1),$$
$$J_\Lambda = \tfrac{5}{2} \quad (5\cos^4\theta - 2\cos^2\theta + 1).$$

[This method to determine the Λ spin was first proposed by Adair (1955). For a discussion see Sakurai (1964) and Tripp (1965).]

(d) State how one might determine the spin of the $\Sigma^\pm$ from the (s-state) capture of negative kaons in hydrogen, $K^- + p \rightarrow \Sigma^\pm + \pi^\mp$.

3.2 The Σ^0-hyperon decays electromagnetically in the mode $\Sigma^0 \rightarrow \Lambda + \gamma$. Show how the relative parity of Σ^0 and Λ determines the multipolarity of the γ-ray emitted. From the polarization vector $\mathbf{\varepsilon}$ of the photon, and the propagation vector $\mathbf{k}$ and spin $\mathbf{\sigma}$ of the Λ, deduce the simplest forms for the matrix element for even or odd relative parity. The experimental determination of the Σ-Λ parity has been based on the analysis of the Dalitz decay $\Sigma \rightarrow \Lambda e^+ e^-$. Which of the parity assignments has the steeper distribution in the invariant mass of the $e^+ e^-$ pair?

3.3 Show that a scalar meson cannot decay to three pseudoscalar mesons in a parity-conserving process.

3.4 The intrinsic parity of the hyperon Ξ^-, of strangeness -2, can in principle be determined from observations on capture in hydrogen from an S-orbit:

$$\Xi^- + p \rightarrow \Lambda + \Lambda.$$

The polarization of the Λ-hyperons can be determined from the asymmetry in the weak decay $\Lambda \to p + \pi^-$ (see Section 6.7). State what is the polarization (if any) of the Λs produced in the above reaction and how the relative polarizations are determined by the Ξ-parity.

3.5 Capture of negative kaons in helium sometimes leads to the formation of a hypernucleus (a nucleus in which a neutron is replaced by a Λ-hyperon) according to the reaction

$$K^- + {}^4\text{He} \to {}^4\text{H}_\Lambda + \pi^0.$$

Study of the decay branching ratios of ${}^4\text{H}_\Lambda$, and the isotropy of decay products establishes that $J({}^4\text{H}_\Lambda) = 0$. Show that this implies negative parity for the K^-, independent of the orbital angular momentum of the state from which the K^- is captured.

3.6 Show that the reaction $\pi^- + d \to n + n + \pi^0$ cannot occur for pions at rest.

3.7 Show that, for pions with zero relative orbital angular momentum, the combination $\pi^+\pi^-$ is an eigenstate of $CP = +1$, and $\pi^+\pi^-\pi^0$ is an eigenstate of $CP = -1$.

3.8 What restrictions does the decay mode $K_1^0 \to 2\pi^0$ place on (a) the kaon spin, (b) the kaon parity?

BIBLIOGRAPHY

Fraser, W. R., *Elementary Particles*, Prentice-Hall, Englewood Cliffs, New Yersey, 1966.

Hamilton, W. D., "Parity violation in electromagnetic and strong interaction processes", *Prog. Nucl. Phys.* **10**, 1 (1969).

Henley, E. M., "Parity and time-reversal invariance in nuclear physics", *Ann. Rev. Nucl. Science* **19**, 367 (1969).

Jackson, J. D., *The Physics of Elementary Particles*, Princeton University Press, Princeton, New Jersey, 1958.

Kemmer, N., J. C. Polkinghorne, and D. Pursey, "Invariance in elementary particle physics", *Rep. Prog. Phys.* **22**, 368 (1959).

Muirhead, A., *The Physics of Elementary Particles*, Pergamon, London, 1965, Chapter 5.

Ramsey, N. F., "Dipole moments and spin rotations of the neutron", Phys. Reports **43**, 409 (1978).

Rowe, E. G., and E. J. Squires, "Present status of C-, P-, and T-invariance", *Rep. Prog. Phys.* **32**, 273 (1969).

Sakurai, J. J., *Invariance Principles and Elementary Particles*, Princeton University Press, Princeton, New Jersey, 1964.

Tadic, D., "Parity non-conservation in nuclei", *Rep. Prog. Phys.* **43**, 67 (1980).

Tripp, R. D., "Spin and parity determination of elementary particles", *Ann. Rev. Nucl. Science* **15**, 325 (1965).

Wick, G. C., "Invariance principles of nuclear physics", *Ann. Rev. Nucl. Science* **8**, 1 (1958).

Williams, W. S., *An Introduction to Elementary Particles*, Academic Press, New York and London, 1971.

Chapter 4

Hadron-Hadron Interactions

In this chapter, we discuss the characteristics of the strong interactions between hadrons. As indicated in Chapter 8, the fundamental strong interactions are considered to take place between the elementary quark constituents of hadrons, and there is indeed a plausible field theory of the interactions between a pair of quarks, called quantum chromodynamics, which is however only calculable for high momentum transfers between the quarks. Hadron-hadron collisions, on the contrary, involve many quark (or antiquark) constituents simultaneously – thus, an insoluble many-body problem – and the individual momentum transfers are generally small.

The situation is somewhat similar to that of collisions between atoms. The fundamental (electromagnetic) interaction between the individual electrons and nuclei is well understood, but the description of atom-atom interactions is very complex because many electrons are involved, and a largely empirical approach to the problem is required. The treatment of hadron-hadron interactions antedates the quark model by many years. One very general approach – the *S*-matrix theory – discusses hadron-hadron collisions in terms of amplitudes and phases of matter waves, in analogy with optics, and using concepts and methods (such as that of resonances) first developed in nuclear structure physics. Another method which has been very successful is to describe hadron-hadron scattering in terms of single-particle exchange – the so-called Regge theory.

The description of hadron-hadron interactions has been of great importance in establishing the existence of the different hadron states, and in measuring their masses and quantum numbers, and this accumulation of data was the essential input required for the quark concept to emerge.

First, we discuss two of the important quantum numbers associated with hadrons: isospin and strangeness.

4.1 ISOSPIN

Heisenberg suggested in 1932 that neutron and proton might be treated as different charge substates of one particle, the nucleon. A nucleon is ascribed a quantum number, isospin, denoted by the symbol I, with a value $I = \frac{1}{2}$, and two substates with I_z, or I_3, equal to $\pm \frac{1}{2}$. The charge is then given by $Q/e = \frac{1}{2} + I_3$, if we assign $I_3 = +\frac{1}{2}$ to the proton and $I_3 = -\frac{1}{2}$ to the neutron. This purely formal description is in complete analogy with that of a particle of ordinary spin $\frac{1}{2}$, with substates $J_z = \pm \frac{1}{2}$ (in units of $h/2\pi$).

Isospin is a useful concept because it is a conserved quantum number in strong interactions.* Consequently these depend on I and not on the third component, I_3. The specifically strong interactions between nucleons, for example, are specified by I and we do not distinguish between neutron and proton – they are degenerate states. Electromagnetic interactions do not conserve isospin, and when we "turn them on" we do provide a quantization axis (the direction of an applied electric field) and, consequently, well-defined and distinguishable eigenvalues of I_3.

It is useful to visualize isospin as a vector **I** in a 3-dimensional "isospin space", in precise analogy with the angular-momentum vector in real space (Fig. 4.1). Conservation of angular momentum is expressed by invariance of the length of this vector under rotations of the coordinate axes. The Cartesian components of angular momentum are denoted by J_x, J_y, J_z, while for isospin the corresponding components are usually labeled I_1, I_2, I_3 – an unfortunate difference of nomenclature.

The earliest evidence for isospin conservation in strong interactions came from the observations of charge symmetry and charge independence of nuclear forces. The former is illustrated by the existence of mirror nuclei. ^{7}Li and ^{7}Be are examples of a mirror pair. The ground states have binding energies differing by only ~ 1.5 MeV, and this can be accounted for by the extra Coulomb energy of

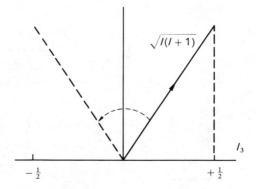

Fig. 4.1 The conservation of isospin I in strong interactions can be interpreted as the invariance, under rotations of the axes, of an isospin vector of length $\sqrt{I(I+1)}$ in a 3-dimensional isospin space. For $I = \frac{1}{2}$, the z-component can have eigenvalues $I_3 = \pm\frac{1}{2}$.

---

* "Isospin" is the word used today, replacing the older usage, "isotopic spin" or "isobaric spin". In some books, isospin is denoted by the symbol T instead of I.

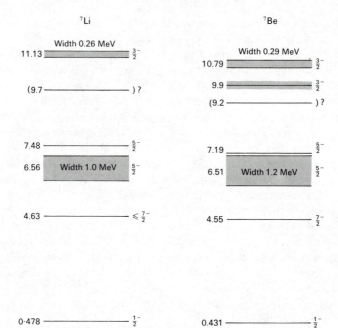

Fig. 4.2 Comparison of the excited states of ^{7}Be and ^{7}Li. The numbers on the left-hand side refer to energies in MeV, and those on the right-hand side to the spin-parity assignments.

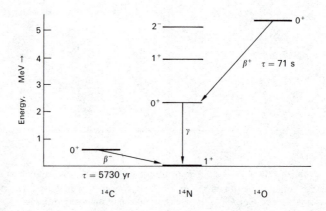

Fig. 4.3 Energy levels of the triad of states ^{14}O, ^{14}N, and ^{14}C. Heavy lines denote the ground states.

^{7}Be. The excited states show a strong similarity (see Fig. 4.2). This can be understood if the configurations of nucleons in the two nuclei are similar, so that corresponding levels have the same energy, width, spin and parity, and the only difference is the interchange of one proton by a neutron. Counting nn, pn, and pp "bonds" in the two cases, it is seen that ^{7}Be has an extra pp bond and ^{7}Li an extra nn bond. Thus, in the same spatial-spin state, the strong nn and pp interactions must be equal – the hypothesis of charge symmetry.

^{7}Be and ^{7}Li form an isospin doublet $(I = \frac{1}{2})$ with substates of $I_3 = (Z - N)/2 = \pm \frac{1}{2}$. An example of an $I = 1$ multiplet with $2I + 1 = 3$ substates is shown in Fig. 4.3. These nuclei consist of the ground states of ^{14}C and ^{14}O (with $I_3 = -1$ and $+1$ respectively) and the first excited state of ^{14}N ($I_3 = 0$). The binding-energy differences between these states, all of spin-parity $J^P = 0^+$, can be ascribed to Coulomb-energy differences. Note that ^{14}N* is produced by the "superallowed" β^+ decay of ^{14}O, to be contrasted with the very slow "forbidden" β^- decay from ^{14}C to the ^{14}N ground state (this last is the famous "carbon-dating" transition). The nuclear configurations consist of a ^{12}C core plus an nn pair (^{14}C), np pair (^{14}N), or pp pair (^{14}O), and the equivalence of their strong interactions implies that, in the same spatial-spin states, the nn, pn, and pp forces are equal – the hypothesis of charge independence.

Other evidence for charge independence comes from the near-equality of the scattering length and potential in pp and pn scattering in the singlet spin state.

At this point, it should be emphasized that the hypothesis of conservation of isospin is a much stronger and more general statement than that nuclear forces are charge independent. The latter is a consequence of the former, but is true only for a pair of similar particles of isospin $\frac{1}{2}$. As we shall see, isospin is conserved in the pion-nucleon interaction, but the interaction is *not* charge-independent. Charge independence follows when, by some symmetry argument, the total isospin of the system is limited to a unique value, and is then equivalent to the statement that, for fixed I, the interaction is independent of I_3.

An example of conservation of isospin is provided by the reaction

$$d + d \rightarrow {}^4\text{He} + \pi^0 \tag{4.1}$$

$$
\begin{array}{ccccc}
I & 0 & 0 & 0 & 1 \\
I_3 & 0 & 0 & 0 & 0
\end{array}
$$

Since no other bound state of two nucleons exists, we can assume that the deuteron has $I = 0$ (see Section 4.2): the same argument applies to ^{4}He. As indicated below, the pion, with three charge states π^+, π^-, π^0, must have $I = 1$. Thus the reaction (4.1) is forbidden as a *strong* (isospin-conserving) reaction. It *can* however proceed as an *electromagnetic* (isospin-violating) reaction, with a correspondingly smaller cross-section.

4.2 ISOSPIN IN THE TWO-NUCLEON SYSTEM

The isospin states of the two-nucleon system can be written down in direct analogy with the combination of two objects of spin $\frac{1}{2}$, as in Section 3.11. Writing the wavefunctions n and p to denote neutron and proton states, we therefore obtain

$$S \begin{cases} \chi(1, 1) = p(1)p(2), & \text{(4.2a)} \\[2mm] \chi(1, 0) = \dfrac{1}{\sqrt{2}}[p(1)n(2) + n(1)p(2)], & \text{(4.2b)} \\[2mm] \chi(1, -1) = n(1)n(2), & \text{(4.2c)} \end{cases}$$

$$A \qquad \chi(0, 0) = \dfrac{1}{\sqrt{2}}[p(1)n(2) - n(1)p(2)], \qquad \text{(4.2d)}$$

where the first three states are members of an $I = 1$ triplet, symmetric under label interchange $1 \leftrightarrow 2$, and the last is an $I = 0$ singlet, which is antisymmetric.

The total wavefunction for a two-nucleon state may be written

$$\psi(\text{total}) = \phi(\text{space})\, \alpha(\text{spin})\, \chi(\text{isospin}) \qquad \text{(4.3)}$$

provided orbital and spin angular momentum can be separately quantized (i.e. the system is nonrelativistic). Applying Eq. (3.37) to a deuteron, which has spin 1, we see that α is symmetric under interchange of the two nucleons. The space wavefunction ϕ has symmetry $(-1)^l$ under interchange. The two nucleons in the deuteron are known to be in an $l = 0$ state (with a few per cent $l = 2$ admixture). Thus ϕ is symmetric, and χ must be antisymmetric in order to satisfy overall antisymmetry of the total wavefunction ψ. From (4.2) is follows that $I = 0$: the deuteron is an isosinglet.

As an example, consider the reactions

$$\begin{array}{lll} \text{(i) } p + p \rightarrow d + \pi^+, & \quad \text{(ii) } p + n \rightarrow d + \pi^0. \\ I \quad\ 1 \qquad\quad 0 \quad\ 1 & \qquad\quad\ 0\,\text{or}\,1 \quad 0 \quad\ 1 \end{array}$$

In each case the final state is of $I = 1$. On the left-hand side, we have a pure $I = 1$ state in reaction (i), but 50% $I = 0$ and 50% $I = 1$ in reaction (ii). Conservation of isospin means that either reaction can only proceed through the $I = 1$ channel. Consequently, $\sigma(\text{ii})/\sigma(\text{i}) = \frac{1}{2}$, as is observed.

4.3 ISOSPIN IN THE PION-NUCLEON SYSTEM

The pion exists in three charge states of roughly the same mass: π^+, π^-, and π^0. Consequently it is assigned $I = 1$, with the charge given by $Q/e = I_3$ simply. This formula is different from that used for the nucleon: $Q/e = I_3 + \frac{1}{2}$. Both can be accommodated in the formula by introducing baryon number:

$$\frac{Q}{e} = I_3 + \frac{B}{2}. \qquad \text{(4.4)}$$

An important application of isospin conservation arises in the strong interactions of nonidentical particles, which will generally consist of mixtures of different isospin states. The classical example of this is pion-nucleon scattering. Since $I_\pi = 1$ and $I_N = \frac{1}{2}$, one can have $I_{\text{total}} = \frac{1}{2}$ or $\frac{3}{2}$. If the strong interactions depend only on I and not on I_3, then the $3 \times 2 = 6$ pion-nucleon scattering processes can all be described in terms of two isospin amplitudes.

Of the six elastic scattering processes,

$$\pi^+ p \to \pi^+ p \tag{4.5a}$$

and

$$\pi^- n \to \pi^- n \tag{4.5b}$$

have $I_3 = \pm \frac{3}{2}$, and are therefore described by a pure $I = \frac{3}{2}$ amplitude. Clearly, at a given bombarding energy, (4.5a) and (4.5b) will have identical cross-sections, since they differ only in the sign of I_3.

The remaining interactions,

$$\pi^- p \to \pi^- p, \tag{4.5c}$$

$$\pi^- p \to \pi^0 n, \tag{4.5d}$$

$$\pi^+ n \to \pi^+ n, \tag{4.5e}$$

$$\pi^+ n \to \pi^0 p, \tag{4.5f}$$

have $I_3 = \pm \frac{1}{2}$ and therefore $I = \frac{1}{2}$ or $\frac{3}{2}$. There is now no principle of exchange symmetry which restricts the isospin to one of the two values, and therefore one has a mixture. The weights of the two amplitudes in the mixture are given by Clebsch-Gordan coefficients (alternatively known as vector-coupling or Wigner coefficients). Their derivation is given in Appendix C. An alternative method, which can be applied to the present case, has been given by Feynman. It not only is elegant, but illuminates the physical meaning of charge independence, and we reproduce it here.

The force between nucleons is charge-independent. This force may be thought of as being due to exchange of virtual particles between the nucleons (just as the electrostatic force between two charges can be thought of in terms of exchange of virtual photons). However, for hadron-hadron interactions, many processes, such as pion exchange, ρ-meson exchange, $K\bar{K}$ exchange, etc., may contribute. It is very plausible to assume that the *contribution* from single pion exchange alone is charge independent—if it were not, charge independence would be an accident depending on chance cancellation of forces due to all possible exchange mechanisms, which individually were not charge-independent. This possibility is discounted as too remote.

The pp, pn, and nn interactions contributed by single pion exchange are represented diagrammatically in Fig. 4.4, where a, b, and c are unknown coupling constants. In the last diagram, the virtual π^- can of course be replaced by a π^+ with arrow reversed—we do not distinguish the two. If the pp, nn, and np forces are equal, we have

$$a^2 = b^2 = ab + c^2.$$

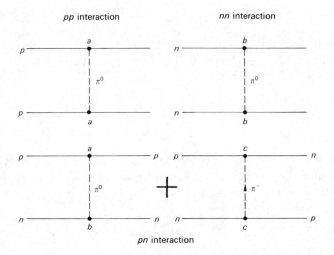

Fig. 4.4

Thus, $a = \pm b$. Furthermore, c cannot be zero, since it is known that the last diagram exists and leads to charge-exchange pn scattering. One must then have

$$b = -a,$$

$$c = \pm\sqrt{2}\,a,$$

where the last sign is arbitrary.

Because of its strong interaction, a nucleon can virtually dissociate into a nucleon plus one or more pions, a hyperon-kaon pair, and so on. The possible virtual states of the neutron involving a nucleon plus one pion are clearly those in the bottom part of Fig. 4.4: we can denote them $|n\pi^0\rangle$ and $|p\pi^-\rangle$ respectively. The initial state (the neutron) has $I = \frac{1}{2}$, $I_3 = -\frac{1}{2}$. Hence the decomposition of this state is

$$|\chi(\tfrac{1}{2}, -\tfrac{1}{2})\rangle = b|n\pi^0\rangle + c|p\pi^-\rangle$$

$$= b(|n\pi^0\rangle \pm \sqrt{2}|p\pi^-\rangle).$$

With the normalization of the states $\langle\chi|\chi\rangle = \int\chi^*\chi\,d\tau = 1$, $\langle n\pi^0|n\pi^0\rangle = \langle p\pi^-|p\pi^-\rangle = 1$ we therefore obtain

$$|\chi(\tfrac{1}{2}, -\tfrac{1}{2})\rangle = \sqrt{\tfrac{1}{3}}\,|n\pi^0\rangle - \sqrt{\tfrac{2}{3}}\,|p\pi^-\rangle. \qquad (4.6)$$

Similarly, the proton state will be

$$|\chi(\tfrac{1}{2}, \tfrac{1}{2})\rangle = -\sqrt{\tfrac{1}{3}}\,|p\pi^0\rangle + \sqrt{\tfrac{2}{3}}\,|n\pi^+\rangle, \qquad (4.7)$$

where the sign of c has been chosen to conform to the usual notation (Condon and Shortley 1951).

To find the coefficients for the $I = \frac{3}{2}$, $I_3 = \pm\frac{1}{2}$ states, we make use of the orthogonality and normalization conditions; for example

$$\langle\chi(\tfrac{3}{2},\tfrac{1}{2})|\chi(\tfrac{1}{2},\tfrac{1}{2})\rangle = 0, \qquad \langle n\pi^0|n\pi^0\rangle = 1, \qquad \langle p\pi^-|n\pi^0\rangle = 0.$$

Thus, if we set

$$|\chi(\tfrac{3}{2},\tfrac{1}{2})\rangle = A|p\pi^0\rangle + B|n\pi^+\rangle, \tag{4.8}$$

$$|\chi(\tfrac{3}{2}, -\tfrac{1}{2})\rangle = C|p\pi^-\rangle + D|n\pi^0\rangle, \tag{4.9}$$

we clearly must have

$$A^2 + B^2 = C^2 + D^2 = 1,$$

and, taking the products of (4.8) with (4.7) and of (4.9) with (4.6),

$$A = B\sqrt{2}, \qquad D = C\sqrt{2}.$$

Thus

$$B = C = \sqrt{\tfrac{1}{3}}, \qquad D = A = \sqrt{\tfrac{2}{3}}, \tag{4.10}$$

again with the usual convention on sign. The resulting list of Clebsch-Gordan coefficients is given in Table 4.1.

Table 4.1 Clebsch-Gordan coefficients in pion-nucleon scattering

Pion	Nucleon	$I = \frac{3}{2}$				$I = \frac{1}{2}$	
		$I_3 = \frac{3}{2}$	$\frac{1}{2}$	$-\frac{1}{2}$	$-\frac{3}{2}$	$\frac{1}{2}$	$-\frac{1}{2}$
π^+	p	1					
π^+	n		$\sqrt{\frac{1}{3}}$			$\sqrt{\frac{2}{3}}$	
π^0	p		$\sqrt{\frac{2}{3}}$			$-\sqrt{\frac{1}{3}}$	
π^0	n			$\sqrt{\frac{2}{3}}$			$\sqrt{\frac{1}{3}}$
π^-	p			$\sqrt{\frac{1}{3}}$			$-\sqrt{\frac{2}{3}}$
π^-	n				1		

We can now calculate the relative cross sections for the following three processes, at a fixed energy:

$$\pi^+ p \to \pi^+ p \qquad \text{(elastic scattering)}, \tag{4.11a}$$

$$\pi^- p \to \pi^- p \qquad \text{(elastic scattering)}, \tag{4.11b}$$

$$\pi^- p \to \pi^0 n \qquad \text{(charge exchange)}. \tag{4.11c}$$

The cross-section is proportional to the square of the matrix element connecting initial and final states, i.e.

$$\sigma \propto \langle \psi_f|H|\psi_i\rangle^2 = M_{if}^2,$$

where H is an isospin operator, having a value $H = H_1$ if it operates on initial and final states of $I = \frac{1}{2}$, and $H = H_3$ for states of $I = \frac{3}{2}$. By conservation of isospin, there is no operator connecting initial and final states of different isospin. Let

$$M_1 = \langle \psi_f(\tfrac{1}{2})|H_1|\psi_i(\tfrac{1}{2})\rangle,$$
$$M_3 = \langle \psi_f(\tfrac{3}{2})|H_3|\psi_i(\tfrac{3}{2})\rangle.$$

The reaction (4.11a) involves a pure state of $I = \frac{3}{2}$, $I_3 = +\frac{3}{2}$. Therefore,

$$\sigma_a = K|M_3|^2,$$

where K is some constant.

Referring to our table, in the reaction (4.11b) we may write

$$|\psi_i\rangle = |\psi_f\rangle = \sqrt{\tfrac{1}{3}}|\chi(\tfrac{3}{2}, -\tfrac{1}{2})\rangle - \sqrt{\tfrac{2}{3}}|\chi(\tfrac{1}{2}, -\tfrac{1}{2})\rangle.$$

Therefore

$$\sigma_b = K\langle \psi_f|H_1 + H_3|\psi_i\rangle^2$$
$$= K|\tfrac{1}{3}M_3 + \tfrac{2}{3}M_1|^2.$$

For the reaction (4.11c), one has

$$|\psi_i\rangle = \sqrt{\tfrac{1}{3}}|\chi(\tfrac{3}{2}, -\tfrac{1}{2})\rangle - \sqrt{\tfrac{2}{3}}|\chi(\tfrac{1}{2}, -\tfrac{1}{2})\rangle,$$
$$|\psi_f\rangle = \sqrt{\tfrac{2}{3}}|\chi(\tfrac{3}{2}, -\tfrac{1}{2})\rangle + \sqrt{\tfrac{1}{3}}|\chi(\tfrac{1}{2}, -\tfrac{1}{2})\rangle,$$

and thus

$$\sigma_c = K|\sqrt{\tfrac{2}{9}}M_3 - \sqrt{\tfrac{2}{9}}M_1|^2.$$

The cross section ratios are then

$$\sigma_a:\sigma_b:\sigma_c = |M_3|^2 : \tfrac{1}{9}|M_3 + 2M_1|^2 : \tfrac{2}{9}|M_3 - M_1|^2. \qquad (4.12)$$

The limiting situations, if one or other isospin amplitude dominates under the experimental conditions, are

$$M_3 \gg M_1, \qquad \sigma_a:\sigma_b:\sigma_c = 9:1:2,$$
$$M_1 \gg M_3, \qquad \sigma_a:\sigma_b:\sigma_c = 0:2:1. \qquad (4.13)$$

Numerous experimental measurements have been made of the total and differential pion-nucleon cross sections. The earliest and simplest experiments measured the attenuation of a collimated, monoenergetic $\pi^{\pm}$-beam in traversing a liquid hydrogen target. Thus, in the sketch of Fig. 4.5, one would measure,

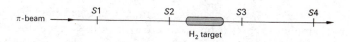

Fig. 4.5 Schematic drawing of measurement of the total pion-proton cross-section.

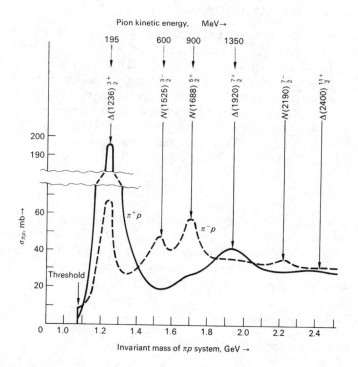

Fig. 4.6 Variation of total cross section for π^+ and π^- mesons on protons, with incident pion energy. The symbol Δ refers to resonances of $I = \frac{3}{2}$; N refers to $I = \frac{1}{2}$. The positions of only a few of the known states, together with their spin-parity assignments, are given.

for a given number of coincidences in the counters $S1$ and $S2$, the change in rate of counters $S3$ and $S4$ with the target both full and empty. The results of such measurements are shown in Fig. 4.6. For both positive and negative pions, there is a strong peak in σ_{total} at a pion kinetic energy of 200 MeV. The ratio $(\sigma_{\pi^+ p}/\sigma_{\pi^- p})_{\text{total}} = 3$, proving that the $I = \frac{3}{2}$ amplitude dominates this region. This bump is referred to as a *resonance*, the $\Delta(1236)$ (1236 MeV being the invariant pion-nucleon mass). The width of this state at half height is 120 MeV. Because the spin-parity turns out to be $J^P = \frac{3}{2}^+$, and $I = \frac{3}{2}$, it is often referred to as the (3, 3) resonance. A discussion of this pion-nucleon state is given in Section 4.6. As Fig. 4.6 indicates, more pion-nucleon resonances are observed – for example, there is an $I = \frac{1}{2}$ resonance at 1525 MeV. In general, in any one region of pion-nucleon invariant mass, several amplitudes will contribute, and one cannot simply interpret a bump in cross section as signifying a unique resonant state. Only for the first $(3, 3)$ resonance is such an interpretation unambiguous.

4.4 STRANGENESS AND ISOSPIN

As explained in Chapter 1, the strange particles were so named because of their long lifetime (for decay via the weak interactions) in contrast with their copious production (in strong interactions). It was argued that a new quantum number, called the strangeness S, was involved. Strangeness would be conserved in the associated strong production of particles of opposite strangeness, but violated in the weak decay of single strange particles into nonstrange particles.

The assignments of isospin and strangeness were obtained as follows. First we note that the Λ-hyperon has no charged counterpart, implying $I_\Lambda = 0$. The values of I and I_3 involved in Λ-decay are as follows:

$$\Lambda \rightarrow p + \pi^-.$$

$$
\begin{array}{cccc}
 & \Lambda & p & \pi^- \\
I & 0 & \frac{1}{2} & 1 \\
I_3 & 0 & \frac{1}{2} & -1
\end{array}
$$

(4.14)

This is a *weak* decay process, and neither I nor I_3 is conserved on the two sides of the equation. From the fact that the Λ-hyperon and neutral kaon have been observed to be produced in association in strong interactions of pions with protons (in the early diffusion-cloud-chamber experiments), we can assign the kaon half-integral isospin. $I_K = \frac{1}{2}$ is the simplest choice:

$$\pi^- + p \rightarrow \Lambda + K^0.$$

$$
\begin{array}{cccc}
 & \pi^- & p & \Lambda & K^0 \\
I & 1 & \frac{1}{2} & 0 & \frac{1}{2} \\
I_3 & -1 & \frac{1}{2} & 0 & -\frac{1}{2}
\end{array}
$$

(4.15)

The correct charge for the neutral kaon is obtained if we set

$$\frac{Q}{e} = I_3 + \tfrac{1}{2},$$

(4.16)

implying that the K^+-meson, of $I_3 = +\frac{1}{2}$, is the charged member of a kaon doublet. This assignment also forbids the strong decay of a kaon according to the scheme $K^+ \rightarrow \pi^+ + \pi^- + \pi^+$; as already seen, this is a weak decay process. The K^--meson does not fit into this scheme. It was therefore necessary to postulate a second K-doublet, with

$$\frac{Q}{e} = I_3 - \tfrac{1}{2},$$

(4.17)

which predicts a second neutral kaon of $I_3 = +\frac{1}{2}$ – called $\bar{K}^0$. K^+ and K^- are considered as particle and antiparticle, as are K^0 and $\bar{K}^0$. The interesting phenomena associated with K^0 and $\bar{K}^0$ are discussed in Section 6.13.

Gell-Mann (1953) and Nishijima (1955) pointed out that the formulae (4.4), (4.16), and (4.17) could be more elegantly expressed by introducing the *strangeness* quantum number, according to the formula

$$\frac{Q}{e} = \frac{B}{2} + \frac{S}{2} + I_3. \tag{4.18}$$

The assignment of strangeness S follows from the isospin assignments. Nucleons and pions clearly must have $S = 0$, the Λ-hyperon $S = -1$, the K^0, K^+ doublet $S = +1$, and the $K^-, \bar{K}^0$ doublet $S = -1$. These assignments are shown in Table 4.2.

An example of conservation of strangeness in the reaction

$$K^- + p \rightarrow \Lambda + \pi^0$$

	K^-	p	Λ	π^0
S	-1	0	-1	0
I_3	$-\frac{1}{2}$	$+\frac{1}{2}$	0	0

is shown in Fig. 4.7.

Table 4.2 Isospin and strangeness assignments for particles decaying by weak or electromagnetic interactions

			I_3				
B	S	I	-1	$-\frac{1}{2}$	0	$+\frac{1}{2}$	$+1$
1	0	$\frac{1}{2}$		n		p	
1	-1	0			Λ		
0	0	1	π^-		π^0		π^+
0	$+1$	$\frac{1}{2}$		K^0		K^+	
0	-1	$\frac{1}{2}$		K^-		$\bar{K}^0$	
1	-1	1	Σ^-		Σ^0		Σ^+
1	-2	$\frac{1}{2}$		Ξ^-		Ξ^0	
1	-3	0			Ω^-		
0	0	0			η		

The Σ-hyperons fit into a charge triplet, $I = 1$. This assignment fits into the observed strong production reactions

$$\pi^\pm + p \rightarrow \Sigma^\pm + K^+$$

	$\pi^\pm$	p	$\Sigma^\pm$	K^+
I	1	$\frac{1}{2}$	1	$\frac{1}{2}$
I_3	± 1	$\frac{1}{2}$	± 1	$\frac{1}{2}$
S	0	0	-1	1

and accounts for the fact that the decay $\Sigma^+ \rightarrow n + \pi^+$ is weak. The predicted neutral member, Σ^0, was not finally identified until 1959. It undergoes the electromagnetic decay mode

$$\Sigma^0 \rightarrow \Lambda + \gamma.$$

	Σ^0	Λ	γ
I	1	0	0
I_3	0	0	0
S	-1	-1	0

Fig. 4.7 Example of the reaction $K^- + p \to \Lambda + \pi^0$ occurring when a K^--meson comes to rest in a hydrogen bubble chamber, at the point A. The neutral pion undergoes Dalitz decay, $\pi^0 \to e^+ e^- \gamma$. The Λ-hyperon decays ($\Lambda \to \pi^- + p$) at the point B. (Courtesy CERN.)

The cascade or Ξ^--hyperon was first observed in the early cosmic-ray work with cloud chambers; the $S = -2$ assignment followed from the production in association with a pair of K^0-mesons, and from the fact that the decay

$$\Xi^- \to \Lambda + \pi^-$$

was weak. The expected neutral counterpart Ξ^0 was detected in 1959. An example of Ξ^--decay is shown in Fig. 4.8.

The existence of the Ω^--baryon of $S = -3$, as well as its mass and decay modes, was predicted before it was observed in 1964, on the basis of unitary symmetry and the quark model (Chapter 5). A picture of the first Ω^--event is given in Fig. 5.3.

The electromagnetic mass splittings between members of isospin multiplets are only easily measurable for the long-lived hadrons, decaying by weak interactions and consequently with very small natural widths. A summary of these mass differences is given in Table 4.3. As expected, they are of order $\Delta m/m \sim \alpha \sim 10^{-2}$. Note that particle and antiparticle must have identical masses, by the *CPT* theorem, and thus $m_{\pi^+} \equiv m_{\pi^-}$, $m_{K^+} \equiv m_{K^-}$. However,

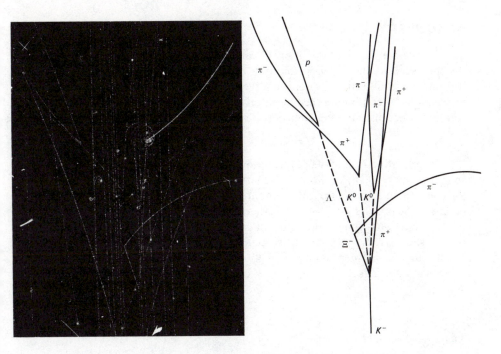

Fig. 4.8 Example of the decay $\Xi^- \to \pi^- + \Lambda$ in a hydrogen bubble chamber. The reaction is produced by an incident 10 GeV/c K^--meson:

$$K^- + p \to \Xi^- + \pi^+ + K^0 + K^0 + \overline{K}^0$$
$$\text{Strangeness } S: \quad -1 \quad 0 \quad -2 \quad 0 \quad +1 \quad +1 \quad -1.$$

The final $\overline{K}^0$ is not observed in this picture. Both K^0-mesons decay in the mode $K^0 \to \pi^+\pi^-$. (Courtesy CERN Information Service.)

Table 4.3 Mass differences in isospin multiplets

	Δm (MeV/c^2)	m_{av} (MeV/c^2)	$10^3\,\Delta m/m$
$n - p$	1.3	939	1.4
$\Sigma^0 - \Sigma^+$	3.1	1190	2.6
$\Sigma^- - \Sigma^0$	4.9	1195	4.1
$\Xi^- - \Xi^0$	6.5	1318	4.9
$K^0 - K^\pm$	4.0	495	8.1
$\pi^\pm - \pi^0$	4.6	140	33

$m_{\Sigma^+} \neq m_{\Sigma^-}$, since Σ^+ and Σ^- are both baryons, rather than baryon and antibaryon.

4.5 G-PARITY

The charge-conjugation operator C can have eigenvalues only for neutral systems, such as the π^0, γ, η, and e^+e^-. It is useful, however, to be able to formulate selection rules for some charged systems; this can be done for strong interactions by combining the operation of charge conjugation with an isospin rotation. For this purpose, consider the operation

$$G = CR = C\exp(i\pi I_2). \tag{4.19}$$

The operation G consists of a rotation R of 180° about the y-axis in isospin space [compare (3.16)], followed by charge conjugation. Applied to a state with a z-component of isospin I_3, this amounts to first flipping $I_3 \to -I_3$ and then reversing the process, $-I_3 \to I_3$. It is therefore plausible that charged states may be eigenfunctions of the G-operator. In order to get the eigenvalues, consider an isospin state $\chi(I, I_3 = 0)$. Under isospin rotations, this state behaves precisely like the angular-momentum wave function $Y_l^{m=0}(\theta, \phi)$ of (3.17) under rotations in ordinary space (see Fig. 3.1). The R-operation $\exp(i\pi L_y)$ implies $\theta \to \pi - \theta$, $\phi \to \pi - \phi$, and hence

$$Y_l^0 \underset{R}{\to} (-1)^l Y_l^0.$$

Therefore,

$$\chi(I, 0) \underset{R}{\to} (-1)^I \chi(I, 0). \tag{4.20}$$

For a state of nucleon and antinucleon, of total spin s and orbital angular momentum l, the effect of the C-operation is to give a factor $(-1)^{l+s}$, just as in the case of positronium. Thus, the effect of the operation $G = CR$ on a neutral $(I_3 = 0)$ nucleon-antinucleon system $|\psi\rangle$ will be

$$G|\psi\rangle = (-1)^{l+s+I}|\psi\rangle. \tag{4.21}$$

Since the strong interactions are invariant under isospin rotations, (4.21) must, in fact, be a general formula, and not limited to the case $I_3 = 0$ for which it was derived. Now suppose the G-operator acts on a pion wave function, $|\pi^+\rangle$. R reverses I_3, thus converting $\pi^+ \to \pi^-$, and C flips the charge back, $\pi^- \to \pi^+$. We may then write

$$G|\pi^+\rangle = \pm|\pi^+\rangle,$$

$$G|\pi^-\rangle = \pm|\pi^-\rangle,$$

$$G|\pi^0\rangle = \pm|\pi^0\rangle.$$

The neutral pion must be an eigenstate of C, with eigenvalue $+1$, since it decays in the mode $\pi^0 \to 2\gamma$. From (4.20), the R-eigenvalue is $(-1)^I = -1$. Thus

$$G|\pi^0\rangle = -|\pi^0\rangle.$$

The eigenvalue of the G-operator is called the G-parity. While the G-parity of the neutral pion is unambiguous, that of the charged pions is not. They are not eigenstates of C, and in the process of charge conjugation, an arbitrary phase appears, which can be chosen at will. For convenience, however, it is the practice to define the phases so that all members of an isospin triplet have the same G-parity as the neutral member. In the present case we can then write

$$G|\pi\rangle = -|\pi\rangle \tag{4.22}$$

provided we define

$$C|\pi^\pm\rangle = -|\pi^\mp\rangle.$$

For details, the reader is referred to Appendix E.

Since the C-operation reverses the sign of the baryon number, it will be apparent that eigenstates of G-parity must have baryon number zero. The charge-conjugation quantum number is multiplicative and isospin additive, so that G-parity is multiplicative. Thus, for a state of n pions,

$$G|\psi(n\pi)\rangle = (-1)^n|\psi(n\pi)\rangle. \tag{4.23}$$

Let us now return to the problem of nucleon-antinucleon annihilation. As an example, consider the annihilation of an antiproton with a neutron into two pions:

$$\bar{p} + n \to \pi^0 + \pi^-. \tag{4.24}$$

Application of the formula

$$Q/e = I_3 + B/2$$

shows that the antiproton has $I_3 = -\frac{1}{2}$ (while the antineutron has $I_3 = +\frac{1}{2}$). The total isospin of a nucleon-antinucleon pair can have the values $I_{total} = 0, 1$. Since, on the left-hand side of (4.24), $I_3 = -1$, the reaction can only proceed through the $I_{total} = 1$ isospin channel. For the two-pion annihilation,

$$G = (-1)^{l+s+I} = +1,$$

so that $l + s$ must be odd. There are two possibilities:

$s = 0$ (*singlet state*). For the choice $l = J = 1$, the parity on the left-hand side of (4.24) is $(-1)^{l+1}$, i.e. even [the extra (-1) factor coming from the opposite intrinsic parity of fermion and antifermion]. The parity of the two-pion $l = 1$ state is $(-1)^l$, or odd. Thus the $^1P_1$-state is forbidden by parity conservation. Equally, $l = 0$ is forbidden, since G-parity conservation requires $l + s$ to be odd. Thus both the singlet states $^1P_1$ and $^1S_0$ are forbidden.

$s = 1$ (*triplet state*). The states $^3P_{0,1,2}$ are forbidden, since $l + s$ must be odd. An S-state is allowed, since $l = 0$, $J = 1$ gives negative parity in both initial and final states. Thus, for values of $l \leqslant 1$, (4.24) can proceed only by annihilation in the $I = 1$, $^3S_1$ state.

We may summarize our discussion of selection rules for both the positronium and nucleon-antinucleon systems as follows:

$$e^+e^- \rightarrow n\gamma, \tag{4.25}$$

$$C\text{-parity} \qquad (-1)^{l+s} \quad (-1)^n$$

$$N\bar{N} \rightarrow n\pi. \tag{4.26}$$

$$G\text{-parity} \quad (-1)^{l+s+I} \quad (-1)^n$$

In both cases, the parity of the initial state is $P = (-1)^{l+1}$.

The concept of G-parity introduces nothing that is not already known from the twin postulates of charge conjugation and isospin invariance; it simply allows some short cuts when we consider selection rules for the decay of meson resonances. The relevant quantum numbers are given in Table 4.4.

Table 4.4

Particle $\begin{pmatrix} \text{mass,} \\ \text{MeV} \end{pmatrix}$	$\pi(140)$	$\rho(770)$	$\omega(783)$	$\phi(1020)$	$f(1270)$	$\eta(549)$	$\eta'(958)$
Spin parity J^P	0^-	1^-	1^-	1^-	2^+	0^-	0^-
Isospin I	1	1	0	0	0	0	0
G-parity	-1	$+1$	-1	-1	$+1$	$+1$	$+1$
Dominant pion decay mode	$-$	2π	3π	3π	2π	3π	5π

Note that the vector mesons ρ, ω, ϕ, and f decay by strong interactions, being resonant states with large widths (3 to 100 MeV), and that the multiplicity of the pion decay modes follows the rule $G = (-1)^n$. On the other hand, the existence of the $\gamma\gamma$-decay mode of the η and η' proves that these mesons decay by electromagnetic transitions, with total widths of 0.9 keV and 0.3 MeV respectively. Since $I = 0$, and $C(2\gamma) = +1$, they must have $G = +1$. The strong decay into two pions is forbidden by parity conservation, leaving the three-pion, G-violating electromagnetic decay as the only possibility.

4.6 DALITZ PLOTS

In a reaction such as

$$\pi + p \rightarrow \pi + \pi + p \tag{4.27}$$

one wants to investigate possible interactions among the final-state particles — for example, the occurrence of resonances (πp, $\pi\pi$) — and thus how the observed distributions in secondary momentum and angle depart from the situation where the matrix element is constant. In the latter case, the distributions will be determined simply from the phase-space factors. The departure of such distributions from phase space can be assessed from a phase-space or *Dalitz* plot (Dalitz 1953, Fabri 1954).

4.6.1 Three-Body Phase Space

As indicated in Eq. (1.28), the number of quantum states available in phase space, per unit normalization volume, is

$$\frac{p^2 \, dp \, d\Omega}{h^3}$$

for a spinless particle of momentum $p \rightarrow p + dp$ inside the solid-angle element $d\Omega$. In a reaction such as (4.27) with three final-state particles, labeled 1, 2, and 3, and fixed initial energy, the number of states will be proportional to

$$p_1^2 \, dp_1 \, p_2^2 \, dp_2 \, d\Omega_1 \, d\Omega_2.$$

In the center-of-momentum system (CMS), $\mathbf{p}_3 = -(\mathbf{p}_1 + \mathbf{p}_2)$ is fixed, so there is no factor for particle 3. If the initial state is unpolarized, the overall orientation in space will be isotropic. The integral over all directions of particle 1 is then $\int d\Omega_1 = 4\pi$, while $d\Omega_2 = 2\pi \, d(\cos\theta_{12})$, where θ_{12} is the angle between particles 1 and 2. Thus the number of states in the element $dp_1 \, dp_2 \, d(\cos\theta_{12})$ is

$$dN = \text{const} \, p_1^2 \, dp_1 \, p_2^2 \, dp_2 \, d(\cos\theta_{12}).$$

The matrix element for the interaction is usually cast in Lorentz-invariant form, by normalizing the wavefunctions of the particles in their individual rest frames. A volume V in the particle rest frame will be Lorentz-contracted to a value Vm/E in a reference frame (in this case the CMS) where the particle has total energy E. The phase-space expression can be made relativistically invariant by including a factor m/E or $1/E$ for each final-state particle (see Appendix D). This then has the form

$$dN = \text{const} \, \frac{p_1^2 \, dp_1 \, p_2^2 \, dp_2 \, d(\cos\theta_{12})}{E_1 E_2 E_3}. \tag{4.28}$$

Using the relations

$$E_1^2 = p_1^2 + m_1^2, \qquad E_2^2 = p_2^2 + m_2^2,$$

$$E_3^2 = p_3^2 + m_3^2 = p_1^2 + p_2^2 + m_3^2 + 2p_1 p_2 \cos\theta,$$

$$E_1 \, dE_1 = p_1 \, dp_1, \qquad E_2 \, dE_2 = p_2 \, dp_2,$$

$$(E_3 \, dE_3)_{p_1, p_2 \text{ fixed}} = p_1 p_2 \, d(\cos\theta),$$

one obtains

$$dN = \text{const} \, \frac{E_1 \, dE_1 \, E_2 \, dE_2 \, E_3 \, dE_3}{E_1 E_2 E_3}.$$

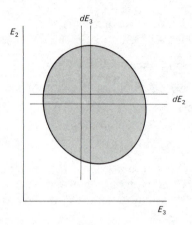

Fig. 4.9

The density of final states is obtained by dividing by dE_f, where $E_f = E_1 + E_2 + E_3$ is the total energy and, for E_1 and E_2 fixed, $dE_f = dE_3$. So

$$\rho = \frac{dN}{dE_f} = \text{const } dE_1 \, dE_2. \tag{4.29}$$

Thus, three-body phase space, by itself, predicts that if we plot E_2 and E_1 (or E_2 and E_3) along orthogonal axes, the density of events per unit area (i.e. per unit of $dE_2 \, dE_1$) should be uniform (Fig. 4.9). In the sketch the curve indicates the kinematically allowed region. Such a plot is called a Dalitz plot. Clearly, if $|M|^2$ depends on the momenta and angles of the particles, then one will not get a uniform Dalitz plot:

$$\rho = |M(E_1, E_2)|^2 \, dE_1 \, dE_2, \tag{4.30}$$

so that the density at any point is a measure of the square of the matrix element.

4.6.2 $K_{\pi 3}$-Decay

The original Dalitz plot was made for the decay

$$K^+ \rightarrow \pi^+ + \pi^+ + \pi^-, \qquad Q = 75 \text{ MeV} = \text{total kinetic energy}.$$

In this case, the three particles have equal mass and are nonrelativistic (or very nearly so). Let $\mathbf{p}_i$ and T_i be the momenta and kinetic energies of the pions ($i = 1$, 2, 3). One can then make the construction shown in Fig. 4.10. ABC is an equilateral triangle of height Q. Then, if one plots T_1, T_2, T_3 along axes normal to the sides of the triangle, all events (represented by points P) will lie inside the triangle, since

$$Q = T_1 + T_2 + T_3.$$

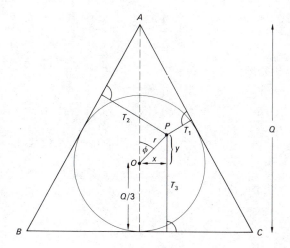

Fig. 4.10

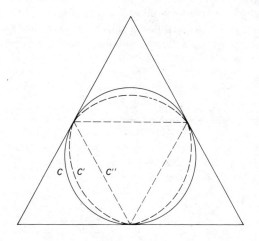

Fig. 4.11

Furthermore, since $\mathbf{p}_1 + \mathbf{p}_2 + \mathbf{p}_3 = 0$, the point P is restricted to lie inside the inscribed circle, center O and radius $Q/3$. In order to prove this, one can express T_1, T_2, T_3 as a function $f(r, \phi, Q)$, where r and ϕ are indicated in Fig. 4.10. Eliminating ϕ, one gets

$$r^2 = \tfrac{4}{9}[Q^2 - 3(T_1 T_2 + T_2 T_3 + T_1 T_3)].$$

Also,

$$|p_1 - p_2| < p_3 < |p_1 + p_2|.$$

Then, using the nonrelativistic relation $p^2 = 2mT$, one obtains

$$|T_3 - T_1 - T_2| < 2\sqrt{T_1 T_2}, \quad \text{etc.,}$$

and, substituting in the expression for r^2, one then finds

$$r^2 < Q^2/9, \tag{4.31}$$

i.e., P must lie within the inscribed circle. This is only true *nonrelativistically*. Even in $K_{\pi 3}$-decay there is a slight correction, so that C becomes distorted to C'. In the *extreme relativistic case* ($p \simeq E \simeq T$), the boundary becomes the inscribed triangle C'' (Fig. 4.11).

Returning to Fig. 4.10, if the Cartesian coordinates of P relative to O are denoted by x and y, then

$$y = T_3 - \frac{Q}{3}, \qquad \frac{dT_3}{dy} = \frac{dE_3}{dy} = -1,$$

$$x = \frac{(T_2 - T_1)}{\sqrt{3}}, \qquad \left(\frac{dT_2}{dx}\right)_{T_1} = \frac{dE_2}{dx} = \sqrt{3}.$$

Then from (4.29) it follows that again the density of points in Fig. 4.10 will be uniform if the matrix element is a constant. The display of the *kinetic energies* along three axes at $120°$ thus results in a simple circular Dalitz plot, for the case of three final-state particles with equal masses and nonrelativistic velocities in the CMS.

Table 4.5

l_-	l_+	
	0	2
0	0^-	2^-
1	1^+	$3^+, 2^+, 1^+$
2	2^-	$4^-, 3^-, 2^-, 1^-, 0^-$

We may now ask what are the departures from phase space for various spin-parity assignments to the $\pi^+ \pi^+ \pi^-$ combination in $K_{\pi 3}$-decay. Let l_+ be the angular momentum of the two *like* pions in the CMS, and let l_- be the angular momentum of the π^- relative to this dipion system. If J is the total angular momentum, then $|l_+ - l_-| \leqslant J \leqslant |l_+ + l_-|$, since the pions are spinless. Now by symmetry, the orbital angular momentum l_+ of two identical pions must be even. We therefore obtain the assignments shown in Table 4.5 for the spin-parity J^P of the three-pion state, for l_+ or $l_- \leqslant 2$.

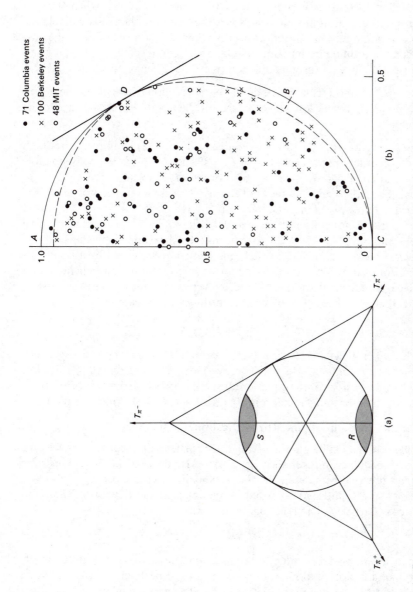

Fig. 4.12 (a) Expected regions of depletion, R and S, in the $K \to 3\pi$ Dalitz plot, for a nonzero kaon spin, J_K. (b) Observed distribution of events in $K \to 3\pi$, showing uniform population, and thus $J_K = 0$. All events are plotted in one semicircle. The dashed curve is the boundary using relativistic kinematics, and departs slightly from a circle. (After Orear *et al.* 1956.)

If $J_K \neq 0$, then l_+ and/or l_- must be nonzero. We may consider the following possibilities:

(a) $l_+ \geqslant 2$.   Here we expect the matrix element to vanish when the two positive pions are mutually at rest, since they cannot then possess relative angular momentum. This situation occurs when T_- is a maximum, the two positive pions recoiling in the same direction with equal velocity. Thus, for any assignment $l_+ > 0$, we expect a depletion of events in the region S of Fig. 4.12.

(b) $l_- \geqslant 1$.   In this case, there should be a depletion in density in the region R, corresponding to very low kinetic energy of the negative pion, relative to the dipion pair.

The observed Dalitz plot in $K \to 3\pi$ decay (Fig. 4.12) is uniform to a high degree of accuracy, proving that $l_+ = l_- = 0$ and $J_K = 0$ is the only possible assignment. (The decay $K^0 \to \pi^+ \pi^-$ also proves J_K is even.)

It must be emphasized that the above analysis gives no information about the kaon *parity*. Three pions in a relative S-state have parity $(-1)^3 = -1$; since, however, parity is not conserved in the weak decay $K \to 3\pi$, this tells us nothing about the initial state. The kaon parity may be determined by the study of *hypernuclei*. These are nuclei in which one neutron is replaced by a bound Λ-hyperon. For example, when negative kaons are brought to rest in helium, and undergo capture from an atomic S-state, a few per cent of the events give rise to bound, rather than free, Λ-hyperons, according to the reaction

$$K^- + {}^4\text{He} \to {}^4\text{H}_\Lambda + \pi^0,$$

where the hypernucleus ${}^4\text{H}_\Lambda$ consists of a triton (${}^3\text{H}$) plus a bound Λ-particle. Detailed measurements of its (weak) decay modes establish that $J_{{}^4\text{H}_\Lambda} = 0$. Thus $J = l = 0$ on both sides of the equation. If the Λ-parity is defined to be positive, like that of the nucleon, the kaon must then have $J^P = 0^-$, like the pion.

4.6.3 Dalitz Plots Involving Three Dissimilar Particles

As an example of a Dalitz plot involving three particles of unequal mass, we take the case of the hyperon resonance $Y_1^*(1385)$ — also denoted $\Sigma(1385)$. This state was first observed by Alston *et al.* in 1960 in interactions of 1.15-GeV/c K^--mesons in a liquid-hydrogen bubble chamber at the Lawrence Radiation Laboratory, Berkeley. The reaction studied was

$$K^- + p \to \pi^+ + \pi^- + \Lambda. \tag{4.32}$$

The kinetic energies of π^+- and π^--mesons, as computed in the overall center-of-mass frame, are plotted along the x- and y-axes respectively. If Q is the total available kinetic energy of the three final-state particles in this frame, then $T_\Lambda = Q - (T_{\pi^+} + T_{\pi^-})$, and lines of constant T_Λ are thus inclined at 45° to the axes, as shown in Fig. 4.13. Momentum-energy conservation keeps the points representing individual events inside the distorted ellipse shown. If there are no

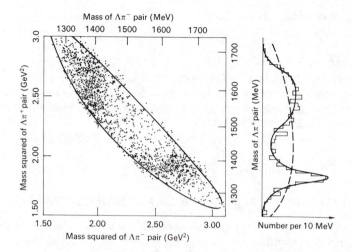

Fig. 4.13 Dalitz plot of the $\Lambda\pi^+\pi^-$ events from reaction (4.32), as measured by Shafer *et al.* (1963), for 1.22-GeV/c incident momentum. The effective $\Lambda\pi^+$ mass spectrum is shown at right. The dashed curve is that expected for a phase-space distribution (ordinate equal to the interval in $M^2_{\Lambda\pi^-}$ within the Dalitz-plot boundary), while the full curve corresponds to a Breit-Wigner resonance expression fitted to the $\Lambda\pi^+$ and $\Lambda\pi^-$ systems.

strong correlations in the final state (4.32), the density of points should be uniform, as usual.

Figure 4.13 shows clearly the strong departures from uniform density, as evidenced by horizontal and vertical bands on the plot, corresponding to favored values of T_{π^-} and T_{π^+}. This means that the reaction (4.32) is then proceeding as a two-body reaction, the two bodies consisting of one pion ($\pi 1$) and a resonant state of the Λ with the other pion ($\pi 2$), with (more or less) unique mass $M_{\Lambda\pi} = 1385$ MeV. Thus, if p is the CMS momentum of each, the total CMS energy will be

$$W = \sqrt{M^2_{\Lambda\pi} + p^2} + \sqrt{m^2_\pi + p^2}, \tag{4.33}$$

so that, for fixed $M_{\Lambda\pi}$, the momentum p is unique. Obviously, if there is a broad resonance, the quantity T_π will have a spread in values, and thus form a band rather than a line on the Dalitz plot. Note that *both* vertical and horizontal bands are found, since either π^+ or π^- can resonate with the Λ-hyperon. Thus

$$Y^*_1(1385) \rightarrow \Lambda + \pi^\pm,$$
$$I \quad\quad 1 \quad\quad\quad 0 \quad 1$$

and this resonance must have isospin 1, since it decays by strong interaction and isospin is conserved.

In the Dalitz plot it is usual to display $M^2_{\Lambda\pi^-}$ instead of T_{π^+} along the x-axis (and $M^2_{\Lambda\pi^+}$ along the y-axis), so that one can read off the $\Lambda\pi$ invariant mass directly. It is readily shown that

$$M^2_{\Lambda\pi^-} = W^2 + m^2_\pi - 2WE_{\pi^+} = a + bT_{\pi^+}, \qquad (4.34)$$

so that $M^2_{\Lambda\pi^-}$ is proportional to T_{π^+} for fixed W, the total CMS energy. Figure 4.13 also includes the $\Lambda\pi^+$ mass spectrum. The pure phase-space distribution (4.29) is shown, together with a curve obtained by assuming that either $\Lambda\pi^+$ or $\Lambda\pi^-$ may resonate, the resonance being described by the Breit-Wigner formula (4.51) with a width $\Gamma = 40$ MeV.

Finally, the spin-parity of the $Y^*_1(1385)$ has been determined to be $J^P = \frac{3}{2}^+$; this conclusion is arrived at by analysis of the polarization in the Λ-decay, relative to the production plane of reaction (4.32).

4.7 WAVE-OPTICAL DISCUSSION OF HADRON SCATTERING

Consider a beam of particles to be represented by a plane wave traveling in the z-direction and incident on a spinless target particle, or scattering center. Such an incident wave, which we take as of unit amplitude, is represented by

$$\psi_i = e^{ikz},$$

where $k = 1/\lambda$ and $2\pi\lambda$ is the de Broglie wavelength, and where the time dependence $e^{-i\omega t}$ has been omitted for brevity. A plane wave can be represented as a superposition of spherical waves, incoming and outgoing. At a radial distance r from the scattering center, such that $kr \gg 1$, the radial dependence of these spherical waves has the form $e^{\pm ikr}/kr$, so that the flux through a spherical shell is independent of r, as it must be to conserve probability. The angular dependence is determined by the Legendre polynomials $P_l(\cos\theta)$.

The expansion is $(kr \gg 1)$*

$$\psi_i = e^{ikz} = \frac{i}{2kr}\sum_l (2l + 1)[(-1)^l e^{-ikr} - e^{ikr}]P_l(\cos\theta), \qquad (4.35)$$

where the first term in square brackets denotes the incoming wave and the

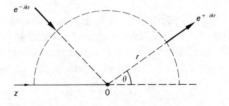

Fig. 4.14

---

* See, for example: F. Mandl, *Quantum Mechanics*, Butterworths, London, 1957, p. 166; L. I. Schiff, *Quantum Mechanics*, McGraw-Hill, New York, 1955, p. 103.

second denotes the outgoing wave as in Fig. 4.14 (the sense is clear if one includes the time-dependent term $e^{-i\omega t}$). In all problems with which we deal here, k is typically 10^{13} cm^{-1}, and r, the distance from the scattering center where the particle wave is observed, is many centimeters. Thus, the asymptotic form (4.35) is appropriate.

The scattering center or potential cannot affect the incoming waves, but can in general alter both phase and amplitude of the outgoing wave. The change of phase of the lth partial wave is denoted by $2\delta_l$, and its amplitude by η_l, where $1 > \eta_l > 0$.

The total wave now has the asymptotic form

$$\psi_{\text{total}} = \frac{i}{2kr} \sum_l (2l+1)[(-1)^l e^{-ikr} - \eta_l e^{2i\delta_l} e^{ikr}] P_l(\cos\theta). \qquad (4.36)$$

Thus, the scattered wave, representing the difference between the outgoing waves with and without the scattering potential, will be

$$\psi_{\text{scatt.}} = \psi_{\text{total}} - \psi_i = \frac{e^{ikr}}{kr} \sum_l (2l+1) \frac{(\eta_l e^{2i\delta_l} - 1)}{2i} P_l(\cos\theta)$$

$$= \frac{e^{ikr}}{r} F(\theta), \qquad (4.37)$$

where the scattering amplitude

$$F(\theta) = \frac{1}{k} \sum_l (2l+1) \left(\frac{\eta_l e^{2i\delta_l} - 1}{2i} \right) P_l(\cos\theta). \qquad (4.38)$$

We note that this corresponds to an *elastically* scattered wave, since the wavenumber k is taken to be the same before and after scattering. This can be true in the laboratory frame only if the scattering center is infinitely massive. In general, the target (scattering) particle will acquire both momentum and energy; thus the quantities k and λ strictly refer to the properties of the wave in the center-of-momentum frame of the incident and target particles, so that they do not change in an elastic collision. The scattered outgoing flux in solid angle $d\Omega$, through a sphere of radius r, is

$$v_0 \psi_{\text{scatt.}} \psi^*_{\text{scatt.}} r^2 \, d\Omega = v_0 |F(\theta)|^2 \, d\Omega, \qquad (4.39)$$

where v_0 is the velocity of the outgoing particles (relative to the scattering center). But (4.39) is, by definition, the product of the scattering cross-section and the incident flux ($= v_i \psi_i \psi_i^* = v_i$). Since $v_i = v_0$ for elastic scattering,

$$v_0 \, d\sigma = v_0 |F(\theta)|^2 \, d\Omega,$$

or

$$\left(\frac{d\sigma}{d\Omega} \right)_{\text{elastic}} = |F(\theta)|^2. \qquad (4.40)$$

The Legendre polynomials P_l obey the orthogonality condition

$$\int P_l P_{l'} \, d\Omega = \frac{4\pi\delta_{l,l'}}{2l+1},$$

where

$$\delta_{l,l'} = 1 \qquad \text{for} \quad l = l',$$
$$= 0 \qquad \text{for} \quad l \neq l'.$$

Thus, the total *elastic* scattering cross section, integrated over angle, is, from (4.38) and (4.40),

$$\sigma_{\text{el.}} = 4\pi\lambda^2 \sum_l (2l+1) \left| \frac{\eta_l e^{2i\delta_l} - 1}{2i} \right|^2. \tag{4.41}$$

When $\eta = 1$, the case for no absorption of the incoming wave, this becomes

$$\sigma_{\text{el.}} = 4\pi\lambda^2 \sum_l (2l+1) \sin^2 \delta_l. \tag{4.42}$$

Obviously, $\sigma_{\text{el.}}$ is zero when $\delta_l = 0$, corresponding to zero scattering potential. If $\eta < 1$, the *reaction* cross section, σ_r, is then obtained from conservation of probability:

$$\sigma_r = \int (|\psi_{\text{in}}|^2 - |\psi_{\text{out}}|^2) r^2 \, d\Omega,$$

where ψ_{in} is the first term of (4.35) and ψ_{out} is the second term of (4.36). This gives

$$\sigma_r = \pi\lambda^2 \sum_l (2l+1)(1 - |\eta_l|^2). \tag{4.43}$$

The *total* cross section will be

$$\sigma_T = \sigma_r + \sigma_{\text{el.}} = \pi\lambda^2 \sum_l (2l+1)2(1 - \eta_l \cos 2\delta_l).$$

Since $P_l(1) = 1$ for all l, (4.38) therefore gives in the *forward direction* $\cos \theta = 1$, $\theta = 0$:

$$\text{Im } F(0) = \frac{1}{2k} \sum_l (2l+1)(1 - \eta_l \cos 2\delta_l).$$

Comparing the last two equations finally gives us the *optical theorem*,

$$\text{Im } F(0) = \frac{k}{4\pi} \sigma_T, \tag{4.44}$$

relating the total cross-section to the imaginary part of the forward elastic scattering amplitude.

The relations derived above describe the various cross-sections $\sigma_{\text{el.}}$, σ_r, and σ_T in terms of the parameters η and δ. They set bounds on the cross-sections imposed by the conservation of probability (often called the unitarity condition). For example, we see from (4.42) that the maximum elastic scattering cross-section for the lth partial wave occurs when $\delta_l = \pi/2$, having the value

$$\sigma_{\text{el.}}^{\text{max}} = 4\pi\lambda^2(2l + 1), \tag{4.45}$$

for $\eta_l = 1$, i.e. the case where there is pure scattering without absorption. Similarly, from (4.43) we obtain the maximum absorption or reaction cross-section by setting $\eta_l = 0$:

$$\sigma_r^{\text{max}} = \pi\lambda^2(2l + 1).$$

This last equation can be reproduced from a simple classical argument. An orbital angular momentum l corresponds to an "impact parameter" b given by $l\hbar = pb$, or $b = l\lambda$. Particles of angular momentum $l \rightarrow l + 1$ therefore impinge on, and are absorbed by, an annular ring of cross sectional area

$$\sigma = \pi(b_{l+1}^2 - b_l^2) = \pi\lambda^2(2l + 1).$$

Note also that, for the case of complete absorption, $\eta = 0$, the elastic cross-section is also $\pi\lambda^2(2l + 1)$ − see Section 4.12.

The quantity

$$f(l) = \frac{\eta_l e^{2i\delta_l} - 1}{2i} = \frac{i}{2} - \frac{i\eta_l}{2}e^{2i\delta_l} \tag{4.46}$$

of (4.38) is the elastic scattering amplitude (for the lth partial wave). It is a

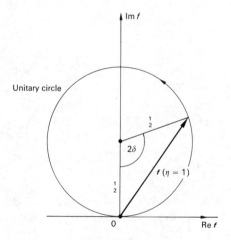

Fig. 4.15 The scattering amplitude **f** plotted as a vector in the complex plane. Causality requires that, as the energy increases, the vector will trace out the circle in an anti-clockwise direction.

complex quantity, and in Fig. 4.15 we show $\mathbf{f}$ plotted as a vector in the complex plane. For $\eta = 1$, the end of the vector describes a circle of radius $\frac{1}{2}$ and center $i/2$, as the phase shift varies from 0 to $\pi/2$. When $\delta = \pi/2$, $\mathbf{f}$ is purely imaginary and has magnitude unity, corresponding to (4.45). If $\eta < 1$, the end of the vector $\mathbf{f}$ lies within the circle shown, sometimes referred to as the *unitary circle*. It is so called because, as $\mathbf{f}$ is defined, its maximum modulus must be unity if probability is conserved — the intensity in a particular outgoing partial wave cannot exceed that in the corresponding incoming wave.

4.8 THE BREIT-WIGNER RESONANCE FORMULA

Let us now make the connection between the preceding wave-optical discussion and the scattering of two elementary particles — one corresponding to the incident wave and the other to the scattering center. To simplify matters let the two particles be spinless. If the elastic scattering amplitude $f(l)$ passes through a maximum for a particular value of l and for a particular CMS wavelength λ, the two particles are said to *resonate*. The resonant state is then characterized by a unique angular momentum or spin $J = l$, a unique parity and isospin, and a mass corresponding to the total CMS energy of the two particles. A criterion of resonance is that the phase shift δ_l of the lth partial wave should pass through $\pi/2$. The cross-section may also be described in terms of the width Γ or lifetime τ of the resonant state, as follows: dropping the subscript l in (4.46), and with $\eta = 1$, we may rewrite f in the form

$$f = \frac{e^{i\delta}(e^{i\delta} - e^{-i\delta})}{2i}$$

$$= e^{i\delta} \sin \delta = \frac{1}{\cot \delta - i}. \tag{4.47}$$

Near resonance $\delta \simeq \pi/2$, so that $\cot \delta \simeq 0$. If E is the total energy of the two-particle state in the CMS, and E_R is the value of E at resonance ($\delta = \pi/2$), then expanding by a Taylor series

$$\cot \delta(E) = \cot \delta(E_R) + (E - E_R)\left[\frac{d}{dE} \cot \delta(E)\right]_{E=E_R} + \cdots$$

$$\simeq -(E - E_R)\frac{2}{\Gamma},$$

where $\cot \delta(E_R) = 0$ and we have defined $2/\Gamma = -[d(\cot \delta(E))/dE]_{E=E_R}$. Neglecting further terms in the series is justified provided $|E - E_R| \simeq \Gamma \ll E_R$. Then from (4.47),

$$f(E) = \frac{1}{\cot \delta - i} = \frac{\Gamma/2}{(E_R - E) - i\Gamma/2}. \tag{4.48}$$

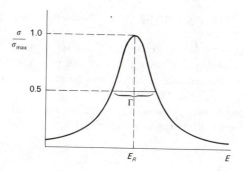

Fig. 4.16 Breit-Wigner resonance curve.

From (4.41) and (4.46), we obtain for the elastic scattering cross section

$$\sigma_{el.}(E) = 4\pi\lambda^2(2l + 1)\frac{\Gamma^2/4}{(E - E_R)^2 + \Gamma^2/4}. \tag{4.49}$$

This is known as the *Breit-Wigner formula.* The resonance curve of $\sigma(E)$ is shown in Fig. 4.16. The width Γ is defined so that the elastic cross section $\sigma_{el.}$ falls by a factor 2 from the peak value when $|E - E_R| = \pm \Gamma/2$.

As pointed out in Chapter 1, the width Γ and lifetime τ of the resonant state are connected by the relation $\tau = \hbar/\Gamma$. The energy dependence of the amplitude (4.49) is simply the Fourier transform of an exponential time pulse, corresponding to the radioactive decay of the resonance. The wavefunction of a nonstationary decaying state of central angular frequency $\omega_R = E_R/\hbar$ and lifetime $\tau = \hbar/\Gamma$ can be written

$$\psi(t) = \psi(0)e^{-i\omega_R t}e^{-t/2\tau}$$

$$= \psi(0)e^{-t(iE_R + \Gamma/2)} \tag{4.50}$$

in units $\hbar = c = 1$. The intensity $I(t) = \psi\psi^* = I(0)e^{-t/\tau}$ obeys the normal exponential law of radioactive decay. The Fourier transform of this expression is

$$g(\omega) = \int_0^\infty \psi(t)e^{i\omega t}\, dt,$$

with $\omega = E/\hbar = E$. The amplitude as a function of E is then

$$\chi(E) = \int \psi(t)e^{iEt}\, dt = \psi(0)\int e^{-t[(\Gamma/2) + iE_R - iE]}\, dt$$

$$= \frac{K}{(E_R - E) - i\Gamma/2},$$

where K is some constant.

The final step consists of making the connection between the probability, integrated over time, that the resonant state will decay with total energy E, and the cross section $\sigma_{el.}(E)$. If the resonance is purely elastic, it is clear from conservation of probability that the cross section — which measures the probability of forming the resonant state — must be proportional to the probability of decay. Thus we may write $\sigma(l, E) \propto \chi^*\chi$, where $\chi = \chi_l(E)$ and l is the value of the orbital angular momentum involved in forming the resonant state in question. Both of these quantities have a maximum when $E = E_R$ and $\delta_l = \pi/2$, so that

$$(\sigma_{el.})_{max} = 4\pi\lambda^2(2l + 1) \quad \text{and} \quad (\chi^*\chi)_{max} = 4K^2/\Gamma^2.$$

Thus, in terms of Γ, measuring the width of a decay process, rather than δ, measuring the phase-shift in a scattering process, we have

$$\sigma_{el.} = 4\pi\lambda^2(2l + 1)\frac{\Gamma^2/4}{(E - E_R)^2 + \Gamma^2/4},$$

as in (4.49).

For a spinless projectile hitting a spinless target, $l = J$, the total angular momentum of the resonant state. Therefore, $(2l + 1) \rightarrow (2J + 1)$. For spin-0 particles (e.g. pions, kaons) incident on nucleons (spin $\frac{1}{2}$), the factor $(2J + 1)$ still applies,* except that now only half the target spin states can contribute. The point is that the spin and parity (J^P) of the resonance are fixed, so that only *one* l-value can contribute, and then only if the target particle has the right spin orientation. For example, the $\Delta(1238)$ πp resonance has $J^P = \frac{3}{2}^+$ and results from a p-wave ($l = 1$) pion-nucleon interaction. Thus the $J = l + \frac{1}{2}$ combination forms this state, not $J = l - \frac{1}{2}$, which has $J^P = \frac{1}{2}^+$ (for $l = 1$). A d-wave interaction ($l = 2$) with $J = l - \frac{1}{2}$ would give the right resonance spin but the wrong parity ($J^P = \frac{3}{2}^-$). For a resonance of angular momentum J formed from a nucleon target by a spinless incident particle, (4.49) becomes

$$\sigma_{el.}(E) = \frac{\pi\lambda^2}{2}\frac{(2J + 1)\Gamma^2}{(E - E_R)^2 + \Gamma^2/4}. \tag{4.51}$$

This is a particular form of the general expression for the cross-section for the formation and decay of a resonance of angular momentum J in the collision of particles of spins s_a and s_b, it being understood that the cross-section is averaged over the spin states of a and b:

$$\sigma_{el.}(E) = \frac{4\pi\lambda^2(2J + 1)\Gamma^2/4}{(2s_a + 1)(2s_b + 1)[(E_R - E)^2 + \Gamma^2/4]}. \tag{4.51a}$$

The spin multiplicity factors follow from considerations of the quantum states available in the reversible reaction $a + b \rightleftharpoons c$, as in (1.31).

---

* Intuitively we see that this is just the spin multiplicity factor for a state J. For a proof, see J. M. Blatt and V. F. Weisskopf, *Theoretical Nuclear Physics*, John Wiley, New York, 1952, p. 423.

In both (4.49) and (4.51) it is assumed that the resonance can only decay *elastically*, i.e. $\pi + n \rightarrow \Delta \rightarrow \pi + n$, but not $\Delta \rightarrow 2\pi + n$. In general, $\Gamma = \Gamma_{\text{el.}} + \Gamma_r$, where $\Gamma_{\text{el.}}$ and Γ_r are *partial widths* for decay in the elastic channel and inelastic channel. Then for $\sigma_{\text{el.}}(E)$, the Γ^2 numerator in (4.51) should be replaced by $\Gamma_{\text{el.}}^2$. The inelastic cross section $\sigma_r(E)$ is the same as (4.51), with the Γ^2 numerator replaced by $\Gamma_{\text{el.}}\Gamma_r$.

The scattering amplitude f, in the case of an elastic resonance ($\eta = 1$), traces out the unitary circle of Fig. 4.15. By applying the concept of causality – the outgoing wave cannot leave the scattering center before the incoming wave arrives – it may be proved that f traces out a circle in the *anticlockwise* direction (for an attractive potential).

4.9 AN EXAMPLE OF A BARYON RESONANCE—THE Δ(1236)

Figure 4.6 shows the $\pi^+ p$ and $\pi^- p$ total cross sections as a function of kinetic energy of the incident pion. There is a very obvious $I = \frac{3}{2}$ resonance at $T_\pi = 195$ MeV, corresponding to a pion-proton mass of 1236 MeV. This is further discussed below. It is designated $P_{33}(1236)$, meaning that it is a p-wave ($l = 1$) pion-nucleon resonance, of $I = \frac{3}{2}$ and $J = \frac{3}{2}$. One can also distinguish other

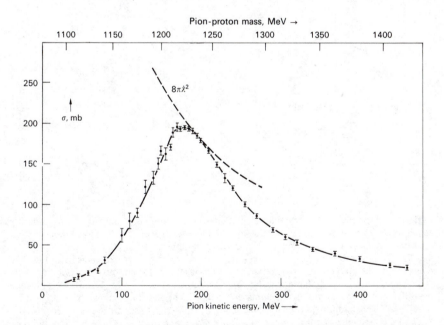

Fig. 4.17 The $\pi^+ p$ total cross section as a function of kinetic energy of the incident pion, or the $\pi^+ p$ mass, in the region of the 1236 MeV, $I = \frac{3}{2}$, $J^P = \frac{3}{2}^+$ resonance. Not all experimental points have been included. The maximum cross section, $8\pi\lambda^2$, allowed by conservation of probability is shown dashed.

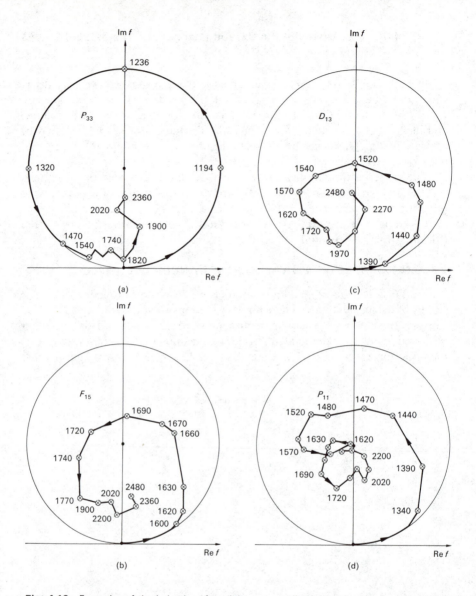

Fig. 4.18 Examples of the behavior of partial-wave amplitudes in pion-nucleon scattering, according to a phase-shift analysis at Saclay (Paris) by Ayed *et al.* (1970). The numbers refer to the center-of-mass energy in MeV. (a) The p-wave amplitude of $I = \frac{3}{2}, J^P = \frac{3}{2}^+$, thus designated P_{33}. The dominant feature is that as the energy increases, the tip of the vector **f** of Fig. 4.15 moves along the unitary circle. The phase shift δ passes $90°$ at 1236 MeV, corresponding to the central resonance mass. Reference to (4.48) shows that the width Γ ($\simeq 120$ MeV) is given by the difference in energy at opposite ends of the diameter parallel to the real axis. Above 1500 MeV, $\eta < 1$, corresponding to the inelastic process $N^* \to N + 2\pi$ as well as $N + \pi$. (b), (c) The F_{15} and D_{13} amplitudes, the former indicating a resonance of $J^P = \frac{5}{2}^+, I = \frac{1}{2}$, of central mass 1690 MeV, and the latter one of $J^P = \frac{3}{2}^-, I = \frac{1}{2}$, and mass 1520 MeV. Note that both are strongly inelastic. (d) The P_{11} amplitude ($I = \frac{1}{2}, J^P = \frac{1}{2}^+$), demonstrating a resonance with these quantum numbers of central mass 1470 MeV. Note that this is *not* apparent as a bump in the cross-section in Fig. 4.6, and only phase-shift analysis is able to reveal its existence.

humps and bumps in σ(total). For example, in the $I = \frac{1}{2}$ channel one can see evidence for the states $D_{13}(1520)$ and $F_{15}(1688)$ as peaks in $\sigma(\pi^- p)$, and $F_{37}(1950)$ as a peak in $\sigma(\pi^+ p)$, i.e. $I = \frac{3}{2}$.

However, it turns out that there are many overlapping pion-nucleon resonances, with at least twenty states of $M_{\pi p} < 2200$ MeV. A mere inspection of the total cross section is therefore misleading. It is necessary to make a sophisticated *phase-shift analysis* of the pion-nucleon elastic scattering data, based on the behavior of the various polynomial coefficients required to fit the angular distributions, as a function of bombarding energy. The best values for the masses, widths, and elasticities of resonant states are fitted to the data by an iterative procedure. Different analyses give somewhat different solutions. Figure 4.18 shows an example of such an analysis for the P_{33}, F_{15}, P_{11}, and D_{13} waves.

For the lowest-lying $\Delta(1236)$ πN resonance, the amplitude is almost purely elastic because of the low mass, and the "tails" of higher-lying resonances may be neglected. From (4.51) we expect $\sigma_{\text{el.}} = 2\pi \lambda^2 (2J + 1)$ at the peak. The limiting value for $J = \frac{3}{2}$, $\sigma_{\text{el.}} = 8\pi\lambda^2$, is shown dashed in Fig. 4.17, clearly proving the $\Delta(1236)$ to be a p-wave resonance of spin-parity $J^P = \frac{3}{2}^+$.

The assignment $J = \frac{3}{2}$ for the spin of the $\Delta(1236)$ may be confirmed from the angular distribution of the $\pi^+ p$ elastic scattering in the region of this resonance. Take the incident pion direction as the quantization axis (z-axis). The angular-momentum wave function of a p-state pion will be $\phi(l, m) = \phi(1, 0)$ since the pion is spinless. For the proton, we have $\alpha(\frac{1}{2}, \pm\frac{1}{2})$ corresponding to the two possible orientations of spin (Fig. 4.19). The product is the state

$$\psi(j, m) = \psi(\tfrac{3}{2}, \pm\tfrac{1}{2}).$$

When the Δ radiates a pion, the remaining proton may or may not have its spin "flipped". From Table 4.1 of Clebsch-Gordan coefficients, we can write for the final-state wave functions α' and ϕ'

$$\psi(\tfrac{3}{2}, \tfrac{1}{2}) = \sqrt{\tfrac{1}{3}}\, \phi'(1, 1)\alpha'(\tfrac{1}{2}, -\tfrac{1}{2}) + \sqrt{\tfrac{2}{3}}\, \phi'(1, 0)\alpha'(\tfrac{1}{2}, \tfrac{1}{2}).$$

Note that, since the scattered pion comes off at some angle θ to the z-direction, it is now possible for its orbital angular momentum to have a finite projection (1 or 0) on the old quantization axis. The ϕ' are simply the spherical harmonics:

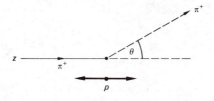

Fig. 4.19

$$\phi'(1,1) = Y_1^1 = -\sqrt{\frac{3}{4\pi}} \sin\theta \frac{e^{i\phi}}{\sqrt{2}},$$

$$\phi'(1,0) = Y_1^0 = \sqrt{\frac{3}{4\pi}} \cos\theta.$$

The angular distribution of the pions is therefore

$$I(\theta) = \psi\psi^* = \tfrac{1}{3}(Y_1^1)^2 + \tfrac{2}{3}(Y_1^0)^2,$$

the cross term being zero, since Y_1^1 and Y_1^0, as well as $\alpha'(\tfrac{1}{2}, -\tfrac{1}{2})$ and $\alpha'(\tfrac{1}{2}, \tfrac{1}{2})$, are orthogonal. Thus

$$I(\theta) \propto \sin^2\theta + 4\cos^2\theta = 1 + 3\cos^2\theta. \tag{4.52}$$

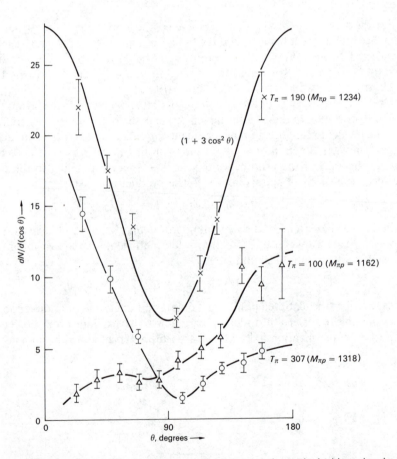

Fig. 4.20 The angular distribution of the scattered pion, relative to the incident pion, in $\pi^+ p$ elastic scattering, as measured in the center-of-mass frame. In the region of the Δ-resonance of mass 1236 MeV ($T_\pi = 190$ MeV), the distribution has the form $1 + 3\cos^2\theta$, as in (4.52).

The differential cross section is plotted in Fig. 4.20 as a function of CMS angle θ for different values of T_π. At resonance, the dependence is in agreement with (4.52).

4.10 CHEW-LOW PLOT

The phase shift due to a scattering potential will depend on the strength and effective range of the interaction. Using methods based on dispersion relations, Chew and Low (1956) derived an effective-range formula for the phase shift applicable to the Δ-resonance region (i.e. for a p-wave πN interaction):

$$\left[\frac{4}{3}\frac{k^3}{m^2\omega}\cot\delta\right]\frac{f^2}{4\pi} = 1 - \frac{\omega}{\omega_r},\tag{4.53}$$

where k is the CMS momentum of the pion, m its rest mass, and ω its total energy plus a recoil correction: $\omega = \sqrt{k^2+m^2}+k^2/2M$. $\omega = \omega_r$ at resonance ($\delta = \pi/2$) and $\hbar/\omega_r = a$, the effective range of the interaction. The quantity $f^2/4\pi$ is the pion-nucleon coupling constant. The value obtained from the linear plot (Fig. 4.21) of Eq. (4.53), due to Barnes et al. (1960), is

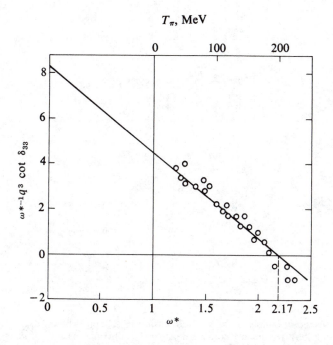

T_π, MeV

Fig. 4.21 Chew-Low plot for the $J = \frac{3}{2}$, $I = \frac{3}{2}$ pion-nucleon phase shift δ [Eq. (4.53)]. The intercept at $\omega/m = 0$ yields the pion-nucleon coupling constant f^2. (From Barnes et al. 1960.)

$$\frac{f^2}{4\pi} = 0.0877 \pm 0.0014 \qquad (4.54)$$

in units $\hbar = c = 1$. In these units $f^2/4\pi$ has the same dimensions as $e^2/4\pi = \frac{1}{137}$. Thus the constant specifying the strong coupling of pions to nucleons is two orders of magnitude larger than the electromagnetic coupling. Other methods of measuring f, for example from nucleon-nucleon scattering via pion exchange, give similar results.

4.11   EXAMPLES OF MESON RESONANCES

4.11.1   A Pion-Pion Resonance – The $\rho(770)$

As indicated in Chapter 6, the ρ-meson is observed as a resonance in e^+e^- annihilation at high energy (i.e. at CMS energy equal to m_ρ), via a virtual photon:

$$e^+ + e^- \rightarrow \gamma \rightarrow \rho \rightarrow \pi^+ + \pi^-.$$

Consequently, the ρ-meson has the same quantum numbers as the photon, that is $J^{PC} = 1^{--}$. Historically, the ρ-meson was first observed, and its quantum numbers first determined, in pion-proton collisions, and we discuss this analysis

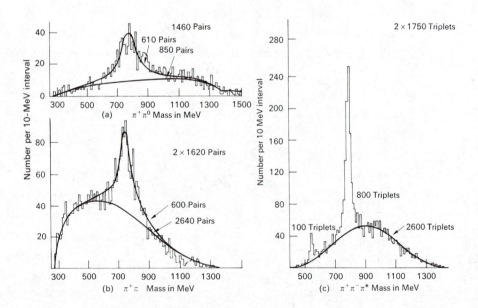

Fig. 4.22   Mass distributions of (a) $\pi^-\pi^+$ pairs from the reaction (4.55a), (b) $\pi^+\pi^-$ pairs from the reaction (4.55b). The smooth curves indicate the distributions expected from phase space. (c) The $\pi^+\pi^-\pi^0$ invariant-mass spectrum from reaction (4.55c). The narrow peak at 785 MeV corresponds to the $\omega \rightarrow 3\pi$ resonance, and that at 550 MeV to $\eta \rightarrow 3\pi$. (From Alff *et al.* 1962.)

as an example of the determination of quantum numbers in hadron-hadron interactions. Figure 4.22 shows results from a study in a hydrogen bubble chamber of the following reactions, using incident pions of momentum 1.6 to 1.9 GeV/c:

$$\pi^+ + p \rightarrow \pi^+ + \pi^0 + p \tag{4.55a}$$

$$\rightarrow \pi^+ + \pi^+ + \pi^- + p \tag{4.55b}$$

$$\rightarrow \pi^+ + \pi^+ + \pi^- + \pi^0 + p. \tag{4.55c}$$

The mass distributions of $\pi^+\pi^0$ pairs from the reaction (4.55a) and $\pi^+\pi^-$ pairs from (4.55b) show a broad peak centered at 765 MeV, with a width about 120 MeV, attributed to the decay $\rho \rightarrow 2\pi$. Experiments have failed to detect a similar peak in the $\pi^+\pi^+$ mode.

These results indicate an isospin assignment $I = 1$ for the ρ-meson. Since we have two identical bosons in the final state, $I = 1$ implies that the spatial wavefunction of the pion pair must be antisymmetric. The simplest possibility is $l = 1$, corresponding to the spin-parity quantum numbers $J^P = 1^-$ for the ρ-meson.

To prove the above spin-parity assignment, we have to rely on observations relating to the production process itself. The easiest way to determine the spin is to consider a special class of "peripheral" events, with small momentum transfer to the nucleon. For this purpose, consider the reactions

$$\pi^- + p \rightarrow \pi^- + \pi^0 + p \tag{4.56a}$$

$$\rightarrow \pi^+ + \pi^- + n. \tag{4.56b}$$

As discussed later, these events can be described by single pion exchange, as indicated diagrammatically in Fig. 4.23(a). If we now look in the rest frame of the ρ-meson [Fig. 4.23(b)], we are essentially observing the collision of the incident π^- with a virtual π^+ of the meson cloud surrounding the nucleon:

$$\pi^- + \pi^+_{(\text{virtual})} \rightarrow \rho \rightarrow \pi^+ + \pi^-.$$

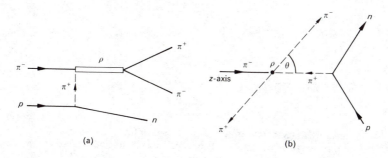

(a)　　　　　　　　　(b)

Fig. 4.23

If we choose the axis of quantization, z, along the incident pion direction, it is clear that in the initial, and hence also the final, state the z-component of angular momentum $m = 0$, since pions have zero spin. Thus, the angular distribution of the scattered pions will be

$$F(\theta) = (Y_l^m)^2 = [P_J^0(\cos\theta)]^2,$$

where J is the ρ-spin and, for $m = 0$, the azimuthal distribution $e^{im\phi}$ is isotropic.

The measured angular distributions for $\pi^-\pi^0$ pairs in (4.56a) is found to have the form

$$F(\theta) = A + B\cos\theta + C\cos^2\theta. \tag{4.57}$$

The fact that no terms higher than $\cos^2\theta$ appear is itself indicative that $J_\rho = 1$. The form (4.57) is interpreted as a coherent superposition of two amplitudes; A_1 due to a $J = 1$ resonance, and A_0 due to an S-wave $\pi\pi$ interaction (of $I = 2$, since

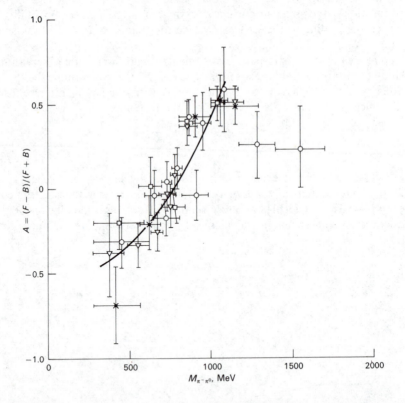

Fig. 4.24 Forward-backward ratio in the pion angular distribution as a function of $M_{\pi^0\pi^-}$ in the reaction (4.56a) as measured by various experiments for incident pion momenta from 2.75 to 6 GeV/c. The forward-backward asymmetry goes to zero at the ρ-mass ($M_{\pi^-\pi^0} = 765$ MeV). (After Baton *et al.* 1965.)

$J = 0$) forming a nonresonant background term. Then

$$F(\theta) = |A_0 + A_1 \cos \theta|^2$$

$$= |A_0|^2 + |A_1|^2 \cos^2 \theta + 2 \operatorname{Re} A_0 A_1^* \cos \theta, \qquad (4.58)$$

where we have used the fact that $P_0^0 = 1$, $P_1^0 = \cos \theta$. Now, as indicated in Fig. 4.15, the amplitude A_1 must become purely imaginary at the ρ-peak, so the interference ($\cos \theta$) term will change sign as one goes through it. This can be seen in terms of the forward-backward ratio of Fig. 4.24.

To summarize, the ρ-meson has $I = 1$, $G = +1$, $J^P = 1^-$. Its production in pion-nucleon interactions may be pictured in terms of collisions of the incident pion with a single virtual pion from the nucleon target.

4.11.2 A Three-Pion Resonance – The $\omega(790)$

Like the ρ-meson, the ω-meson is also produced in $e^+ e^-$ annihilation at high energy and consequently has $J^{PC} = 1^{--}$. Again it was first observed in bubble-chamber experiments in 1961 (Maglic *et al.*) in the analysis of antiproton annihilations of the type

$$p + \bar{p} \to \pi^+ + \pi^+ + \pi^- + \pi^- + \pi^0.$$

In this reaction, one can measure the magnetic curvatures, and hence momenta, of all the charged particles. The neutral pion decays to two γ-rays, which are not normally observed, since the conversion length in hydrogen is some 15 m. However, one can compute the missing energy and momentum and see if they are consistent with the π^0-mass, and in this way some 800 examples of the above reaction were uniquely identified. The invariant masses of all possible three-pion combinations were plotted, as shown in Fig. 4.25. For the $\pi^+ \pi^+ \pi^-$, $\pi^- \pi^- \pi^+$, $\pi^+ \pi^+ \pi^0$, and $\pi^- \pi^- \pi^0$ combinations, the invariant three-pion mass spectrum is consistent with a phase-space distribution. For the $\pi^+ \pi^- \pi^0$ events, however, there is a strong peak at $M_{3\pi} = 790$ MeV, with a width determined in later, more precise experiments to be $\Gamma = 10$ MeV. This departure from phase space proves the existence of a three-pion resonance ω, decaying by strong interactions, so that the reaction is $p\bar{p} \to \pi^+ \pi^- \omega$, $\omega \to \pi^+ \pi^- \pi^0$. Since there is no peak in the singly or doubly charged triplets, the isospin $I_\omega = 0$. Also, since $\omega \to 3\pi$, its G-parity is $(-1)^3 = -1$.

The full spin-parity analysis of the ω-decay Dalitz plot is more complicated than for $K_{\pi 3}$-decay. In the latter case, one could capitalize on the fact that there are two identical π^+-particles in the final state. The first point to remark on is that the Dalitz plot (Fig. 4.26) is not uniformly populated, so that the matrix element describing the process is not constant, and depends on the energies or momenta of the pions. The dependence of the spatial part of the matrix element on these quantities can be denoted by the amplitude

$$A = A(\mathbf{p}_i, E_i), \qquad (4.59)$$

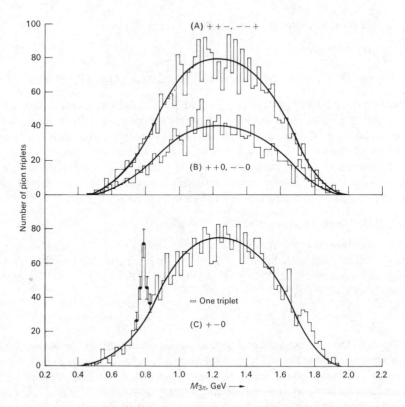

Fig. 4.25 Mass spectra of three-pion combinations selected from events of the type $p + \bar{p} \to \pi^+ + \pi^+ + \pi^- + \pi^- + \pi^0$. (A) Combinations of charge ± 1; (B) combinations of charge ± 2; (C) neutral combinations. The curves refer to the phase-space distribution, i.e. the spectrum expected if there is no strong pion-pion interaction in the final state. (After Maglic *et al.* 1961.)

where $\mathbf{p}_i$ and E_i refer to the ith pion ($i = 1, 2, 3$), and are of course measured in the ω rest frame.

The second point is that, since $I_\omega = 0$ and $I_\pi = 1$, any pair of pions must be in a state $I = 1$, the third having $I = 1$ to give a resultant of zero. As discussed earlier, a pair of like bosons with $I = 1$ must have an antisymmetric space wavefunction. Thus the amplitude A must change sign under interchange of any two of the indices i in (4.59). Thirdly, we consider the properties of the initial (ω) and final (3π) states under space inversion. Since $\omega \to 3\pi$ is a strong decay, the parity of initial and final states must be the same. Thus, if the space amplitude is even under inversion — that is, $A(-\mathbf{p}_i, E_i) = + A(\mathbf{p}_i, E_i)$ — then the final-state parity will be $(+1)(-1)^3 = -1$, where the second factor comes from the intrinsic parity of each pion. On the other hand, if A is odd, the final-state parity — and hence that of the ω-meson — must be even.

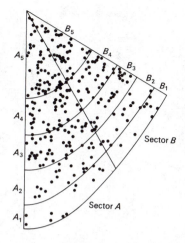

Fig. 4.26 Observed Dalitz-plot distribution for the $\omega \to 3\pi$ decay. Since the plot has sextant symmetry, all events have been displayed in one sextant. (After Stevenson *et al.* 1962.)

We now consider in turn various possible assignments J^P for the spin-parity of the ω-meson, and what they imply for the form of $A(\mathbf{p}_i, E_i)$, which in turn determines the Dalitz-plot population.

(a) $J^P = 0^-$. A particle of zero spin and odd parity is called *pseudoscalar*, since it has the spatial transformation properties of such a quantity. From the above discussion, it follows that A must be a scalar quantity, that is, even under space inversions. The simplest expression with the required properties is

$$A = (E_1 - E_2)(E_2 - E_3)(E_3 - E_1). \tag{4.60}$$

This is clearly a scalar, involves the energies of all three pions, and changes sign under interchange of any two pion labels. It is obviously not the most general form one can write. For example, one could multiply (4.60) by some symmetric scalar function of the energies, or raise it to any odd power. However, one always starts out by trying the simplest matrix elements, in the hope that one of them will work. In any case we are not attempting in the first instance to predict the exact Dalitz-plot population, but rather enquiring about general features, such as where the density is zero or has a maximum value. The most obvious prediction from (4.60) is that the population should vanish at the center of the plot, when $E_1 = E_2 = E_3$, as well as along radial lines separated by 60° (when $E_1 = E_2$, $E_2 = E_3$, or $E_1 = E_3$). After some algebra the population is found to be

$$A^2 = [(E_1 - E_2)(E_2 - E_3)(E_3 - E_1)]^2 \propto r^6 \sin^2 3\phi, \tag{4.61}$$

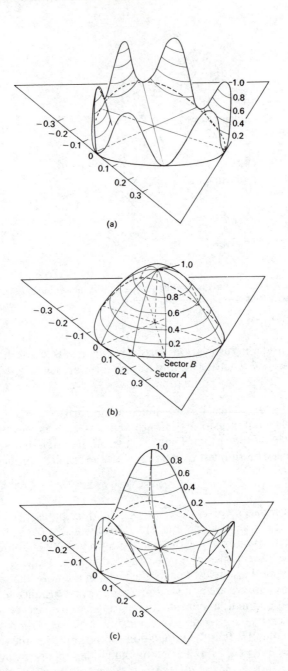

Fig. 4.27 Dalitz-plot densities for the decay into three pions of an ω-meson of spin-parity: (a) 0^- (pseudoscalar), (b) 1^- (vector), (c) 1^+ (axial vector). The density is represented by the height in the three-dimensional diagrams.

where the coordinates r and ϕ are defined in Fig. 4.10. The population has a maximum at the boundary, and has the distribution shown in Fig. 4.27(a). The observed distribution, Fig. 4.26, shows these predictions to be incorrect, so that the ω-meson cannot be pseudoscalar.

(b) $J^P = 0^+$. In this case, the ω-meson would be *scalar*, so that A must be odd under inversion and hence a pseudoscalar. The only simple way to form a pseudoscalar from the pion momentum vectors is to build an axial vector from two of them, and take the scalar product of this with the third pion. Thus, one can form terms of the type $(\mathbf{p}_1 \times \mathbf{p}_2) \cdot \mathbf{p}_3$. However, since $\mathbf{p}_1, \mathbf{p}_2$, and $\mathbf{p}_3$ must be coplanar in order to conserve momentum, all such terms are zero. Thus, the ω-meson cannot be scalar, nor can any particle decaying to three pions, if parity is conserved in the decay.

(c) $J^P = 1^-$. With this spin-parity, the ω-meson would be a *vector* particle, since it is described by an amplitude or wavefunction with the spatial properties of a 3-vector. We therefore require A to transform like a vector, but to be even under space inversion, and thus be an axial vector. From the polar vectors $\mathbf{p}_i$, one constructs the requisite form for A as follows:

$$A = (\mathbf{p}_1 - \mathbf{p}_2) \times \mathbf{p}_3 + (\mathbf{p}_2 - \mathbf{p}_3) \times \mathbf{p}_1 + (\mathbf{p}_3 - \mathbf{p}_1) \times \mathbf{p}_2, \qquad (4.62)$$

which, as before, is antisymmetric under pion label interchange. Since $\mathbf{p}_1 + \mathbf{p}_2 + \mathbf{p}_3 = 0$, this is more simply written

$$A = 6\mathbf{p}_2 \times \mathbf{p}_1. \qquad (4.63)$$

Then A must vanish whenever $\mathbf{p}_1$ and $\mathbf{p}_2$ (and therefore $\mathbf{p}_3$ also) are parallel or antiparallel. This is precisely the condition at the boundary of the Dalitz plot; given the numerical values of p_1 and p_2, p_3 must lie in the interval $|p_1 - p_2| < p_3 < |p_1 + p_2|$, the limits obtaining at the boundary, where all three momenta are collinear. It is clear, therefore, that this assignment predicts a Dalitz-plot density with a maximum at the center and vanishing at the boundary, as in Fig. 4.27(b). The density has the explicit form

$$A^2 = [1 - (1 + B)(r/r_0)^2 - B(r/r_0)^3 \cos 3\phi], \qquad (4.64)$$

where $B = 2(1 - 3m/M)/(1 + 3m/M)^2$, $r_0 = r_{max} = \frac{1}{3}(M - 3m)$, and m and M are the masses of the π- and ω-mesons. Indeed, the right-hand side of (4.64), when set to zero, is the equation of the boundary. The experimental distribution in Fig. 4.26 is consistent with this prediction.

(d) $J^P = 1^+$. To complete the discussion we consider finally the case when the ω is an *axial vector* meson. Then A has the transformation properties of a polar vector. From the $\mathbf{p}_i$ and E_i, such a vector with the required exchange symmetry can be written as

$$A = (\mathbf{p}_1 - \mathbf{p}_2)E_3 + (\mathbf{p}_2 - \mathbf{p}_3)E_1 + (\mathbf{p}_3 - \mathbf{p}_1)E_2. \qquad (4.65)$$

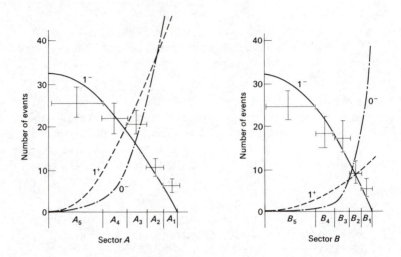

Fig. 4.28 Radial distribution of the Dalitz plot in $\omega \to 3\pi$ decay. The experimental distribution indicates the spin-parity assignment 1^-. The regions A_1–A_5, B_1–B_5 refer to the subdivisions in Fig. 4.26.

In this case, A vanishes at the center of the Dalitz plot, when $E_1 = E_2 = E_3$. Making use of the fact that $E_1 + E_2 + E_3 = M$ and $\mathbf{p}_1 + \mathbf{p}_2 + \mathbf{p}_3 = 0$, (4.65) can also be put in the form

$$A = \mathbf{p}_1(M - 3E_2) - \mathbf{p}_2(M - 3E_1),$$

which goes to zero along three radial lines parallel to the energy axes, that is, when $\mathbf{p}_1 = \mathbf{p}_2$ and $E_1 = E_2$. In terms of r, ϕ the density has the form

$$A^2 = r^2[1 - \tfrac{1}{2}(1 - 3m/M) - (r/r_0)\cos 3\phi], \qquad (4.66)$$

and is shown in Fig. 4.27(c). As Fig. 4.26 shows, this is also inconsistent with experiment.

To summarize therefore, the scalar assignment 0^+ is forbidden by parity conservation, and the pseudoscalar 0^- and axial-vector 1^+ because they predict zero density at the center of the plot. Only the vector assignment 1^- agrees with experiment. Figure 4.28 summarizes the results on the radial density distribution, integrated over the azimuthal angle ϕ.

4.12 TOTAL AND ELASTIC CROSS-SECTIONS; THE BLACK-DISC MODEL

Figure 4.29 shows total and elastic pp cross-sections at high energy, and Fig. 4.30 the total cross-section for various particles on protons. The elastic cross-section in Fig. 4.29 accounts for only a small fraction of the total cross-section at

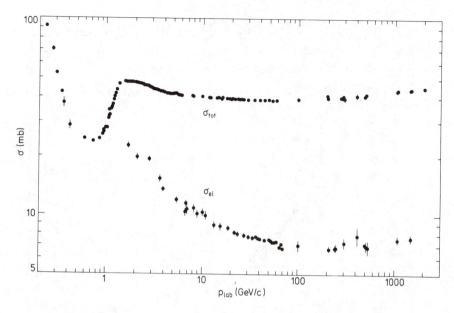

Fig. 4.29 *pp* elastic and total cross-sections as a function of energy.

high energy. The total cross-sections are constant within 10% above 5 GeV, with a slow decrease followed by an increase at the upper energies. The magnitude of the cross-section varies with the type of particle, but is in the region of 20 to 40 mb. If one equates this value to a "geometrical" cross-section πR^2, one obtains $R \sim 10^{-13}$ cm = 1 fm as the "range" of the strong interaction. Secondly, especially in the lower energy range, one observes a large difference in the $\bar{p}p$ and pp cross sections, and this is not unexpected in view of the larger number of isospin channels open for the nucleon-antinucleon process, as well as the higher available energy following annihilation. Similar remarks apply in comparing $\pi^- p$ with $\pi^+ p$, and $K^- p$ with $K^+ p$. At very high energies there is a prediction from quantum field theory, known as the Pomerančuk theorem, that the cross-sections should become the same for particle and antiparticle, and moreover isospin-independent. Thus, cross-sections for $\pi^- p$ and $\pi^+ p$ ($\equiv \pi^- n$ from charge independence) should become equal on both counts. The trends in the data (Fig. 4.31) support this.

The simplest possible model of absorption and scattering is that of a totally absorbing black disc of well-defined radius R. Setting $\eta_l = 0$ in (4.41) and (4.43) for this case, one obtains

$$\sigma_{\text{elastic}} = \pi \lambda^2 \sum (2l + 1) = \pi R^2,$$

$$\sigma_{\text{inelastic}} = \pi \lambda^2 \sum (2l + 1) = \pi R^2,$$

$$\sigma_{\text{total}} = 2\pi \lambda^2 \sum (2l + 1) = 2\pi R^2,$$

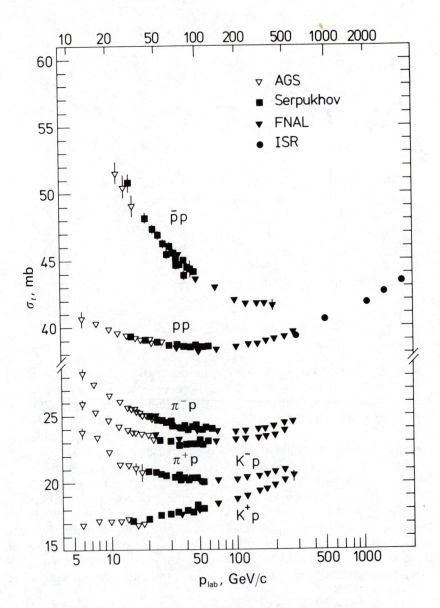

Fig. 4.30 Total cross-sections for various particles on proton targets.

where λ is the de Broglie wavelength of the colliding particles in their center-of-momentum system. In this model, the angular momentum contributed by the incident wave, l, varies from 0 to $l_{max} = R/\lambda$. For example, for $R = 1$ fm, and 20 GeV/c incident momentum, $\lambda = 0.01$ fm and $l_{max} = 100$. Then

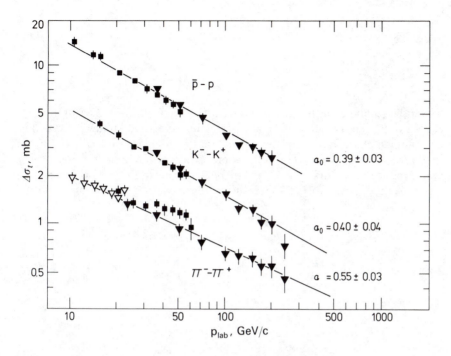

Fig. 4.31 Difference of antiparticle and particle cross-sections on protons as a function of energy.

$$\sum (2l + 1) = (l_{\max} + 1)^2 \simeq l_{\max}^2 = R^2/\lambdabar^2.$$

As expected, the inelastic or reaction or absorption cross-section is simply the geometrical area of the disc, πR^2. We note that $\sigma_{\text{elastic}} = \sigma_{\text{inelastic}}$ and represents the diffraction or shadow scattering, which is familiar in optics when a plane wave is interrupted by a completely absorbing obstacle. The angular distribution of the elastic scattering is the Fourier transform of the spatial distribution of the obstacle. For the lth partial wave, it is represented by the polynomial $P_l(\cos\theta)$. The sum over Legendre polynomials can be approximated, for small scattering angles, by a Bessel function of first order. Actually, it is more useful to discuss the differential elastic cross-section in terms of the momentum transfer q, rather than the angular deflection θ. Referring to Fig. 4.37, we have

$$q = 2k \sin \frac{\theta}{2}, \tag{4.67}$$

where, with units $\hbar = c = 1$, we have $k = 1/\lambdabar = p$, the CMS momentum of each of the colliding particles. Then it may be shown that, in the small-angle approximation,

$$\frac{d\sigma_{\text{el.}} \text{ (black disc)}}{dq^2} = \pi R^4 \left| \frac{J_1(Rq)}{Rq} \right|^2 \tag{4.68}$$

$$\simeq \frac{\pi R^4}{4} \exp\left(-\frac{R^2 q^2}{4}\right), \tag{4.69}$$

where the second expression is accurate, with $R = 1$ fm, for values of $q^2 < 0.2$ $(\text{GeV}/c)^2$. For larger values of Rq, the Bessel function $J_1(Rq)$ undergoes maxima and minima characteristic of the diffraction phenomenon. For $R = 1$ fm, the first zero is at $q^2 \sim 0.6$ $(\text{GeV}/c)^2$. Some experimental data on $\pi^\pm p$ elastic scattering are shown in Fig. 4.32, and on pp scattering in Fig. 4.33. There is clear diffractive-type structure in these elastic-scattering data. The position of the first

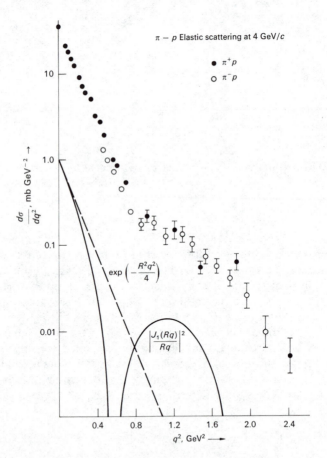

Fig. 4.32 Elastic scattering cross-section $d\sigma/dq^2$ for $\pi^\pm$ on protons, at 4-GeV/c incident momentum. The curve shows the distribution (Bessel function) expected for the black-disc model, with $R = 1$ fm. Data from Coffin et al. (1966).

minimum or change of slope occurs at a value of q^2 close to that expected for an absorbing hadronic "disc" of radius of order 1 fm. It is also noticeable that in the pp scattering, the minimum in q^2 decreases slowly with increasing energy, corresponding to a slow (logarithmic) increase in the effective size of the disc. A similar dilation effect is seen in the total cross-sections at very high energy (Fig. 4.30).

For small values of q^2 ($q^2 < 2$ GeV/c^2) all elastic cross-sections seem to be well fitted by the empirical formula

$$\frac{d\sigma}{dq^2} \bigg/ \frac{d\sigma(0)}{dq^2} = \exp[-(Aq^2 - Bq^4)], \qquad (4.70)$$

A and B being both positive [i.e., when log $(d\sigma/dq^2)$ is plotted against q^2, the curve is concave upward]. Equation (4.70) reproduces the first term of (4.69), which dominates at small q^2, with $R = \sqrt{4A}$. For $\pi^\pm p$ scattering, $A \simeq 8$ (GeV/c)$^{-2}$—corresponding to $R = 1$ fm. The parameter A is weakly energy-dependent and at medium energies, may decrease or increase with E,

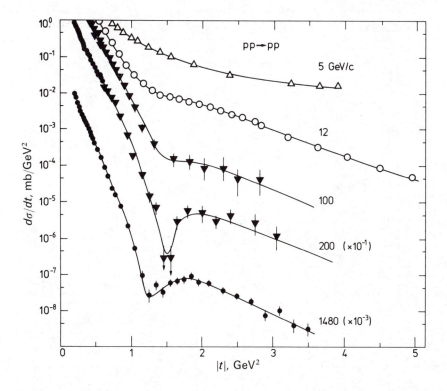

Fig. 4.33 Differential cross-section for elastic pp scattering as a function of the square of the momentum transfer, $|t| = q^2$.

depending on the particles involved. At high enough energies (above 100 GeV) however, A increases slowly with E in $\pi^{\pm}p$, $K^{\pm}p$, pp and $\bar{p}p$ elastic scattering. Even more significant is the discrepancy between the expected ratio $\sigma_{\text{el.}}/\sigma_{\text{total}} = 0.5$ and that observed. In all cases, $\sigma_{\text{el.}}/\sigma_{\text{total}} < 0.5$ and falls off with increasing energy.

To summarize, therefore, the classical optical model of absorption and scattering by an opaque disc of radius ~ 1 fm describes some of the gross features of the observations, but fails to account in detail for the shape of the elastic scattering at large momentum transfer, the value of $\sigma_{\text{el.}}/\sigma_{\text{total}}$, and the dependence on bombarding energy. We may also remark that such a model does not incorporate spin, and thus is unable to account for transverse polarization in the scattering process (typically of order 10%). Somewhat better fits to the data can be obtained from Regge-pole theory.

4.13   PARTICLE PRODUCTION AT HIGH ENERGIES

As is clear from Fig. 4.29, inelastic processes dominate high-energy hadron collisions, which are characterized by multiple production of secondary mesons (and baryon-antibaryon pairs) as for example in Fig. 1.1. The main features are:

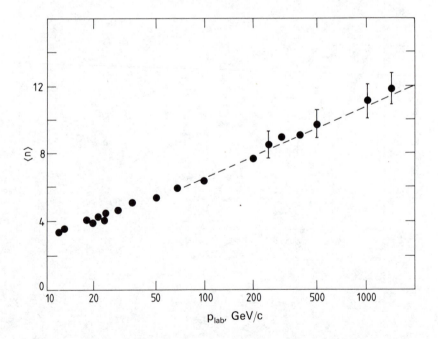

Fig. 4.34 Mean multiplicity of charged particles produced in pp collisions, showing a slow (logarithmic) increase with incident proton momentum.

(a) The average multiplicity increases logarithmically with energy (Fig. 4.34).

(b) The transverse component of momentum of a secondary (relative to the beam direction) is limited, with $\langle p_T \rangle \simeq 0.4 \, \text{GeV}/c$, independent of the incident energy. Thus "soft" collisions dominate.

(c) The longitudinal component of momentum has a roughly exponential distribution in terms of the Feynman variable $x_F = p_{||}/p_{||}(\text{max})$, and exhibits "scaling" i.e. is nearly independent of collision energy. Thus, small $p_{||}$-values are dominant, the bulk of the secondaries being emitted in a so-called "central" region in the CMS. In addition, a few high-energy particles form a "diffractive" component at small angle in the forward and backward directions, as sketched in Fig. 4.35.

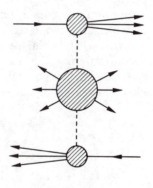

Fig. 4.35

Pions are the most prolifically produced secondaries in high-energy collisions, with kaons and baryon-antibaryon pairs less frequent by one and two orders of magnitude respectively.

4.14 ONE-PARTICLE-EXCHANGE (OPE) MODEL

We have already noted that at low energy ($M_{\pi p} < 3$ GeV), pion-nucleon scattering is dominated by the formation of resonances, but that above this limit, the cross-sections have a smooth energy dependence, and are characterized by a pronounced forward peaking in the elastic channel. In this situation, a more useful description of a two-body $\rightarrow$ two-body process is to consider the interaction mediated by a single virtual particle exchange (for example, a pion) in analogy with the forward-peaked Coulomb scattering mediated by single photon exchange (Fig. 4.36). Associated with a virtual boson of free mass m will be a propagator term entering the scattering amplitude of the form

$$f(q^2) \sim \frac{g^2}{m^2 + q^2}.\tag{4.71}$$

Since q^2 is positive for a scattering process, this term is positive and finite. However, it would "blow up" (i.e. has a singularity or pole) in the *unphysical* region $q^2 = -m^2$, and we note that the smaller the value of m, the larger will be the amplitude at finite, positive q^2. So we may expect pion exchange to dominate other possible boson exchanges for reactions such as

$$\pi^- + p \rightarrow \rho^- + p$$

at small values of q^2. Indeed, the one-pion-exchange model provides a fair description of elastic scattering for terms of high l, that is, for "distant" collisions. The data on pp elastic scattering at 100 MeV has been analysed by fitting the angular distributions with a set of arbitrary and smoothly varying phase shifts for the lower partial waves only, and then relying on the OPE model to supply the scattering amplitude for the remaining terms of high l. The analysis gave a value for $f^2/\hbar c = (g^2/\hbar c)(m_\pi/2m_p)^2$ in agreement with the value (4.54) from low-energy pion-nucleon scattering. Often, π-exchange is disallowed; for example, in the charge-exchange scattering $\pi^- p \rightarrow \pi^0 n$, the exchanged particle must have even G-parity and the simplest possibility is a ρ-meson. In such reactions, the single-particle-exchange model gives predictions in serious disagreement with the data.

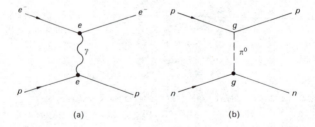

Fig. 4.36

4.15   THE REGGE-POLE MODEL

A basic problem in the OPE model is that the poles in the scattering amplitude, which are assumed to dominate the scene, correspond to the exchange of particles carrying fixed angular momentum J. Thus, in the pion exchange discussed above, $J = 1$ (p-wave). As a consequence, it is found that at high incident energies the amplitude contains a factor E^J, so that it always "blows up" if J is large enough. A possible way out of this difficulty was suggested by Regge in 1959, who proposed to treat the angular momentum as a continuous, complex variable – although, clearly, physically observable states must have integral or half-integral angular momentum. Although Regge worked in nonrelativistic potential theory, his ideas were quickly applied to high-energy particle physics by Chew *et al.* (1962). These objects with complex angular momentum are referred to as *Regge poles*.

First let us consider the general collision process $a + b \rightarrow c + d$, where a, b, c, and d are hadronic states. These particles have four-momenta denoted by p_a, p_b, p_c, and p_d respectively [Fig. 4.37(a)]. The four-momentum $p_a = (\mathbf{p}_a, iE_a)$, where $\mathbf{p}$ is the three-momentum and E the total energy of the particle.

There are two independent Lorentz-invariant quantities one can form from these four-vectors (apart from the particle rest mass $p^2 = -m^2$). These are denoted by the so-called Mandelstam variables

$$s = -(p_a + p_b)^2 = -(p_c + p_d)^2 = E^2, \tag{4.72}$$

$$t = -(p_a - p_c)^2 = -(p_b - p_d)^2 = -q^2. \tag{4.73}$$

In the CMS of the collision, a and b have equal and opposite three-momentum, so that $s = E^2$, the square of the total CMS energy [Fig. 4.37(b)]. t is the square of the four-momentum transfer between a and c, or b and d. For an *elastic collision*, we note that in the CMS no energy is transferred and $|\mathbf{k}| = |\mathbf{k}'|$, so that

$$t = -(\mathbf{k} - \mathbf{k}')^2 = -2k^2(1 - \cos\theta), \tag{4.74}$$

where θ is the scattering angle in the CMS. Note that $t = -q^2$ is negative for a scattering process. We emphasize that s and t are invariants and therefore have the same values in both laboratory and center-of-mass systems. It is also possible to consider the crossed momentum transfer, defined by $u = -(p_a - p_d)^2$. It is left as an exercise to show that

$$s + t + u = m_a^2 + m_b^2 + m_c^2 + m_d^2,$$

so that u is not an independent quantity.

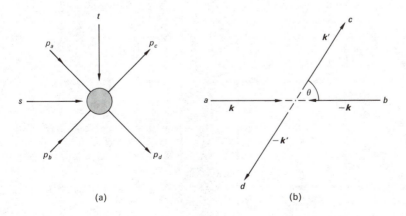

(a)   (b)

Fig. 4.37

When we are interested in the properties of the intermediate state formed from a and b (for example, a resonance of a and b), it is natural to describe the partial-wave scattering amplitude in terms of s-channel quantities, i.e. as $f(l, E)$ where l and E are the angular momentum and energy of the combination a and b. At low energies (< 1 GeV) we know that pion-nucleon elastic scattering is dominated by s-channel resonances. However, at higher energies (above 10 GeV) the cross-section varies very smoothly with energy and such a description is not appropriate. Rather, we think of the scattering as dominated by exchange of poles in the momentum-transfer or t-channel (just as for electron-proton scattering due to photon exchange). Since t is negative, these exchanged particles are outside the physical region for the reaction $a + b \rightarrow c + d$. But now suppose that we replace b and c by their antiparticles $\bar{b}$ and $\bar{c}$ and reverse their momenta (Fig. 4.38).

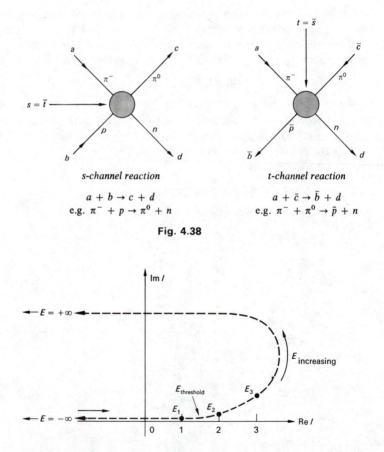

s-channel reaction

$a + b \rightarrow c + d$

e.g. $\pi^- + p \rightarrow \pi^0 + n$

t-channel reaction

$a + \bar{c} \rightarrow \bar{b} + d$

e.g. $\pi^- + \pi^0 \rightarrow \bar{p} + n$

Fig. 4.38

Fig. 4.39 Typical behavior of a Regge trajectory in the complex angular-momentum plane.

Then in (4.72) and (4.73), s and t change sign. t is now the energy variable for the t-channel reaction $a + \bar{c} \rightarrow \bar{b} + d$, and we call it $\bar{s}$. s is negative; it is the momentum-transfer variable for the t-channel reaction, and we call it $\bar{t}$. The reaction $a + \bar{c} \rightarrow \bar{b} + d$ is called the *crossed reaction* to $a + b \rightarrow c + d$. The point is that the t-channel exchange poles of the original reaction now become $\bar{s}$-channel resonances, in the physical region of the crossed reaction. The important *principle of crossing symmetry* states that both reactions are described by one and the same amplitude; we can write this as

$$\bar{F}(\bar{s}, \bar{t}) = F(t = \bar{s}, s = \bar{t}). \tag{4.75}$$

Obviously, this principle requires that the function F can be continued into the unphysical regions in the s, t plane. For example, $\bar{s} > 0$ for the t-channel corresponds to unphysical values of $t > 0$ or, from (4.74), $\cos \theta > 1$ in the s-channel. Crossing symmetry is of vital importance in seeking to describe high-energy processes in the Regge picture.

Let us now turn to the s-channel, and consider the behavior of the amplitude $f(l, E)$ according to the Regge hypothesis. [The index l, for the lth partial wave contributing to the scattering, implies the usual angular dependence $P_l(\cos \theta)$.] The angular momentum is written as a function of energy, $\alpha(E)$, with real and imaginary parts: $l = \operatorname{Re} \alpha(E)$. The trajectory described by α as E varies is called a Regge trajectory. Figure 4.39 shows an example of the form such a trajectory might have in the complex angular-momentum plane.

The trajectory starts off along the negative real axis, and, depending on the strength of the scattering potential, may cross over the origin along the positive real axis as E increases. The particles a and b may possess a bound state of total energy E_1 and $l = 1$, say. This corresponds to $\operatorname{Im} \alpha(E) = 0$ and $\operatorname{Re} \alpha(E) = n$, an integer. Such a bound state or pole in $\alpha(E)$ will necessarily occur at $E < E_{\text{threshold}}$, where $E_{\text{threshold}} = m_a + m_b$. Increasing E still further through $E_{\text{threshold}}$, the trajectory now leaves the real axis, $\alpha(E)$ acquiring a positive imaginary part. Whenever the trajectory passes $\operatorname{Re} \alpha(E) = l = n$, an integer, one can have an unbound state of energy E_n and angular momentum n, which is identified as an s-channel *resonance*. This can be seen as follows. If $\operatorname{Im} \alpha \ll \operatorname{Re} \alpha$, we may write a Taylor expansion for $E \simeq E_n$ of the form

$$\alpha(E) = \operatorname{Re} \alpha(E) + i \operatorname{Im} \alpha(E)$$

$$\simeq n + (E - E_n) \left[\frac{d}{dE} \operatorname{Re} \alpha(E) \right]_{E = E_n} + i \operatorname{Im} \alpha(E).$$

Setting

$$\varepsilon = \frac{d}{dE} (\operatorname{Re} \alpha(E))_{E = E_n} \quad \text{and} \quad \Gamma = \frac{2}{\varepsilon} \operatorname{Im} \alpha(E)$$

gives

$$\alpha(E) = n + \varepsilon[(E - E_n) + i\Gamma/2].$$

Whenever l and E have values corresponding to a resonance, the amplitude $f(l, E)$ must have a singularity or *pole*. We express this by writing

$$f(l, E) = \frac{R(E)}{l - \alpha(E)}, \tag{4.76}$$

where $R(E)$ is some function (the residue function). Thus

$$f(l = n, E) = -\frac{R}{\varepsilon} \frac{1}{(E - E_n) + i\Gamma/2}, \tag{4.77}$$

which is the usual Breit-Wigner formula for the amplitude near a resonance of angular momentum n and energy E_n. Since Im $\alpha(E)$ and Γ are both positive, we require the slope ε of Re $\alpha(E)$ versus E to be positive. A corollary is therefore that when the trajectory turns over and heads to $E \to +\infty$, as it eventually must, there will be no further resonant states. Thus, the same Regge trajectory

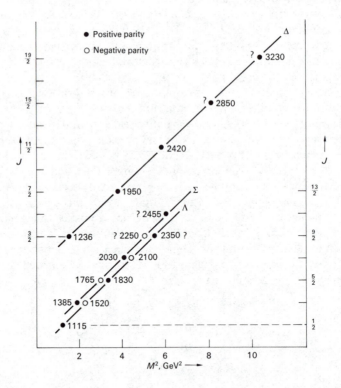

Fig. 4.40 Chew-Frautschi plots of fermion Regge trajectories. The trajectory marked Δ consists of the sequence $I = \frac{3}{2}$, $S = 0$, and $J^P = \frac{3}{2}^+, \frac{7}{2}^+, \frac{11}{2}^+, \ldots$; that marked Λ, of the sequence $I = 0$, $S = -1$, $J^P = \frac{1}{2}^+, \frac{3}{2}^-, \frac{5}{2}^+, \ldots$; and that marked Σ, of the sequence $I = 1$, $S = -1$, $J^P = \frac{3}{2}^+, \frac{5}{2}^-, \frac{7}{2}^+, \ldots$. Resonances for which the spin-parity is not firmly established are indicated by a question mark.

connects together bound states and resonances, and so to speak interpolates between the integral (or half-integral) values of $\mathrm{Re}\,\alpha(E)$ which these states possess. The number of bound states or resonances on a particular trajectory will depend on the "strength" of the interaction potential.

What other properties do Regge trajectories have, and what evidence is there for them? If we seek to describe a process in these terms, it is clear that all the quantum numbers except angular momentum ought to be the same for all poles on a given trajectory. Thus, the parity, baryon number, isospin, strangeness, and G-parity are defined by conservation laws applying to the particular reaction, independent of how we choose to describe it. Thus $\pi^+ + p \rightarrow$ anything can only proceed via the $I = \frac{3}{2}$, $B = 1$, $S = 0$ channel. The requirement that Regge poles should also describe exchange forces in the t-channel has the effect that the consecutive poles must be separated by two units of angular momentum ($\Delta l = 2$), and this fixes the *parity* of the trajectory. Consider, for example, the scattering of two pions. By Bose symmetry, states of even (odd) isospin must be associated with even (odd) l, so that, given the isospin channel, all two-pion poles must be separated along the trajectory by $\Delta l = 2$. Physically, we can observe the l- or J-values of resonances as a function of the mass $E = M$. This corresponds to a trajectory in the l, E plane, i.e. the variation of $\mathrm{Re}\,\alpha(E)$ with E as we move along the curve of Fig. 4.39. Such a plot is called a Chew-Frautschi diagram. Figure 4.40 gives some examples of fermion trajectories. For example, the so-called Δ-trajectory links together baryon states of $I = \frac{3}{2}$, $B = 1$, $S = 0$, and positive parity. The trajectory in the l, E plane is found to be a straight line if J is plotted against M^2, with slope $dJ/dM^2 = 0.9\,\mathrm{GeV}^{-2}$. The Σ- and Λ-trajectories have a similar slope. Regge theory cannot account for the linearity of these trajectories; a possible explanation, in terms of quantum chromodynamics, is given in Section 4.17 below.

4.16 APPLICATION OF REGGE POLES TO HIGH-ENERGY REACTIONS

Perhaps the most important application of the Regge-pole model is to the study of t-channel exchange processes in high-energy reactions, where s is large and t is small. Let us consider the elastic scattering process $\pi^+ + p \rightarrow \pi^+ + p$. From (4.38) and (4.46) we may write for the scattering amplitude summed over all contributing partial waves

$$F(\theta) = F(E, \cos\theta) = \frac{1}{k}\sum_l (2l+1)f(l, E)P_l(\cos\theta), \qquad (4.78)$$

where $E = \sqrt{s}$ is the total energy in the CMS.

In the Regge-pole model, this partial-wave expansion may be transformed, by a method due to Watson and Sommerfeld, to a sum over Regge poles, plus a background integral, which we shall ignore hereafter. The new form is

$$F(E, \cos\theta) = F(s, t) = \frac{1}{k}\sum_i \frac{\beta_i(s)}{\sin\pi\alpha_i(s)}[P_{\alpha_i(s)}(-\cos\theta)] + \mathrm{B.I.} \qquad (4.79)$$

This result follows from the properties of certain contour integrals in the complex angular-momentum plane, which we shall not discuss here. Here

$$\beta_i(s) = -\pi[2\alpha_i(s) + 1]R_i(s), \tag{4.80}$$

where $R_i(s)$ is defined in (4.76) as the residue at the pole. The denominator term $1/\sin \pi l$ has poles at $l = 0, \pm 1, \pm 2, \ldots$, with residue $(-1)^l/\pi$. Furthermore, $P_l(-\cos\theta) = (-1)^l P_l(\cos\theta)$. Using these facts, (4.78) and (4.79) are seen to be entirely equivalent. The angle θ, of course, refers to the scattering angle in the s-channel. For the elastic scattering of particles of mass m and M, (4.74) gives after a little algebra

$$\cos\theta = 1 + \frac{t}{2k^2} = 1 + \frac{2ts}{-4m^2s + (s - M^2 + m^2)^2}. \tag{4.81}$$

As mentioned previously, the description (4.79) in terms of s-channel poles is not particularly useful for studying the processes at high s and small t. We are interested in *exchange* (t-channel) poles, as in the OPE model. Let us instead go to the crossed (t-channel) reaction, i.e. $\pi^+\pi^- \to p\bar{p}$. Then

$$\bar{F}(\bar{s}, \bar{t}) = F(t, s) = \frac{1}{k}\sum \frac{\beta_i(t)}{\sin \pi\alpha_i(t)}[P_{\alpha_i(t)}(-\cos\theta_t)], \tag{4.82}$$

where

$$\cos\theta_t = 1 + \frac{2st}{-4m^2t + (t - M^2 + m^2)^2} \tag{4.83}$$

is now in the unphysical region. Thus, when t is small and $s \to \infty$, $\cos\theta_t \propto (-s)$, and it may be shown that $P_{\alpha(t)}(-\cos\theta_t) \propto (s/s_0)^{\alpha(t)}$. If we assume that one Regge pole dominates everything, we have

$$F(s, t) \propto \frac{1}{k}\beta_1(t)\left(\frac{s}{s_0}\right)^{\alpha_1(t)}, \tag{4.84}$$

where s_0 is a parameter with dimensions (energy)^2. The cross-section then has the form, from (4.40) and (4.74),

$$\frac{d\sigma}{dt} = \frac{d\sigma}{d\Omega}\frac{d\Omega}{dt} = [F(s, t)]^2 \frac{\pi}{k^2}.$$

In the limit $s \to \infty$, $s \simeq 4k^2$, since the masses of the particles involved may be neglected [refer to Fig. 4.37(b)]. Then

$$\frac{d\sigma(\text{elastic})}{dt} = D_1(t)\left(\frac{s}{s_0}\right)^{2\alpha_1(t) - 2}, \tag{4.85}$$

where D is some function of t only. From the optical theorem (4.44) we know that the total cross-section

$$\sigma_T = \frac{4\pi}{k} \operatorname{Im} F(t = 0).$$

Thus, if σ_T is to be nearly constant at high s, one needs $\alpha_1(0) = 1$ in (4.84). This then is one property required of the trajectory of a single exchanged (t-channel) Regge pole if it is to account for elastic scattering at high energy. Of course, this one trajectory ought to account for all elastic scattering phenomena, and it is not difficult to see that, since no quantum numbers (strangeness, isospin, baryon number, etc.) apart from angular momentum may be exchanged in such processes, this trajectory must have the quantum numbers of the vacuum. It is therefore called the vacuum trajectory. The implication is that all other Regge trajectories must have $\alpha(t = 0) < 1$, so that the vacuum trajectory

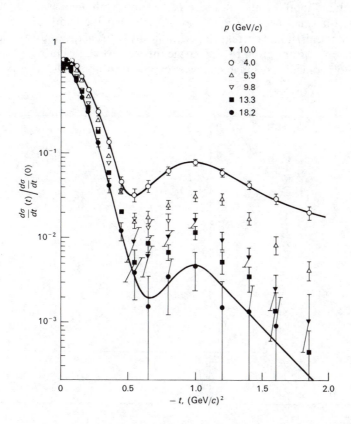

Fig. 4.41 Early observations on the differential cross-section for elastic charge exchange, $\pi^- p \to \pi^0 n$, compiled by Sonderegger *et al.* (1966). The shrinkage of the t-distribution with increasing incident momentum p yields $\alpha_\rho(t) = 0.6 + 0.9t$ for the ρ-exchange trajectory.

can dominate elastic scattering in the high-energy limit, as in (4.85). Since exchange of a vacuum pole implies, for example, that $\sigma_{\pi^+ p}$(total) $= \sigma_{\pi^- p}$(total) as predicted by the Pomerančuk theorem, this trajectory is also called the Pomerančuk trajectory. It is often denoted α_P.

In (4.85) the term containing s must dominate both the s- and the t-dependence at large values of s. If b is the (positive) slope at small t, and the trajectory is linear in this small region, then

$$\alpha_P(t) = 1 + bt,$$

so that

$$\frac{d\sigma(\text{elastic})}{dt} = \left(\frac{d\sigma}{dt}\right)_{t=0} \exp\left[2b\log\left(\frac{s}{s_0}\right)\right]t. \tag{4.86}$$

Comparing with (4.70) and with $q^2 = -t$, we see that this correctly predicts the exponential falloff of the elastic cross-section with increasing momentum transfer, for small negative t. It also predicts that the width of the forward diffraction peak should decrease logarithmically with energy (unfortunately, the scale factor s_0 is not given by the Regge theory). At sufficiently high energies, where Pomerančuk exchange could be expected to dominate, $\pi^\pm p, pp$ and $\bar{p}p$

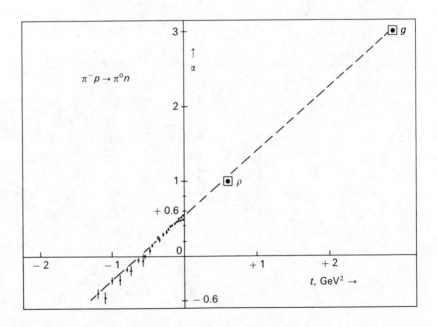

Fig. 4.42   The ρ-trajectory. For $t < 0$, the data come from the analysis of $\pi^- p$ charge-exchange scattering. The line drawn through it extrapolates fairly well through the resonances or poles of the ρ- and g-mesons in the region $t > 0$.

elastic scattering all show a uniform shrinkage (see for example Fig. 4.33). This prediction was one of the great successes of the Regge model.

Let us now look at pion-nucleon charge-exchange scattering, $\pi^- p \to \pi^0 n$. The exchanged particle must have G-parity $+1$, $J = 1$, spin-parity 1^-, 3^-, $5^-, \ldots$, and it is therefore identified with the ρ-trajectory; this is the only one which can obviously contribute. The formula (4.84) applies, with $\alpha_1 = \alpha_\rho$. The data (Fig. 4.41) appear to show a shrinkage, and from its magnitude one can compute $\alpha_\rho(t)$, with the results shown in Fig. 4.42. A line of the form

$$\alpha_\rho(t) = 0.55 + 0.9t \tag{4.87}$$

fits the charge-exchange data for $t < 0$ and extrapolates fairly well to the ρ ($J^P = 1^-$) and g ($J^P = 3^-$) mesons lying on the ρ-trajectory at $t > 0$. The region $t > 0$ is the physical region for the crossed channel $\pi^- \pi^0 \to \bar{p}n$. Note that the slope of the trajectory (0.9 GeV^{-2}) is the same as that of the baryon trajectories, a point to be discussed below.

We have touched here on a few of the features of the Regge-pole model. Single Regge exchange is unable to account for all the features of hadron-hadron scattering, particularly multiparticle production and more complex descriptions (triple Regge exchange) have been developed. These involve more arbitrary parameters, with some corresponding loss in predictive power. While one cannot regard the Regge model as in any sense a fundamental theory of hadronic interactions, it does give a unified description of a broad mass of data at small momentum transfers which would otherwise be uncorrelated.

4.17 THE STRING MODEL OF REGGE TRAJECTORIES

The linearity of the Regge trajectories is not predicted by the above theory, but it can be (at least qualitatively) accounted for in terms of a field theory of interquark forces, called quantum chromodynamics, which will be discussed in Chapter 8. In this theory, there is a so-called color field between the quarks, the color force being carried by the vector bosons of the strong interaction between quarks, called gluons. The characteristic feature of the gluons is their strong self-interaction. In analogy with the electric lines of force between two charges, as in Fig. 4.43(a), we can imagine that quarks are held together by color lines of force as in Fig. 4.43(b), but the gluon-gluon interaction pulls these together into the form of a tube or string (recall our discussion in terms of elastic bands in Chapter 1).

Suppose k is the energy density per unit length of such a string, and that it connects together two massless quarks as in Fig. 4.43(c). The angular momentum of the quark pair will then be equal to the total angular momentum of the gluon tube, and we can calculate this if we assume that the ends of the tube rotate with velocity $v = c$. Then the local velocity at radius r will be

$$\frac{v}{c} = \frac{r}{r_0},$$

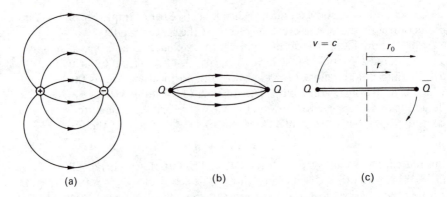

Fig. 4.43 (a) Electric lines of force between two charges. (b) Color lines of force between quarks are pulled together into a tube or string, because of the strong self-interaction between the gluons, which are the carriers of the color field. (c) String model used in calculating the relation between angular momentum and mass of a hadron.

where r_0 is half the length of the string. The total mass is then (relativistically)

$$E = Mc^2 = 2 \int_0^{r_0} \frac{k \, dr}{\sqrt{1 - v^2/c^2}} = k r_0 \pi,$$

and its orbital angular momentum will be

$$J = \frac{2}{\hbar c^2} \int_0^{r_0} \frac{k r v \, dr}{\sqrt{1 - v^2/c^2}} = \frac{k r_0^2 \pi}{2 \hbar c}.$$

Eliminating r_0 between these equations gives the observed relation between the angular momentum and energy of a hadron state:

$$J = \alpha' E^2 + \text{const}, \tag{4.88}$$

where in our model

$$\alpha' = \frac{1}{2 \pi k \hbar c}.$$

From the slope of the ρ-trajectory, for example, we have

$$\alpha' = 0.9 \, \text{GeV}^{-2}$$

or

$$k = 0.2 \, \text{GeV}^2 \qquad (\hbar = c = 1). \tag{4.89}$$

This number also comes from consideration of the sizes of hadrons. A typical hadron mass is about 1 GeV, and its radius, as measured in electron scattering (see Chapter 7) is about 1 fm, so the linear energy density will be $k \sim 1 \, \text{GeV} \, \text{fm}^{-1}$

or $0.2 \, \text{GeV}^2$. The value of k is an important input in discussions of the interquark potential, where the confining part of the potential is of the form kr, as discussed in Chapter 5.

PROBLEMS

4.1 Find a relation between the total cross-sections (at a given energy) for the reactions

$$\pi^- p \to K^0 \Sigma^0,$$

$$\pi^- p \to K^+ \Sigma^-,$$

$$\pi^+ p \to K^+ \Sigma^+.$$

4.2 At a given center-of-mass energy, what is the ratio of the cross-sections for $p + d \to {}^3\text{He} + \pi^0$ and $p + d \to {}^3\text{H} + \pi^+$?

4.3 A hypernucleus is one in which a neutron is replaced by a bound Λ-hyperon. ${}^4\text{He}_\Lambda$ and ${}^4\text{H}_\Lambda$ are a doublet of mirror hypernuclei. Deduce the ratio of the reaction rates

$$K^- + {}^4\text{He} \to {}^4\text{He}_\Lambda + \pi^-$$

$$\to {}^4\text{H}_\Lambda + \pi^0.$$

4.4 State which of the following combinations can or cannot exist in a state of $I = 1$, and give the reasons:

(a) $\pi^0 \pi^0$,
(b) $\pi^+ \pi^-$,
(c) $\pi^+ \pi^+$,
(d) $\Sigma^0 \pi^0$,
(e) $\Lambda \pi^0$.

4.5 In which isospin states can (a) $\pi^+ \pi^- \pi^0$, (b) $\pi^0 \pi^0 \pi^0$ exist? [*Hint*: First write down the isospin functions for a pair, e.g. $\pi^0 \pi^0$, and then combine with the third pion. Refer to Table III in the Appendix for any Clebsch-Gordan coefficients required.]

4.6 Deduce through which isospin channels the following reactions may proceed:

(a) $K^+ + p \to \Sigma^0 + \pi^0$,
(b) $K^- + p \to \Sigma^+ + \pi^-$.

Find the ratio of cross sections for (a) and (b), assuming that one or other channel dominates.

4.7 Write down the quantum numbers (G, I, J^P) of the S- and P-states of the $p\bar{p}$ system able to decay to (a) $\pi^+ \pi^-$, (b) $\pi^0 \pi^0$, (c) $\pi^0 \pi^0 \pi^0$. In annihilations at rest, the process $p\bar{p} \to 2\pi^0$ does not appear to occur; what do you conclude from this fact?

4.8 The $A1$ meson, of $I = 1$, is considered to be a resonant state of a ρ-meson ($I = 1$) and pion ($I = 1$). Thus decay $A1 \to \rho + \pi$ is dominant. Find the expected branching ratio

$$\frac{A1 \to \pi^0 \pi^0 \pi^+}{A1 \to \pi^+ \pi^- \pi^+}.$$

4.9 The ω-meson has G-parity -1, and the ρ-meson has G-parity $+1$. The ω and ρ have the same spin and parity (1^-). The ρ-meson has central mass 775 MeV and is a broad state with a width $\Gamma \simeq 120$ MeV, overlapping the ω-state (783 MeV, $\Gamma \sim 10$ MeV). Would you expect the ω- and ρ-states to interfere, and what qualitative effects would any interference have on the $\pi^+\pi^-$ and $\pi^+\pi^-\pi^0$ mass spectra in reactions where both ω and ρ can be produced?

4.10 As shown in Chapter 6, the neutral kaons decay from the states K_1^0 and K_2^0, of CP eigenvalues $+1$ and -1 respectively. If, as we believe, $p\bar{p}$ annihilation at rest takes place from an atomic S-state only, show that $p\bar{p} \rightarrow K_1^0 + K_2^0$ occurs, but that $p\bar{p} \rightarrow 2K_1^0$ or $p\bar{p} \rightarrow 2K_2^0$ does not.

4.11 Show that, in $K \rightarrow 3\pi$ decay, the relativistic factor $E_1 E_2 E_3$ is constant within a range of $\pm 1\%$ over different regions of the Dalitz plot.

4.12 In the decay of a resonance of mass M into three particles of momenta $\mathbf{p}_1$, $\mathbf{p}_2$, and $\mathbf{p}_3$, show that the boundary of the Dalitz plot is given by the condition $|\mathbf{p}_1| + |\mathbf{p}_2| - |\mathbf{p}_3| = 0$. Deduce the equation of the boundary in terms of E_1 and E_2 in the case where the three decay products have the same rest mass m. Show that if $m \ll M$, the boundary of the Dalitz plot becomes the inscribed triangle of Fig. 4.11.

4.13 In a reaction $A + B \rightarrow C + D + E$, at a fixed bombarding energy, the quantities m_{CD}^2 and m_{DE}^2 are displayed along the x- and y-axes in a Dalitz plot (m_{CD} is the mass of particles C and D, etc.). If θ is the direction of particle E with respect to either C or D in the CD center-of-mass frame, show that when C and D resonate at a fixed mass, we have $m_{DE}^2 = \alpha - \beta \cos \theta$, where α and β are constants.

4.14 In the following reaction in hydrogen,

$$\pi^- + p \rightarrow X^- + p,$$

a boson resonance X is observed with mass 2.4 GeV. The incident pion momentum is 12 GeV/c. Calculate the maximum angle of emission of the recoil proton with respect to the beam direction, and its momentum. Calculate also the angle and momentum of the proton when the four-momentum transfer is a maximum, and compute q_{max}^2.

4.15 The Breit-Wigner formula (4.49) describes a resonance of width $\Gamma \ll E_0$, the peak energy. For a broad resonance, this formula is not exact, since the phase-space available for two-body decay changes appreciably as one goes through the resonance. Show that for an S-wave resonance this effect may be accounted for by replacing Γ with $\Gamma_0 \cdot (pE_0/E p_0)$, where p is the CMS momentum of either particle, and Γ_0, p_0, and E_0 refer to the resonance peak. What additional factors would you expect for a resonance decaying to two particles with orbital angular momentum l?

4.16 Show that:

(a) $\rho \rightarrow \eta + \pi$ is forbidden as a strong decay,
(b) $\omega \rightarrow \eta + \pi$ is forbidden as a strong or electromagnetic decay.

BIBLIOGRAPHY

Amaldi, U., M. Jacob, and G. Matthiae, "Diffraction of matter waves", *Ann. Rev. Nucl. Science* **26**, 385 (1976).

Blatt, J., and V. F. Weisskopf, *Theoretical Nuclear Physics*, John Wiley, New York, 1952.
Bøggild, H., and T. Ferbel, "Inclusive reactions", *Ann. Rev. Nucl. Science* **24**, 451 (1974).
Dalitz, R. H., "Strange particle resonant states", *Ann. Rev. Nucl. Science* **13**, 339 (1963).
Foa, L., "High energy hadron physics", *Riv. Nuovo Cim.* **3**, 283 (1973).
Källen, G., *Elementary Particle Physics*, Addison-Wesley, Reading, Mass., 1964.

Chapter 5

Static Quark Model of Hadrons

5.1 INTRODUCTION

In this chapter, we discuss the evidence from the regularities observed among the hadrons for the hypothesis that hadrons are built from fundamental fermions called *quarks*. Several types or flavors of quark are required, characterized by quantum numbers which are conserved in the strong interactions. These conserved quantities lead to invariance principles applying to the hadron states and to hadron-hadron interactions. Actually some of these invariance principles were discovered long before quarks were invented – the most notable being isospin invariance. The regularities and patterns among the hadron states were also first interpreted in terms of an approximate higher symmetry, called unitary symmetry. However, these descriptions have been superseded by the hypothesis of quark constituents, which goes much further than symmetry principles alone and receives very strong support from lepton-hadron scattering experiments, discussed in Chapter 7, and from the study of the level spectrum of meson states formed from heavy quarks, discussed in Section 5.11.

5.2 BARYON STATES AND THE QUARK MODEL: THE $\frac{3}{2}^+$ DECUPLET AND $\frac{1}{2}^+$ OCTET

5.2.1 The Baryon Decuplet and Quark Quantum Numbers

Scores of hadron states are observed in accelerator experiments; the values of mass and spin-parity J^P of the lower-lying baryon states are depicted in Fig. 5.1 (see also Table IV at the end of the book). Most of these baryons are very short-lived, decaying by strong interactions (e.g. $\Delta^{++} \rightarrow \pi^+ + p$) with widths ~ 100 MeV and lifetimes $\tau \sim 10^{-23}$ seconds. Obviously, these myriad states cannot all be fundamental. In a sense, they can be regarded as excited states of the nucleon, much like the Balmer series of excited states of a hydrogen atom. We note that the energy levels of the hydrogen atom are described quantum-mechanically as the eigenstates of the constituent proton and electron in their mutual Coulomb potential, and it is obvious that we are seeking a similar description of the baryon in terms of more elementary constituents.

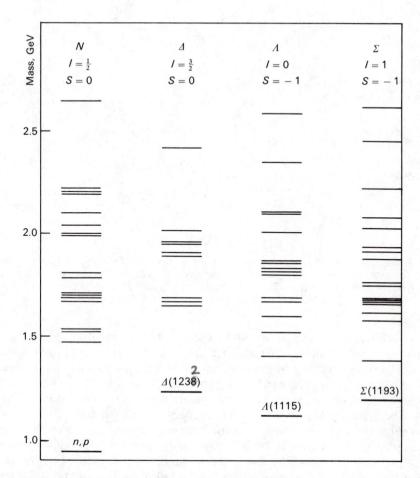

Fig. 5.1 Energy levels (masses) of the $S = 0$ and $S = -1$ baryon states. All except the lowest three states decay by strong interactions, with lifetimes of order 10^{-23} sec, and with the emission of mesons, for example $\Delta(1232) \to p + \pi$. The $\Lambda(1115)$ and $\Sigma(1193)$ decay by weak ($\Delta S = 1$) or electromagnetic ($\Delta S = 0$, $\Delta I = 1$) transitions.

It is found that the baryon states can be arranged systematically in multiplets or families of a fixed spin and parity, J^P, for all members. The most famous one is the baryon decuplet of $J^P = \frac{3}{2}^+$, shown in Fig. 5.2, where we plot the strangeness S against the third component of isospin, I_3, for each of the 10 members. Working downwards, these consist of an $S = 0$, $I = \frac{3}{2}$ isospin quadruplet, the $\Delta(1232)$, existing in the charge substates $\Delta^{++}, \Delta^+, \Delta^0, \Delta^-$. The number 1232 in parentheses indicates the central resonance mass in MeV. Next come an $I = 1$ isospin triplet of $S = -1$, the $\Sigma(1384)$; an $S = -2$, $I = \frac{1}{2}$ isospin doublet, the $\Xi(1533)$; and finally an $I = 0$ singlet of $S = -3$, the $\Omega^-(1672)$. The members of each isospin multiplet have essentially the same central

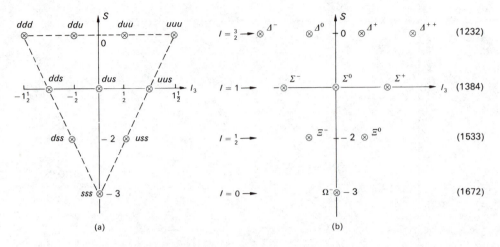

Fig. 5.2 (a) Quark label assignments in the baryon decuplet. Except for the states at the apices, each site label corresponds to the first term only of (5.3). (b) The observed decuplet of baryon states of spin-parity $\frac{3}{2}^+$. The mean mass of each isospin multiplet is given in brackets.

mass, differing only by a few MeV, characteristic of electromagnetic mass splittings in isospin multiplets. The states of different strangeness differ considerably in mass, but the mass difference for each increment of strangeness is roughly the same. This surely cannot be an accident; indeed, the Ω^- baryon was predicted on this basis three years before it was observed (see Fig. 5.3).

The regularities such as that in the decuplet can be accounted for by postulating three types of fermion constituent in a baryon, called *quarks* with the quantum numbers shown in Table 5.1. The quark hypothesis was put forward in 1964 by Gell-Mann and by Zweig. These quarks consist of an $S = 0$ isospin doublet, labeled u and d [standing for $I_3 = +\frac{1}{2}$ (up) and $I_3 = -\frac{1}{2}$ (down)], and a $S = -1$ isosinglet, labelled s (for strange). The assignments u, d, s are called the *flavor* of the quark. Baryons are assumed to be composed of three quarks, and the most democratic assignment is that each has a fractional baryon number, $B = \frac{1}{3}$. From the relation (4.18), i.e.

$$Q/e = \frac{1}{2}(B + S) + I_3, \tag{5.1}$$

where the combination $Y = B + S$ is called the hypercharge, it follows that quarks must also carry fractional charges of $\frac{2}{3}$ and $-\frac{1}{3}$. The appropriate combinations of quarks indicated in Fig. 5.2 can then account for the quantum numbers I, I_3, S (or Y) of the members of the decuplet, and of course their electric charges. The progressive increase in mass of the decuplet members with decreasing S can then be simply ascribed to a difference in mass of s, as compared with u and d, quarks in the region $m_s - m_{u,d} \simeq 150$ MeV. The masses of u and d quarks are expected to be nearly equal, since any difference must be of

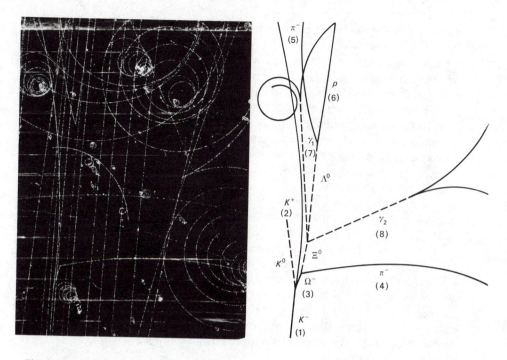

Fig. 5.3 The first Ω^- event (Barnes *et al.* 1964). (Courtesy Brookhaven National Laboratory.) It depicts the following chain of events:

$$K^- + p \to \Omega^- + K^+ + K^0$$

$$\quad \hookrightarrow \Xi^0 + \pi^- \;(\Delta S = 1 \text{ weak decay})$$

$$\qquad \hookrightarrow \pi^0 + \Lambda \;(\Delta S = 1 \text{ weak decay})$$

$$\qquad\qquad \hookrightarrow \pi^- + p \;(\Delta S = 1 \text{ weak decay})$$

$$\qquad\quad \hookrightarrow \gamma \;\;+\;\; \gamma \;\;(\text{e.m. decay})$$

$$\qquad\qquad\quad \downarrow \qquad \downarrow$$

$$\qquad\qquad\; e^+e^- \quad e^+e^-.$$

TABLE 5.1   Quark quantum numbers (as of 1964)[a]

Flavor	B	J	I	I_3	S	Q/e
u	$\frac{1}{3}$	$\frac{1}{2}$	$\frac{1}{2}$	$+\frac{1}{2}$	0	$+\frac{2}{3}$
d	$\frac{1}{3}$	$\frac{1}{2}$	$\frac{1}{2}$	$-\frac{1}{2}$	0	$-\frac{1}{3}$
s	$\frac{1}{3}$	$\frac{1}{2}$	0	0	-1	$-\frac{1}{3}$

[a] Antiquarks $\bar{u}$, $\bar{d}$, and $\bar{s}$ have the signs of B, I_3, S, and Q/e reversed.

the order of the electromagnetic mass differences among the members of an isospin multiplet (see Table 4.3).

Since, as we shall see, quarks are not normally observed as free particles, it is assumed that they are confined inside a hadron by suitable binding forces. The quark behaves in this force field approximately like a free particle with a momentum $\sim r_0^{-1}$, where r_0 is the range of the potential, and an effective mass m^* which is reduced from the value it would have for a free particle by an amount determined by the nature of the potential (e.g. whether it is vector or scalar). The usual assumption made in quark-model calculations is that the motion of the quarks inside the potential can be treated as nonrelativistic, implying $m^* > r_0^{-1}$. Fits to baryon masses, in order to extract m^*, show that this assumption is a marginal one, but seems to work. It is for this reason that the best tests of the quark model apply to systems containing the more recently discovered heavy quarks, c and b, discussed in Section 5.11 below.

It turns out that most, if not all, the baryon states in Fig. 5.1 can be accounted for in terms of the u, d, s quark combinations, three at a time, introducing relative orbital angular momentum l of the quarks as necessary to account for the states of high J. This separate quantization of orbital and spin angular momentum of the quarks is valid only in the approximation of nonrelativistic motion. The meson states ($B = 0$), to be described below, can be described in terms of quark-antiquark combinations. Thus the rule is

$$\text{baryon} = QQQ,$$

$$\text{meson} = Q\bar{Q}. \tag{5.2}$$

We now have to discuss the decuplet in more detail. Given that a baryon is to consist of 3 quarks chosen from any 3 flavors, 27 combinations are possible: so what is special about only 10 of them? We have to introduce some symmetry principle peculiar to members of a particular multiplet. Such symmetry will concern quark spin as well as quark flavor. First, with regard to flavor, we can require that the flavor part of the baryon wavefunction should have a definite symmetry under interchange of any pair of quarks. The corner states uuu, ddd and sss of Fig. 5.2 are clearly symmetric under interchange, so it is natural to require the same symmetry for the other states. We can write the wavefunctions as follows:

Δ: $ddd, (ddu + udd + dud)/\sqrt{3}, (duu + udu + uud)/\sqrt{3}, uuu,$

Σ: $(dds + sdd + dsd)/\sqrt{3}, (dsu + uds + sud \qquad (uus + suu + usu)/\sqrt{3}$

$\qquad\qquad\qquad\qquad + sdu + dus + usd)/\sqrt{6},$

Ξ: $(dss + sds + ssd)/\sqrt{3}, (sus + ssu + uss)/\sqrt{3},$

Ω: $\qquad\qquad\qquad\qquad\qquad sss, \tag{5.3}$

where the top row is constructed by replacing u-quarks with d-quarks as we go from right to left, by applying the isospin shift operators successively (see Appendix C). Thus

$$I^-(uuu) = [I^-(u)](uu) + (u)[I^-(u)](u) + (uu)[I^-(u)]$$
$$= duu + udu + uud.$$

Similarly, successive rows are constructed by replacing u-quarks with s-quarks. Note that, like the corner states, the remaining 7 wavefunctions are symmetric under interchange of any two quarks.

These 10 states are the *only* completely symmetric combinations we can make. Of the remaining 17 of the total of 27, one is completely antisymmetric. The wavefunction

$$(dsu + uds + sud - usd - sdu - dus) \tag{5.4}$$

describes this state, changing sign under interchange of any two quarks, for example $1 \leftrightarrow 2$ or $2 \leftrightarrow 3$. It is easily constructed by adding an s-quark to an antisymmetric u, d combination, i.e. $(ud - du)s$, plus cyclic permutations thereof. This leaves 16 states, which turn out to consist of two octets of mixed symmetry.

In the first studies of hadron families, the emphasis was placed on the symmetry rather than the quark constitution, and indeed the quarks were thought of as convenient mathematical fictions, the base states of the symmetry. The various multiplets were described as (irreducible) representations of the symmetry group SU(3), standing for the special unitary group in 3 dimensions, i.e. with 3 degrees of freedom, corresponding to the 3 quark flavors u, s, d. Exact unitary symmetry would imply invariance under transformations in "unitary spin space", analogous to the SU(2) transformations in "isospin space". The inequality of the masses of the decuplet members indicates that unitary symmetry is in fact badly broken, although still recognizable. The above combinations were written symbolically in the form

$$3 \otimes 3 \otimes 3 = 10 \oplus 8 \oplus 8 \oplus 1. \tag{5.5}$$

$$\begin{array}{ccc} \uparrow & \smile & \uparrow \\ S & \text{mixed} & A \end{array}$$

The actual quark combinations, as written down in (5.3) for example, are indeed most readily described by resorting to group-theoretical methods. However, we shall not discuss these here.

5.2.2 Quark Spin and Color: The Baryon Decuplet and Octet

To proceed further, and to solve the mystery of the octets of mixed symmetry in (5.5), we now incorporate the half-integer spins of the quark constituents. Since the members of the decuplet (5.3) consist of the spin-$\frac{3}{2}$ baryons of lowest mass, we assume that the quarks sit in the spatially symmetric ground state ($l = 0$). The value $J = \frac{3}{2}$ is then obtained by having the quarks in a symmetric spin state, with spins "parallel", as in $\Delta^{++} = u\uparrow u\uparrow u\uparrow$, for example. Hence the $\frac{3}{2}^+$ decuplet is characterized by *symmetry of the three-quark wavefunction in both flavor and spin, as well as space*. This clearly violates the Pauli principle, that two or more fermions may not exist in the same quantum state. This was a problem years ago: subsequently it turned out that another degree of freedom, called *color*, was necessary for other reasons. It is postulated that quarks exist in three colors – say red, green, blue – and that baryons and mesons built from quarks have zero net color, that is, they are *color singlets*. If so, Δ^{++} consists of one red, one green, and one blue u-quark, which makes them nonidentical. "Color" is perhaps an unfortunate name; it is simple a notation for a new property of quarks, quite separate from the flavor quantum number. The three colors specify the "strong charges" of the quarks, in just the same way that the signs $+$ and $-$ specify their electric charges. The evidence for color comes from a number of sources. For example, the predicted rate of decay $\pi^0 \to 2\gamma$ is found to be proportional to the square of the number of colors N_c, and comparison with experiment gives $N_c = 2.98 \pm 0.11$; and the cross-section ratio

$$\sigma(e^+e^- \to \text{hadrons})/\sigma(e^+e^- \to \mu^+\mu^-)$$

at high energy is proportional to N_c and requires a value of 3 within about 10% (see Section 7.8).

If spin is included together with flavor, the symmetry group is called SU(6), since there are now 6 base states, and the irreducible representations are given by the decomposition

$$6 \otimes 6 \otimes 6 = 20 \oplus 70 \oplus 70 \oplus 56. \tag{5.6}$$

$$\begin{array}{cccc} & \uparrow & \underbrace{\quad} & \uparrow \\ & A & \text{mixed} & S \end{array}$$

The SU(6)-symmetric 56-plet consists of SU(3) multiplets:

$$56 = (2, 8) + (4, 10), \tag{5.7}$$

$$\begin{array}{cc} \uparrow & \uparrow \\ \tfrac{1}{2}^+ \text{octet} & \tfrac{3}{2}^+ \text{decuplet} \end{array}$$

where the first number in brackets refers to the spin multiplicity $(2J + 1)$, the second to the SU(3) representation. We have already seen that the decuplet members are described by wavefunctions which are symmetric with respect to space, spin, and flavor, so that they are overall symmetric. In exact SU(6) we do not distinguish between flavor and spin, and it is plausible therefore that the octet states in (5.7) have wavefunctions which are symmetric under *simultaneous* interchange of flavor and spin of any quark pair, although not, as in the decuplet, under each separately. These states are identified with the members of the lowest-lying baryon states of $J^P = \tfrac{1}{2}^+$, the baryon octet which includes the proton and neutron as members.

To construct the baryon octet wavefunctions we first start with a proton (uud) and put two quarks in a spin singlet state (see Section 3.11)

$$\frac{1}{\sqrt{2}}(\uparrow\downarrow - \downarrow\uparrow)$$

which is antisymmetric. To make the overall state symmetric, we need a flavor-antisymmetric combination of u and d (since u and u cannot do it), which is the isosinglet (see Section 4.2)

$$\frac{1}{\sqrt{2}}(ud - du).$$

We then add the third quark u, with spin up, to obtain

$$(u\uparrow d\downarrow - u\downarrow d\uparrow - d\uparrow u\downarrow + d\downarrow u\uparrow)u\uparrow.$$

Although the expression in brackets is symmetric under interchange of the first and second quarks (flavor and spin), the whole expression has to be symmetrized by making a cyclic permutation (12 terms in all), giving finally

$$\phi(P, J_z = +\tfrac{1}{2}) = \frac{1}{\sqrt{18}}[2u\uparrow u\uparrow d\downarrow + 2d\downarrow u\uparrow u\uparrow + 2u\uparrow d\downarrow u\uparrow$$

$$- u\downarrow d\uparrow u\uparrow - u\uparrow u\downarrow d\uparrow - u\downarrow u\uparrow d\uparrow$$

$$- d\uparrow u\downarrow u\uparrow - u\uparrow d\uparrow u\downarrow - d\uparrow u\uparrow u\downarrow]. \qquad (5.8)$$

The other members of the baryon octet of $J^P = \tfrac{1}{2}^+$ can be worked out in similar fashion. This octet is depicted in Fig. 5.4, where the flavor combinations have been indicated as uud, ssu, etc., but it should be understood that these are abbreviations for the properly symmetrized expressions. The eight members consist of the n and $p(939)$ nucleon isospin doublet ($I = \tfrac{1}{2}$, $S = 0$), the $\Sigma(1193)$

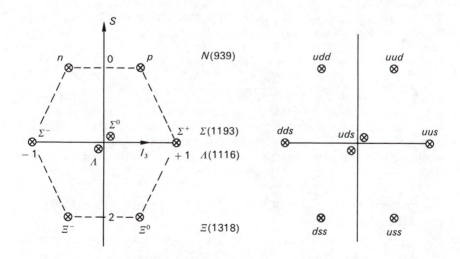

Fig. 5.4 The baryon octet of spin-parity $\tfrac{1}{2}^+$. The observed states are given on the left, and quark flavor assignments on the right.

isotriplet ($I = 1$, $S = -1$), the $\Xi(1318)$ isodoublet ($I = \frac{1}{2}, S = -2$), and the $\Lambda(1116)$ isosinglet ($I = 0$, $S = -1$).

The hypothesis that baryon masses differ because of differences in the strange quark content gave for the decuplet

$$\underset{\text{152 MeV}}{\Sigma(1384) - \Delta(1232)} = \underset{\text{149 MeV}}{\Xi(1533) - \Sigma(1384)} = \underset{\text{139 MeV}}{\Omega^-(1672) - \Xi(1533)}. \quad (5.9)$$

In the $\frac{1}{2}^+$ octet, the same hypothesis gives:

$$\underset{\text{1193 MeV}}{M_\Sigma} = \underset{\text{1116 MeV}}{M_\Lambda},$$

$$\underset{\text{177 MeV}}{M_\Lambda - M_N} = \underset{\text{203 MeV}}{M_\Xi - M_\Lambda}. \quad (5.10)$$

There is rather poor agreement for each of the three equalities above and a large discrepancy between mass differences in decuplet and octet. Historically, mass relations in multiplets were first obtained by making certain assumptions about the nature of symmetry breaking in SU(3), namely, that the mass terms should be dominated by the effects of certain operators, predicting a specific (and quadratic) dependence of mass on the hypercharge $Y = B + S$ and the isospin I. These assumptions were enshrined in the Gell-Mann–Okubo mass formula for baryons:

$$M = M_0 + M_1 Y + M_2 \left[I(I + 1) - \frac{Y^2}{4} \right]. \quad (5.11)$$

In the decuplet of Fig. 5.2, the states are related by $Y = B + S = 2(I - 1)$, so that $M = A + BY$ and we obtain the linear S-dependence of mass as in (5.9). For the baryon octet, (5.11) predicts

$$\underset{\text{4541 MeV}}{3M_\Lambda + M_\Sigma} = \underset{\text{4514 MeV}}{2M_N + 2M_\Xi} \quad (5.12)$$

correct to 1% accuracy. However, there are no clear physical reasons for the assumptions on which (5.11) is based, and they certainly do not explain the great difference in the M_0-values for octet and decuplet states. A more quantitative description of baryon and meson masses is obtained in terms of hyperfine-splitting effects of the interquark interactions, and these are discussed in Section 5.8 below.

5.3 QUARK-ANTIQUARK COMBINATIONS: THE PSEUDOSCALAR MESONS

As indicated in (5.2), the states observed in nature consist of three-quark combinations (the baryons) and quark-antiquark combinations (the mesons). Restricting ourselves to three flavors, we expect families of mesons containing $3^2 = 9$ states, or nonets. Given spin $\frac{1}{2}$ for quarks and antiquarks, we might expect both spin triplet ($\uparrow\uparrow$) states of $J = 1$ (the vector mesons) and spin singlet ($\downarrow\uparrow$) states of $J = 0$ (the pseudoscalar mesons).

In discussing the baryon multiplets, emphasis was placed on the quark-exchange symmetry. But now we are dealing with quarks and antiquarks – thus the interchange $u \to \bar{u}$, for example. It is necessary therefore to consider the effect of charge conjugation applied to quark wavefunctions. If the baryon number B is conserved, there is no actual physical process $Q \leftrightarrow \bar{Q}$: thus as a result of the operation of charge conjugation, or particle-antiparticle conjugation, an arbitrary and unobservable phase occurs. We can write $|u\rangle \overset{C}{\to} |\bar{u}\rangle$ or $|u\rangle \overset{C}{\to} |\bar{u}\rangle e^{i\phi}$. The phase ϕ is generally chosen according to the Condon-Shortley convention, which actually introduces a minus sign in some places. In the following table, arrows denote the C-operation:

	Nucleons		Quarks		
I_3	Particle	Antiparticle	Particle	Antiparticle	(5.13)
$+\frac{1}{2}$	$\|p\rangle$	$+\|\bar{n}\rangle$	$\|u\rangle$	$+\|\bar{d}\rangle$	
$-\frac{1}{2}$	$\|n\rangle$	$-\|\bar{p}\rangle$	$\|d\rangle$	$-\|\bar{u}\rangle$	

Assuming that our quark-antiquark combinations are in the $l = 0$ singlet spin state (Section 3.11) and recalling the opposite intrinsic parity of fermion and antifermion (Section 3.6), the quantum numbers will be $B = 0$, $J^P = 0^-$. These correspond to *pseudoscalar mesons*, so-called because the wavefunctions have $J = 0$, have odd parity, and change sign under spatial inversion.

With only u and d quarks and antiquarks we can make $2^2 = 4$ combinations as follows:

I	I_3	Wavefunction	Q/e		
1	1	$u\bar{d} = \pi^+$	$+1$		
1	-1	$-\bar{u}d = \pi^-$	-1	isospin triplet	(5.14)
1	0	$\sqrt{\frac{1}{2}}(d\bar{d} - u\bar{u}) = \pi^0$	0		
0	0	$\sqrt{\frac{1}{2}}(d\bar{d} + u\bar{u}) = \eta$	0	isospin singlet	

These assignments are justified by applying the isospin shift operators $I^\pm$, in exact analogy with the angular-momentum operators $J^\pm$ [see Appendix C]:

$$I^\pm|\Psi(I, I_3)\rangle = \sqrt{I(I+1) - I_3(I_3 \pm 1)}\,|\Psi(I, I_3 \pm 1)\rangle. \qquad (5.15)$$

Thus applying to single quark states:

$$I^+|d\rangle \to |u\rangle, \qquad I^+|\bar{u}\rangle \to |-\bar{d}\rangle, \qquad I^+|u\rangle = I^+|\bar{d}\rangle = 0. \qquad (5.16)$$

Furthermore,

$$I^-|\Psi(1,1)\rangle = I^+|\Psi(1,-1)\rangle = \sqrt{2}|\Psi(1,0)\rangle,$$

$$I^+|\Psi(1,0)\rangle = \sqrt{2}|\Psi(1,1)\rangle, \qquad I^+\Psi|1,1\rangle = I^-\Psi|1,-1\rangle = 0. \quad (5.17)$$

Applying these results to quark-antiquark combinations in (5.14), we obtain, for example,

$$I^+|\pi^-\rangle = I^+|-d\bar{u}\rangle = |-u\bar{u}+d\bar{d}\rangle = \sqrt{2}|\pi^0\rangle,$$

$$I^+|\pi^0\rangle = I^+\frac{|d\bar{d}-u\bar{u}\rangle}{\sqrt{2}} = \frac{|u\bar{d}+0-0+u\bar{d}\rangle}{\sqrt{2}} = \sqrt{2}|u\bar{d}\rangle = \sqrt{2}|\pi^+\rangle. \quad (5.18)$$

The $I = 1$ combinations in (5.14) are thus identified with π^+, π^-, π^0, the lowest-mass pseudoscalar mesons. The fourth combination in (5.14) has the property

$$I^{\pm}|\eta\rangle = I^{\pm}\frac{|d\bar{d}+u\bar{u}\rangle}{\sqrt{2}} = \frac{|u\bar{d}-u\bar{d}\rangle}{\sqrt{2}} = 0. \quad (5.19)$$

Thus $|\eta\rangle$ is an isospin singlet, which does not transform under an isospin transformation into any other state; it is orthogonal to the $I = 1$ combinations, so that $\langle\eta|\pi^0\rangle = 0$, for example. With the phase convention under C-conjugation which we have adopted, the singlet state is *symmetric* in quark labels $(d \to u, \bar{d} \to \bar{u})$, whereas the π^+, π^-, π^0 states all change sign. The singlet is identified with the η-meson, of mass 550 MeV.

TABLE 5.2 Pseudoscalar-meson states as quark-antiquark combinations

	I	I_3	S	Meson	Quark combination	Decay	Mass, MeV
SU(3) octet	1	1	0	π^+	$u\bar{d}$	$\pi^{\pm} \to \mu\nu$	140
	1	-1	0	π^-	$d\bar{u}$		
	1	0	0	π^0	$(d\bar{d}-u\bar{u})/\sqrt{2}$	$\pi^0 \to 2\gamma$	135
	$\frac{1}{2}$	$\frac{1}{2}$	$+1$	K^+	$u\bar{s}$	$K^+ \to \mu\nu$	494
	$\frac{1}{2}$	$-\frac{1}{2}$	$+1$	K^0	$d\bar{s}$	$K^0 \to \pi^+\pi^-$	498
	$\frac{1}{2}$	$-\frac{1}{2}$	-1	K^-	$\bar{u}s$	$K^- \to \mu\nu$	494
	$\frac{1}{2}$	$\frac{1}{2}$	-1	$\bar{K}^0$	$\bar{d}s$	$\bar{K}^0 \to \pi^+\pi^-$	498
	0	0	0	η_8	$(d\bar{d}+u\bar{u}-2s\bar{s})/\sqrt{6}$	$\eta \to 2\gamma$	549
SU(3) singlet	0	0	0	η_0	$(d\bar{d}+u\bar{u}+s\bar{s})/\sqrt{3}$	$\eta' \to \eta\pi\pi$ $\to 2\gamma$	958

Introduction of s-quarks in addition to u and d gives us a total of $3^2 = 9$ states, which are listed in Table 5.2, together with their assignment to the pseudoscalar mesons. Transformations from strange to nonstrange mesons are effected simply by replacing u and d quarks with s quarks. The square-root factors are inserted where appropriate so that all states are normalized to unity. The unitary SU(3) singlet state $|\eta_0\rangle$ is written down as the symmetric $Q\bar{Q}$ combination, while the eighth member state of the octet, $|\eta_8\rangle$, becomes $(d\bar{d} + u\bar{u} - 2s\bar{s})/\sqrt{6}$, orthogonal to $|\eta_0\rangle$.

On the basis of exact unitary symmetry, we expect the decomposition

$$3 \otimes 3 = 8 \oplus 1 \tag{5.20}$$

with the SU(3) singlet quite distinct from the octet members. The linear octet mass formula (5.12) used for the baryons is not satisfied for mesons; it is found that it works if (mass)^2 is used instead of mass:

$$2(M_{K^0}^2 + M_{\bar{K}^0}^2) = \underset{0.988\,\text{GeV}^2}{4M_{K^0}^2} = \underset{0.924\,\text{GeV}^2}{M_{\pi^0}^2 + 3M_{\eta}^2}. \tag{5.21}$$

However, the actual states η and η' observed in nature appear to be linear combinations of the wavefunctions η_8 and η_0 written down in the last two lines of Table 5.2. On the basis of mass formulae, the mixing angle is $\theta \simeq 11°$. This matter is discussed more fully below, for the vector-meson nonet.

5.4 THE VECTOR MESONS

The spin triplet combinations ($\uparrow\uparrow$) of Q and $\bar{Q}$ in s-states give us the vector mesons $(J^P = 1^-)$ (see Fig. 5.5). In this case the octet-singlet mixing angle is large $(\theta \simeq 35°)$. Formally we can write

$$\phi = \phi_0 \sin\theta - \phi_8 \cos\theta,$$
$$\omega = \phi_8 \sin\theta + \phi_0 \cos\theta, \tag{5.22}$$

where ϕ, ω denote the physical vector-meson states and ϕ_0, ϕ_8 the singlet and octet states, respectively, of $I = S = 0$. Assuming that the matrix element of the Hamiltonian between the states yields the (mass)^2, i.e. $M_\phi^2 = \langle\phi|H|\phi\rangle$ etc., we obtain from (5.22)

$$M_\phi^2 = M_0^2 \sin^2\theta + M_8^2 \cos^2\theta - 2M_{08}^2 \sin\theta\cos\theta, \tag{5.23a}$$
$$M_\omega^2 = M_8^2 \sin^2\theta + M_0^2 \cos^2\theta + 2M_{08}^2 \sin\theta\cos\theta, \tag{5.23b}$$

in an obvious notation. Further, since ϕ and ω are orthogonal, we obtain

$$M_{\phi\omega}^2 = 0 = (M_0^2 - M_8^2)\sin\theta\cos\theta + M_{08}^2(\sin^2\theta - \cos^2\theta). \tag{5.24}$$

Eliminating M_{08} and M_0 between these three equations gives

$$\tan^2\theta = \left(\frac{M_\phi^2 - M_8^2}{M_8^2 - M_\omega^2}\right). \tag{5.25}$$

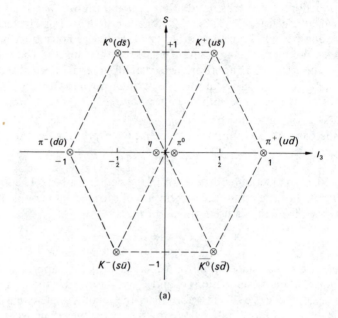

(a)

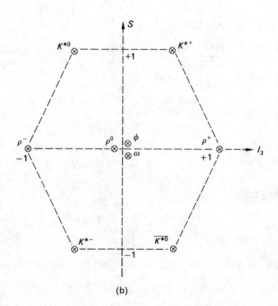

(b)

Fig. 5.5 (a) The lowest-lying pseudoscalar-meson states ($J^P = 0^-$). Quark flavor assignments are indicated. (b) The vector-meson nonet ($J^P = 1^-$). The quark assignments are the same as in (a).

TABLE 5.3 Vector-meson nonet

State	I	Y	Mass, MeV	Dominant decay mode
ρ	1	0	776	$\rho \to 2\pi$
K^*	$\frac{1}{2}$	± 1	892	$K^* \to K\pi$
ω	0	0	783	$\omega \to 3\pi$
ϕ	0	0	1019	$\phi \to K\bar{K}$

Using the analogue of (5.21), we have

$$M_8^2 = \tfrac{1}{3}(4M_{K^*}^2 - M_\rho^2), \tag{5.26}$$

so that the observed masses (Table 5.3) give $\theta \simeq 40°$. For the particular case $\sin\theta = 1/\sqrt{3}$, $\theta \simeq 35°$, Eq. (5.22) would give

$$\phi = \frac{1}{\sqrt{3}}(\phi_0 - \sqrt{2}\,\phi_8),$$

$$\omega = \frac{1}{\sqrt{3}}(\phi_8 + \sqrt{2}\,\phi_0),$$

where from Table 5.2

$$\phi_0 = (d\bar{d} + u\bar{u} + s\bar{s})/\sqrt{3},$$

$$\phi_8 = (d\bar{d} + u\bar{u} - 2s\bar{s})/\sqrt{6}, \tag{5.27}$$

so that

$$\phi = s\bar{s},$$

$$\omega = (u\bar{u} + d\bar{d})/\sqrt{2}. \tag{5.28}$$

In this case of "ideal mixing" — which is almost true in practice — ϕ is composed entirely of s-quarks, and ω of u and d. We note that these simple expressions predict similar masses for ω and ρ as well as a larger mass for the ϕ — as is observed. Even more importantly, they allow some understanding about the decay modes of ω and ϕ. These are

$$\left.\begin{array}{l} \phi(1020) \to K^+K^- \\ \quad\to K^0\bar{K}^0 \end{array}\right\} 84\% \qquad \left.\begin{array}{l} \omega(783) \to \pi^+\pi^-\pi^0 \quad 90\% \\ \quad\to \pi^+\pi^- \\ \quad\to \pi^0\gamma \end{array}\right\} 10\%$$

$$\phi \to \pi^+\pi^-\pi^0 \quad 15\%$$

The phase-space factors favor 3π decay of the ϕ, since the Q-value is 600 MeV, compared with $Q = 24$ MeV for $K\bar{K}$ decay. Yet the $K\bar{K}$ decay is dominant. This must be somehow connected with the $s\bar{s}$ composition of the ϕ-meson as in (5.28). To understand this we can draw quark flow diagrams (Fig. 5.6). Diagram

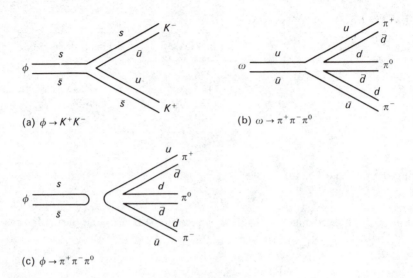

(a) $\phi \to K^+ K^-$

(b) $\omega \to \pi^+ \pi^- \pi^0$

(c) $\phi \to \pi^+ \pi^- \pi^0$

Fig. 5.6 Quark flow diagrams for ϕ-and ω-decay.

(c) is for $\phi \to 3\pi$, involves unconnected quark lines, and is suppressed. This suppression is known as the Zweig rule and is especially important in understanding the narrow widths of the more massive mesons (ψ, Υ) built from b and c quarks (see Section 5.11). In the context of a field theory of quark interactions (quantum chromodynamics) the Zweig suppression is viewed in terms of a multigluon intermediate state (see Section 8.10).

5.5 LEPTONIC DECAYS OF VECTOR MESONS

As an illustration of the quark-antiquark assignments for the vector mesons mentioned above,

$$\rho^0 = \frac{1}{\sqrt{2}}(u\bar{u} - d\bar{d}),$$

$$\omega^0 = \frac{1}{\sqrt{2}}(u\bar{u} + d\bar{d}), \tag{5.29}$$

$$\phi^0 = s\bar{s},$$

we consider their leptonic decays $V \to l^+ l^-$ (where $l = e, \mu$), which are assumed to proceed via exchange of a single virtual photon (see Fig. 5.7). The partial width for decay is given by the Van Royen-Weisskopf formula (1967):

$$\Gamma(V \to l^+ l^-) = \frac{16\pi\alpha^2 Q^2}{M_V^2}|\psi(0)|^2, \tag{5.30}$$

Fig. 5.7

where $Q^2 = |\sum a_i Q_i|^2$ is the squared sum of the charges of the quarks, in the meson, $\psi(0)$ is the amplitude of the $Q\bar{Q}$ wavefunction at the origin, and M_V is the meson mass. Apart from numerical factors, the form of this expression follows in straightforward fashion. The square of the propagator of the single photon exchanged introduces a factor q^{-4} (where $|q|^2 = M_V^2$), and the phase-space factor for the two-body final state, a factor q^2. The factor α^2 results from the coupling of the photon to the quarks and to the leptons. As in Rutherford scattering, the cross-section will also be proportional to the square of the quark charges. Finally, $|\psi(0)|^2$ is the probability that the quark and antiquark will interact with the photon at a point in space-time (the origin of their relative coordinates). In the expression for the phase space, we have assumed $M_V^2 \gg m_l^2$.

Since ρ, ω, ϕ have similar masses, we expect $|\psi(0)|^2/M_V^2$ to be practically constant, and thus $\Gamma_{e^+e^-} \propto Q^2$ (see also Fig. 5.17). The Q^2 factors are, from (5.29),

$$\rho^0: \quad \left[\frac{1}{\sqrt{2}}\left(\frac{2}{3} - \left(-\frac{1}{3}\right)\right)\right]^2 = \frac{1}{2},$$

$$\omega^0: \quad \left[\frac{1}{\sqrt{2}}\left(\frac{2}{3} - \frac{1}{3}\right)\right]^2 = \frac{1}{18},$$

$$\phi^0: \quad \left(\frac{1}{3}\right)^2 = \frac{1}{9}.$$

The expected and observed leptonic widths are in the ratio

$$\Gamma(\rho^0):\Gamma(\omega^0):\Gamma(\phi^0) = \begin{array}{ll} 9:1:2 & \text{predicted,} \\ 8.8 \pm 2.6:1:1.70 \pm 0.41 & \text{observed.} \end{array} \quad (5.31)$$

This result is a test both of the quark assignments in the vector mesons and of the quark charges.

5.6 DRELL-YAN PRODUCTION OF LEPTON PAIRS BY PIONS ON ISOSCALAR TARGETS

Another test of the quark charge assignments is provided by the process of lepton pair production by pions on nucleons. As shown in Fig. 5.8, we imagine

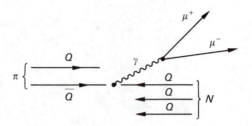

Fig. 5.8 The Drell-Yan mechanism for production of lepton pairs is viewed as the fusion of quark and antiquark to a virtual photon, decaying to the pair.

this process to proceed by annihilation of an antiquark from the pion with a quark from the nucleon, to form a virtual photon transforming to a muon pair. Again, the cross-section is proportional to the squares of the quark charges. For π^- $(= \bar{u}d)$ on an isoscalar ^{12}C nucleus $(= 18u + 18d)$ we expect annihilation of $u\bar{u}$; thus

$$\sigma(\pi^- C \to \mu^+ \mu^- + \cdots) \propto 18Q_u^2 = 18(\tfrac{4}{9})$$

while for incident π^+ $(= u\bar{d})$,

$$\sigma(\pi^+ C \to \mu^+ \mu^- + \cdots) \propto 18Q_d^2 = 18(\tfrac{1}{9}).$$

The cross-section ratio $\sigma(\pi^- C)/\sigma(\pi^+ C)$ is indeed $4:1$ in the region well away from any heavy meson resonances (e. g. $\psi \to \mu^+ \mu^-$).

5.7 PION-NUCLEON CROSS-SECTIONS

The quark model makes predictions on the relative magnitudes of high-energy hadron-hadron cross-sections, interpreting these as the additive effects of the scattering amplitudes between the individual constituent quarks. The individual collisions between pairs of quarks are treated as quasi-independent of the other constituents. Then the total amplitude $f(0)$ for forward elastic scattering will be the sum of the amplitudes for each quark pair. The optical theorem (4.44) tells us that the total hadron-hadron cross-section $\sigma_T \propto \text{Im} f(0)$. The real part of the amplitude $f(0)$ is assumed to be small — and this is confirmed, for example, by looking at the Coulomb interference in the forward direction in pp elastic scattering. We also need to use the Pomerančuk theorem (Section 4.12) which implies that at high energy, $\sigma(QQ) = \sigma(Q\bar{Q})$, as well as isospin invariance for u- and d-quark cross-sections: $\sigma(uu) = \sigma(dd) = \sigma(ud)$. Putting all these things together, it then follows, from simple quark counting, that we expect a ratio

$$\frac{\sigma(\pi N)}{\sigma(NN)} = \frac{2}{3}.$$

At $E_\pi = 60$ GeV, $\sigma(\pi^+ p)$ and $\sigma(\pi^- p)$ are both equal to 24 mb, while $\sigma(pp) = \sigma(pn) = 38$ mb in an equivalent energy range (see Fig. 4.30). This ratio is indeed approximately $\frac{2}{3}$. For further examples, see Problem 5.4.

5.8 MASS RELATIONS AND HYPERFINE INTERACTIONS

So far, mass differences in hadron multiplets have been simply ascribed to those between the constituent u-, d-, and s-quarks. These alone cannot account for the observed differences, especially those between baryon octet and decuplet members of the same quark constitution. The discrepancy has to be attributed to the effects of quark-quark interactions, for which there is a theory called quantum chromodynamics, discussed in Chapter 8. Here, we shall only sketch out briefly how the most important new contribution, the hyperfine splitting of the hadron energy levels, can be described in terms of the color force between quarks.

First, let us recall that, in the energy-level diagram of the hydrogen atom, each level of given n, l, and j quantum numbers is split into two very close hyperfine levels, due to the interactions of the magnetic moments of the constituent proton and electron. In the ground ($1S$) state, the transition between the two states gives rise to the famous 1420-MHz (21-cm) radio-frequency spin-flip line. Now consider two charged point fermions with magnetic dipole moments $\boldsymbol{\mu}_i$ and $\boldsymbol{\mu}_j$ separated by distance r_{ij}. The interaction energy is proportional to $\boldsymbol{\mu}_i \cdot \boldsymbol{\mu}_j / r_{ij}^3$. For two particles in a relative S-state, the interaction when averaged over all directions in space is zero except at the origin. The dipole moment has the form expected for a Dirac pointlike fermion:

$$\boldsymbol{\mu}_i = \frac{e_i}{2m_i} \boldsymbol{\sigma}_i \tag{5.32}$$

in units $\hbar = c = 1$, where e_i and m_i are the electric charge and mass of the particle, $\boldsymbol{\sigma}_i$ its spin-vector ($\sigma_i^2 = 1$). The interaction energy due to this dipole-dipole interaction is then*

$$\Delta E = \frac{2\pi}{3} \frac{e_i e_j}{m_i m_j} |\psi(0)|^2 \boldsymbol{\sigma}_i \cdot \boldsymbol{\sigma}_j, \tag{5.33}$$

where $\psi(0)$ is the wavefunction of our two-particle system at the origin ($r_{ij} = 0$). The numerical factor arises from the angular integration for the S-state wavefunction.

Turning now to quarks, the normal magnetic interaction associated with the electric charge and spin of the quarks is small on the scale of hadron masses; it is of order of the electromagnetic mass differences (~ 1 MeV). But quarks carry a strong color charge, and at small interquark separation the color potential is

---

* See for example J. D. Bjorken and S. D. Drell, *Relativistic Quantum Mechanics*, McGraw-Hill, New York, 1964, p. 58.

assumed to have the same $(1/r)$ form as the Coulomb potential [see (5.47)]. Associated with the color charges of spinning quarks is a color magnetic interaction, of the same form as (5.33) but with electric charges replaced by color charges. The numerical coefficient in the expression for ΔE depends on whether the interaction is between a quark pair or a quark-antiquark pair, and on the eigenvalues of certain operators associated with the color symmetry. The actual expressions are (see Appendix G)

$$\Delta E(Q\bar{Q}) = \frac{8\pi\alpha_s}{9m_i m_j}|\psi(0)|^2 \boldsymbol{\sigma}_i \cdot \boldsymbol{\sigma}_j, \tag{5.34}$$

$$\Delta E(QQ) = \frac{4\pi\alpha_s}{9m_i m_j}|\psi(0)|^2 \boldsymbol{\sigma}_i \cdot \boldsymbol{\sigma}_j, \tag{5.35}$$

where α_s is the strong coupling constant (Section 8.9.3) and equal to the square of the color charge, in analogy with the fine structure constant of electromagnetism, $\alpha = e^2/4\pi\hbar c$. The product of the Pauli vectors $\boldsymbol{\sigma}_i, \boldsymbol{\sigma}_j$ depends in magnitude and sign on the relative quark spin orientations, just as for two bar magnets the force depends on orientation. Denoting the spin vectors of the quarks by $\mathbf{s}_i, \mathbf{s}_j$ (where $s_z = \pm\frac{1}{2}$) and the total spin by $\mathbf{S} = \mathbf{s}_i + \mathbf{s}_j$, one obtains

$$\boldsymbol{\sigma}_i \cdot \boldsymbol{\sigma}_j = 4\mathbf{s}_i \cdot \mathbf{s}_j = 2[S(S+1) - s_i(s_i+1) - s_j(s_j+1)]$$

$$= \begin{cases} +1 & \text{for} \quad S = 1, \\ -3 & \text{for} \quad S = 0. \end{cases} \tag{5.36}$$

Turning to baryons, consisting of three quarks, we have to sum (5.36) over the quark spins, to obtain, with $\mathbf{S} = \mathbf{s}_i + \mathbf{s}_j + \mathbf{s}_k$,

$$\sum \boldsymbol{\sigma}_i \cdot \boldsymbol{\sigma}_j = 4\sum \mathbf{s}_i \cdot \mathbf{s}_j = 2[S(S+1) - 3s(s+1)]$$

$$= \begin{cases} +3 & \text{for} \quad S = \frac{3}{2}, \\ -3 & \text{for} \quad S = \frac{1}{2}. \end{cases} \tag{5.37}$$

This formula will be satisfactory for the nucleon N- or Δ-states, where the three (u and d) quark masses in the denominator of (5.35) are equal. For a $\Sigma^+(1193)$ hyperon (uus) in the octet, the different mass of s and u quarks has to be taken into account in summing (5.36) over the three-quark state. The like pair is $u\uparrow u\uparrow$, in a triplet spin state with $\boldsymbol{\sigma}_u \cdot \boldsymbol{\sigma}_u = 1$, so that from (5.37), $2\boldsymbol{\sigma}_u \cdot \boldsymbol{\sigma}_s = \sum \boldsymbol{\sigma}_i \cdot \boldsymbol{\sigma}_j - \boldsymbol{\sigma}_u \cdot \boldsymbol{\sigma}_u = -4$. Thus we find, for example,

$$(\Delta E)_\Delta = +\frac{3}{m_u^2}K,$$

$$(\Delta E)_N = -\frac{3}{m_u^2}K,$$

$$(\Delta E)_\Sigma = \left(\frac{1}{m_u^2} - \frac{4}{m_u m_s}\right)K. \tag{5.38}$$

where $K = 4\pi\alpha_s|\psi(0)|^2/9$. From the eight isomultiplets of baryons in the decuplet and octet, we are thus able to fit the four unknown parameters, K, m_u, m_s, and the constant M_0 of Eq. (5.11). The resulting mass values are compared with those observed in Table 5.4, which yields

$$m_n(= m_u = m_d) = 363 \text{ MeV},$$

$$m_s = 538 \text{ MeV}, \tag{5.39}$$

$$K/m_n^2 = 50 \text{ MeV}.$$

TABLE 5.4 Masses of baryons predicted from hyperfine-splitting effects (from Rosner 1980)

Baryon and mass (MeV)	Quark composition (n denotes u or d)	$\Delta E/K$	Predicted mass, MeV
$N(939)$	$3n$	$-3/m_n^2$	939
$\Lambda(1116)$	$2n, 1s$	$-3/m_n^2$	1114
$\Sigma(1193)$	$2n, 1s$	$1/m_n^2 - 4/(m_n m_s)$	1179
$\Xi(1318)$	$1n, 2s$	$1/m_s^2 - 4/(m_n m_s)$	1327
$\Delta(1232)$	$3n$	$3/m_n^2$	1239
$\Sigma(1384)$	$2n, 1s$	$1/m_n^2 + 2/(m_n m_s)$	1381
$\Xi(1533)$	$1n, 2s$	$1/m_s^2 + 2/(m_n m_s)$	1529
$\Omega(1672)$	$3s$	$3/m_s^2$	1682

In all cases, agreement is within 1% or better. In particular, the 300-MeV mass difference between Δ and N is successfully accounted for. For the pseudoscalar- and vector-meson states, discussed in Section 5.3, a similar treatment is possible. In particular, the large mass difference between the $\pi(140)$ and $\rho(776)$, representing singlet and triplet spin combinations of u, d quarks and antiquarks, is ascribed to the hyperfine interaction. We note from (5.34) and (5.35) that the coefficient for mesons is twice that for baryons; this is a crucial prediction of the theory of the interquark color field. Again, the predicted masses agree to within $\sim 1\%$ with the observed values. The fitted m_n, m_s values are slightly less than for the baryons, while K/m_n^2 as defined above is 80 MeV (cf. 50-MeV for baryons). However, it is known that the rms radius of the charge distribution of mesons is smaller ($R_0 \simeq 0.6$ fm) than it is for baryons ($R_0 \simeq 0.8$ fm), so that $|\psi(0)|^2$, which is proportional to $1/R_0^3$, will be almost a factor 2 larger. So the expectedly larger hyperfine splitting in the mesons, compared with baryons, seems to be verified in the data.

5.9 ELECTROMAGNETIC MASS DIFFERENCES AND ISOSPIN SYMMETRY

The actual mass of a charged hadron can be thought of as made up of two components. First there is a sort of "bare" mass m originating from that of the quark constituents and from their strong mutual interactions, of the type described in the previous section. Secondly, there will be a contribution Δm due to the electric charge of the hadron — basically equal to the work required to put the charge on the previously uncharged particle. If all baryons in the $\frac{1}{2}^{+}$ octet have similar charge distributions, then we might expect similar values of Δm for similar charges:

$$\Delta m_p = \Delta m_{\Sigma^+},$$

$$\Delta m_{\Sigma^-} = \Delta m_{\Xi^-},$$

$$\Delta m_{\Xi^0} = \Delta m_n.$$

Adding the "bare" masses to each side and summing these equations gives

$$m_p + m_{\Sigma^-} + m_{\Xi^0} = m_{\Sigma^+} + m_{\Xi^-} + m_n$$

or

$$\underset{-1.3\,\text{MeV}}{(m_p - m_n)} = \underset{-8.0\,\text{MeV}}{(m_{\Sigma^+} - m_{\Sigma^-})} + \underset{+6.4\,\text{MeV}}{(m_{\Xi^-} - m_{\Xi^0})}. \qquad (5.40)$$

$$-1.6\,\text{MeV}$$

This formula was due originally to Coleman and Glashow and is well verified within the errors. The individual mass differences are associated with isospin symmetry breaking, which has already been discussed in Section 4.4 (see Table 4.3). In the context of the quark model, there are several distinct effects to consider in accounting for the mass differences:

(a) *Difference in masses of the u- and d-quarks.* The sign of each term in (5.40) indicates $m_d > m_u$.

(b) *Coulomb energy difference associated with the electrical energy between pairs of quarks.* This will be of order $e^2/R_0 = (e^2/\hbar c)(\hbar c/R_0)$ where R_0 is the size of a baryon. With $\hbar c = 197$ MeV fm, $R_0 \simeq 0.8$ fm, we have $e^2/R_0 \simeq 2$ MeV.

(c) *Magnetic energy difference associated with the magnetic-moment (hyperfine) interaction between quark pairs.* From Eq. (5.33) this will be of order of magnitude $(e\hbar/mc)^2(1/R_0)^3$, where m is the quark mass and $|\psi(0)|^2 \simeq R_0^{-3}$. Since $\hbar/mc$ is of order R_0, the magnetic energy is also in the region of e^2/R_0, i.e. 1 or 2 MeV.

Fitting the exact forms of these terms to the numbers in (5.40), it is found that $m_d - m_u = 2$ MeV. While, given the values of the spins, charges, and radii of baryons, any model must predict Coulomb and magnetic energy differences of the above magnitudes, the closeness in mass of the u- and d-quarks could not have been foreseen. Thus the property of approximate isospin invariance of the interactions between hadrons and in atomic nuclei can be associated with the near equality of m_u and m_d, and of no more fundamental significance. At the present time, the origin of quark masses and the above near equality is however not understood.

5.10 BARYON MAGNETIC MOMENTS

From the symmetry properties of the three-quark wavefunction in the baryon octet, it is a fairly straightforward matter to compute the magnetic moments of the various members, by assuming that they are equal to the vector sums of the quark moments. If the quarks behave as pointlike Dirac particles, each will have a magnetic dipole moment as in (5.32). For the proton, uud, we have already noted that the two u-quarks will be in a symmetric (triplet) spin state described by a spin function $\chi(J = 1; m = 0, \pm 1)$, while the third ($d$) quark can be denoted by $\phi(J = \frac{1}{2}, m = \pm \frac{1}{2})$. The total-angular-momentum function for a spin-up proton will be $\psi(J = \frac{1}{2}, m = \frac{1}{2})$, and the Clebsch-Gordan coefficients for combining $J = 1$ and $J = \frac{1}{2}$ give us (see Appendix C, and Table III)

$$\psi(\tfrac{1}{2}, \tfrac{1}{2}) = \sqrt{\tfrac{2}{3}}\,\chi(1, 1)\phi(\tfrac{1}{2}, -\tfrac{1}{2}) - \sqrt{\tfrac{1}{3}}\,\chi(1, 0)\phi(\tfrac{1}{2}, \tfrac{1}{2}). \tag{5.41}$$

For the first combination, the moment will be $\mu_u + \mu_u - \mu_d$, and for the second, just μ_d. Hence we get for the proton moment

$$\mu_p = \tfrac{2}{3}(2\mu_u - \mu_d) + \tfrac{1}{3}\mu_d = \tfrac{4}{3}\mu_u - \tfrac{1}{3}\mu_d. \tag{5.42}$$

The result for the neutron is the same, with the labels u and d interchanged. For Σ^+ one replaces μ_d by μ_s in (5.42), and for Σ^-, μ_d by μ_s and μ_u by μ_d. For the Λ-hyperon, which is a uds combination with $I = 0$, the pair u and d must be

TABLE 5.5 Magnetic moments of "stable" baryons in the $J^P = \frac{1}{2}^+$ octet

Baryon	Magnetic moment in quark model	Prediction, n.m.	Observed, n.m.
p	$\frac{4}{3}\mu_u - \frac{1}{3}\mu_d$	2.79	2.793
n	$\frac{4}{3}\mu_d - \frac{1}{3}\mu_u$	-1.86	-1.913
Λ	μ_s	-0.58	-0.614 ± 0.005
Σ^+	$\frac{4}{3}\mu_u - \frac{1}{3}\mu_s$	2.68	2.33 ± 0.13
Σ^0	$\frac{2}{3}(\mu_u + \mu_d) - \frac{1}{3}\mu_s$	0.82	$-$
Σ^-	$\frac{4}{3}\mu_d - \frac{1}{3}\mu_s$	-1.05	-1.41 ± 0.25
Ξ^0	$\frac{4}{3}\mu_s - \frac{1}{3}\mu_u$	-1.40	-1.20 ± 0.06
Ξ^-	$\frac{4}{3}\mu_s - \frac{1}{3}\mu_d$	-0.47	-1.85 ± 0.75

in an $I = 0$ (that is, antisymmetric) isospin state. So they must be in an antisymmetric spin state ($J = 0$). Hence the u and d in the Λ make no contribution to the moment, and $\mu_\Lambda = \mu_s$. Values of μ for Σ^0, Ξ^0, and Ξ^- can be worked out similarly. Table 5.5 shows the values observed and predicted, using the values of quark mass deduced from Table 5.4 to evaluate the quark moments using (5.39). All numbers are given in terms of the nuclear magneton (n.m.) $\mu = eh/2Mc$, where M is the proton mass.

The predicted p, n, and Λ moments agree very well with experiment. In particular, the expected ratio $\mu_n/\mu_p = -\frac{2}{3}$ is in impressive agreement with the observed value -0.685. There are however considerable discrepancies for the Σ and Ξ states, and these underline a basic weakness of the static or nonrelativistic quark model. In later chapters, we shall see that a more sophisticated picture of baryons (or mesons) has them consisting of the "valence" quarks of the static model, plus neutral vector gluons (the mediating vector bosons of the color field) some of which can transform to quark-antiquark pairs. The $Q\bar{Q}$ pair content of a baryon, although small, can contribute extra currents and thus affect the magnetic moment. At the present time, however, no comprehensive dynamical theory of quark interactions, applicable to the static baryon states, exists.

A brief discussion of the measurement of hyperon magnetic moments is appropriate here. For illustration we shall discuss that of the Λ-hyperon. The Λ-hyperons can be produced in hadron-hadron collisions, for example in the reaction

$$\pi^- + p \to \Lambda^0 + K^0,$$

and are found to be polarized with spins normal to the ΛK production plane, i.e. along the vector $(\mathbf{p}_K \times \mathbf{p}_\Lambda)$. The sense and degree of polarization can be found by

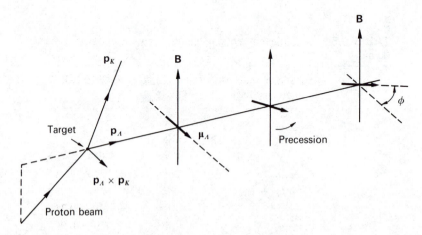

Fig. 5.9 Sketch of precession of spin of Λ-hyperons produced in association with neutral kaons, in a magnetic field. Initially, the hyperons are polarized normal to the production plane.

observing the up-down asymmetry in the subsequent parity-violating weak decay $\Lambda \to p + \pi^-$ (see Section 6.7). Suppose that the Λ traverses a magnetic field **B** perpendicular to $\mathbf{p}_\Lambda$ and in the production plane, as in Fig. 5.9. The Larmor precession frequency is

$$\nu_L = \mu_\Lambda B/h,$$

and the angle of rotation is therefore

$$\phi = \mu_\Lambda \int \frac{B\,dl}{h}$$

proportional to the line integral of the field, and is numerically $18° \cdot (\mu_\Lambda/\mu_N)$ for $\int B\,dl = 1$ tesla metre, where μ_N is the nuclear magneton.

The first experiments on μ_Λ were done with nuclear-emulsion detectors and an incident pion momentum of $\simeq 1.05$ GeV/c, where the transverse polarization was very high ($\simeq 100\%$), but the path length in the field was limited by the lifetime to $c\tau_\Lambda \sim 10$ cm. Thus, appreciable precession angles could be obtained only with the use of pulsed fields of order 20 tesla. The more recent experiments have employed high-energy beams of 300-GeV incident protons, which produce more intense secondary hyperon beams of much higher momentum ($\simeq 100$–250 GeV/c). Thus the useful path length is increased due to relativistic time dilation by 2 orders of magnitude, and d.c. spectrometer magnets can be employed. The decay products are recorded in multiwire proportional chambers. The greatly reduced polarization ($\simeq 8\%$ instead of $\simeq 100\%$) at high energy is more than compensated by the huge increase in hyperon intensities available – the most recent experiment involving a sample of 3×10^6 Λ-particles.

5.11 HEAVY-MESON SPECTROSCOPY AND THE QUARK MODEL

Over the past six years, our quark Table 5.1 has become enlarged by the discovery of two more flavors, c for charm and b for bottom, associated with much more massive quarks (see Table 5.6).

TABLE 5.6 Quark quantum numbers (as of 1977): $Q/e = I_3 + \frac{1}{2}(B + S + C + B^*)$[a]

Flavor	I	I_3	S	C	B^*	Q/e
u	$\frac{1}{2}$	$\frac{1}{2}$	0	0	0	$+\frac{2}{3}$
d	$\frac{1}{2}$	$-\frac{1}{2}$	0	0	0	$-\frac{1}{3}$
s	0	0	-1	0	0	$-\frac{1}{3}$
c	0	0	0	1	0	$+\frac{2}{3}$
b	0	0	0	0	-1	$-\frac{1}{3}$

[a] B denotes baryon number, which is $\frac{1}{3}$ for all quarks. B^* denotes the bottom or beauty quantum number.

It is a characteristic of the strong "color" forces between quarks that these heavier constituents involve somewhat weaker interactions at smaller distances. The series of heavy-meson states ψ, Υ formed from $c\bar{c}$ and $b\bar{b}$, respectively, have narrow widths and present a spectrum of discrete spectroscopic levels similar to (but on a quite different energy scale from) that of positronium (e^+e^-). The level sequence and spacing can be calculated using simple potential models and is in astonishingly close agreement with what is observed. These heavy-meson spectra indeed present the clearest evidence yet for the existence of basic quark and antiquark constituents of hadrons (see Fig. 5.15).

5.11.1   Charmonium Levels

The ψ series of meson resonances was first observed in 1974 in e^+e^- collisions at SLAC (Stanford), using the e^+e^- collider SPEAR (Augustin *et al.* 1974); and the lowest-lying state, called ψ or J/ψ, was simultaneously observed in experiments at the Brookhaven AGS in collisions of 28-GeV protons on a beryllium target (Aubert *et al.* 1974), leading to a massive e^+e^- pair:

SLAC: $$e^+e^- \rightarrow \psi \rightarrow \text{hadrons} \tag{5.43}$$
$$\rightarrow e^+e^-, \mu^+\mu^-$$

BNL: $$p + Be \rightarrow \psi/J + \text{anything.} \tag{5.44}$$
$$\rightsquigarrow e^+e^-$$

The original data on the reaction (5.43) are shown in Fig. 5.10, and on (5.44) in Fig. 5.11. In both cases, a sharp resonance ψ is observed, peaking at a mass of 3.1-GeV. In (5.44), massive electron pairs were detected by means of a magnet spectrometer and detectors downstream of the target, electrons and positrons being recorded in coincidence at large angles on either side of the incident proton beam axis. In the e^+e^- experiment, the reaction rate in the beam intersection region was measured as the beam energies were increased in small steps. In addition to the particle ψ, a second resonance ψ' of mass 3.7 GeV was also found in this first SPEAR experiment (Abrams *et al.* 1974).

The observed widths of the peaks in Fig. 5.10 are dominated by the experimental resolution, on the secondary-electron momentum in the BNL experiment and on the circulating-beam momentum in the SLAC experiment. The true width of the ψ is much smaller and can be determined from the total reaction rate and the leptonic branching ratio, both of which have been measured. Recalling the Breit-Wigner formula (4.51) for the formation of a resonance of spin J from two particles of spin s_1 and s_2, we can write

$$\sigma(E)_{e^+e^- \rightarrow \psi \rightarrow e^+e^-} = \frac{4\pi\lambda^2(2J+1)\Gamma_{e^+e^-}^2/4}{(2s_1+1)(2s_2+1)[(E-E_R)^2 + \Gamma^2/4]}, \tag{5.45}$$

where λ is the de Broglie wavelength of the e^+ and e^- in the CMS, E is the CMS energy, E_R is the energy at the resonance peak, Γ is the total width of the

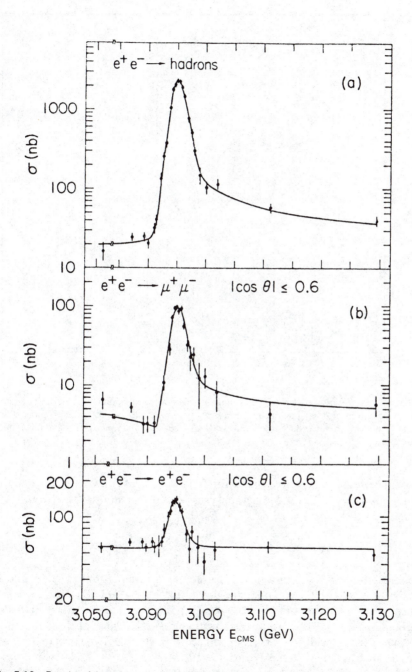

Fig. 5.10 Results of Augustin *et al.* (1974) showing the observation of the ψ/J resonance of mass 3.1 GeV, produced in e^+e^- annihilation at the SPEAR storage ring, SLAC.

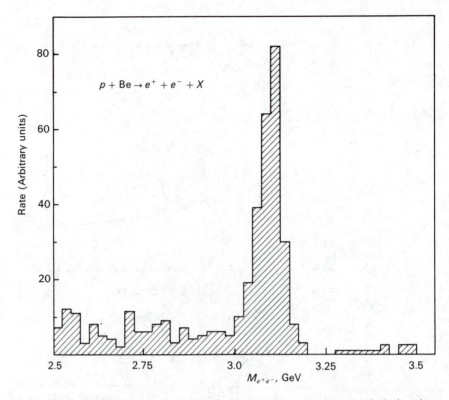

Fig. 5.11   Results of Aubert *et al.* (1974) indicating the narrow resonance ψ/J in the invariant-mass distribution of e^+e^- pairs produced in inclusive reactions of protons with a beryllium target. The experiment was carried out with the 28-GeV AGS at Brookhaven National Laboratory.

resonance, and $\Gamma_{e^+e^-}$ is its partial width for $\psi \to e^+e^-$. With $s_1 = s_2 = \frac{1}{2}$ and the assumption $J = 1$, the total integrated cross-section is readily found from (5.45) using the substitution $\tan\theta = 2(E - E_R)/\Gamma$:

$$\int_0^\infty \sigma(E)\,dE = \frac{3\pi^2}{2}\lambdabar^2\left(\frac{\Gamma_{e^+e^-}}{\Gamma}\right)^2\Gamma. \tag{5.46}$$

The integrated cross-section in Fig. 5.10(c) must be equal to $\int\sigma(E)\,dE$, and is numerically 900 nb MeV. The branching ratio $\Gamma_{e^+e^-}/\Gamma = 0.07$, and $\lambdabar = \hbar c/pc$, where $pc = 1500$ MeV and $\hbar c = 200$ MeV fm. Inserting these numbers in (5.46), we obtain $\Gamma = 0.067$ MeV for the true width of the ψ, which is much smaller than the experimental width, of order several MeV. In comparison with other vector mesons such as the $\rho(776\text{ MeV})$ with $\Gamma = 100$ MeV and $\omega(784\text{ MeV})$ with $\Gamma = 11$ MeV, the $\psi(3100\text{ MeV})$ has an extremely small width, and the purely electromagnetic decay $\psi \to e^+e^-$ competes with that into hadrons. Note that the

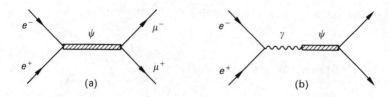

Fig. 5.12

partial width $\Gamma(\psi \to e^+ e^-) = 4$ keV and is not so different from that of the other vector mesons. For example, $\Gamma(\omega \to e^+ e^-) = 0.8$ keV and $\Gamma(\phi \to e^+ e^-) = 1.6$ keV.

The assumption $J^P = 1^-$, i.e. the vector nature of the ψ-particle, is justified by observing the shape of the resonance curve in Fig. 5.10(b). It has the characteristic dispersionlike appearance characteristic of two interfering amplitudes: these are due to the direct channel [Fig. 5.12(a)] and the production of ψ via an intermediate virtual photon [Fig. 5.12(b)]. The interference between these diagrams is proof that ψ must have the same quantum numbers as the photon. The isospin assignment $I = 0$ is based on the characteristics of hadronic decays. Since the ψ decays predominantly to an odd number of pions, its G-parity is -1. Since from (4.21), we have $G = (-1)^{J+I}$, it follows that I must be even. The $I = 0$ assignment is confirmed by observations on the decay mode $\psi \to \rho\pi$: the various charge states $\rho^+\pi^-, \rho^0\pi^0, \rho^-\pi^+$ are found to be equally populated. Reference to the Clebsch-Gordan coefficients (Table III, Appendix) for combining two states of $I = 1$ then shows $I = 0$ is the correct assignment for ψ.

In summary, some properties of the particles ψ and ψ' are listed in Table 5.7. An example of the decay $\psi' \to \psi + \pi^+\pi^-, \psi \to e^+ e^-$ is shown in Fig. 5.13.

The extreme narrowness of the ψ and ψ' states indicated that there was no possibility of understanding them in terms of $u, d,$ and s (and $\bar{u}, \bar{d},$ and $\bar{s}$) quarks. A new type of quark had in fact been postulated some years before by Glashow, Iliopoulos, and Maiani (1970), in connection with the nonexistence of strangeness-changing neutral weak currents (see Section 6.11 below). This carried a new quantum number, C for *charm*, which, like strangeness, would be conserved in strong and electromagnetic interactions. The large masses of the ψ, ψ' mesons imply that, if they contain such charmed quarks, these in turn must be massive. It is indeed postulated that ψ, ψ' consist of the vector combination of $c\bar{c}$, called *charmonium*, just as the ρ^0 consists of $u\bar{u}$ and $d\bar{d}$. Other combinations with a net charm number, e.g. $c\bar{d}$, form the so-called charmed mesons, which had been observed previously in neutrino experiments (but not clearly identified) and were soon to be catalogued in SLAC experiments. With the exception of the lowest-lying D-meson, of mass 1870 MeV, which decays weakly in a $\Delta C = 1$ transition, all are broad states, just like the ρ and ω. Since $M_\psi < 2M_D$, the decay of the ψ into meson states with $C = +1$ and -1 is energetically impossible; it follows that ψ must decay into states containing only u, d, s quarks and antiquarks.

TABLE 5.7   Charmonium states and decay modes

State	Mass, MeV	J^P, I	Γ, MeV	Branching ratio	
$J/\psi(3100)$	3097 ± 1	1^-, 0	0.063	Hadrons	86%
				[mostly $(2n + 1)\pi$]	
				$e^+ e^-$	7%
				$\mu^+ \mu^-$	7%
$\psi(3700)$	3685 ± 1	1^-, 0	0.228	$\psi + 2\pi$	50%
				$\chi + \gamma$	21%
				$e^+ e^-$	0.9%
				$\mu^+ \mu^-$	0.9%

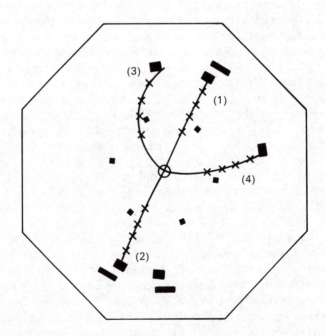

Fig. 5.13 Example of the decay $\psi'(3.7) \rightarrow \psi(3.1) + \pi^+ + \pi^-$ observed in a spark chamber detector. The $\psi(3.1)$ decays to $e^+ + e^-$. Tracks (3) and (4) are due to the relatively low-energy (150-MeV) pions, and (1) and (2) to the 1.5-GeV electrons. The magnetic field and the SPEAR beam pipe are normal to the plane of the figure. The trajectory shown for each particle is the best fit through the sparks, indicated by crosses. [From G. S. Abrams *et al.*, *Phys. Rev. Letters* **34**, 1181 (1975).]

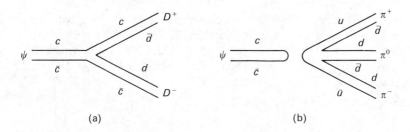

Fig. 5.14 Quark diagrams for charmonium decay. The diagram (a) is favored but forbidden by energy conservation for charmonium states $\psi(3.1)$ and $\psi'(3.7)$ with masses below threshold $2M_D = 3.75$ GeV. The "Zweig-forbidden" diagram (b) is therefore the only one allowed for hadronic decay.

We recall that the small branching ratio for the decay $\phi \to \pi^+ \pi^- \pi^0$ in comparison with $\phi \to K\bar{K}$ (see Fig. 5.6) was ascribed to the so-called Zweig rule, where unconnected lines in the quark flow diagram lead to a suppression of the decay amplitude. We can draw similar diagrams for ψ-decay (Fig. 5.14).

According to this rule, the decay $\psi \to D\bar{D}$ is preferred but is not allowed by energy conservation, leaving the decay into noncharmed mesons as in (b), which is much more strongly suppressed than in ϕ-decay. We shall return to the problem of the ψ width in the last chapter, on quantum chromodynamics (Section 8.10).

The narrowness of the ψ, ψ' states provides us with excellent tests of quark-model spectroscopy. If the ψ is really to be described as a quark-antiquark combination in nonrelativistic motion in a potential well, one should observe singlet and triplet spin states and $s, p, d, \ldots$ states of orbital motion, just as in positronium or in the hydrogen atom. A rather complete system of such spectroscopic levels is in fact found (see Fig. 5.15). The solution of the positronium or hydrogen level structure depends on knowing the form of the Coulomb potential. The potential for charmonium is unknown. The most plausible theory, quantum chromodynamics, involves a color force between quarks mediated by massless vector bosons called gluons. In analogy with the electrical interaction mediated by massless vector photons, it is assumed that the potential must have the $1/r$ form at small distance. At large distances however, the potential must increase with r, so that the quarks are confined inside the hadron. Thus, a plausible potential is

$$V = -\frac{k_1}{r} + k_2 r, \tag{5.47}$$

where k_1 and k_2 are arbitrary constants. Here, we have to identify k_2 with k, the "string constant" that was given a value 0.2 GeV^{-2} to fit the Regge trajectories in (4.89). Then $k_1 \simeq 0.5$ can be fixed from the $2\,^3S_1$–$1\,^3S_1$ level separation, as discussed below. The ordering of the expected levels in the sequence

$1s, 1p, 2s, 1d, 2p$ does not in fact depend on the values of k_1 and k_2 within a broad range. The observed states are shown in Fig. 5.15(a), the $1p$ states being denoted by χ. As in the case of positronium or hydrogen, the three p substates are separated by relativistic (spin-orbit) effects. These χ-states are produced in radiative decays of the $2\,^3S_1$ states $\psi'(3685)$. Like the ψ, the ψ' has $I = 0$ (evidenced by the decay $\psi' \to \eta\psi$) and $J^{PC} = 1^{--}$, since interference between the

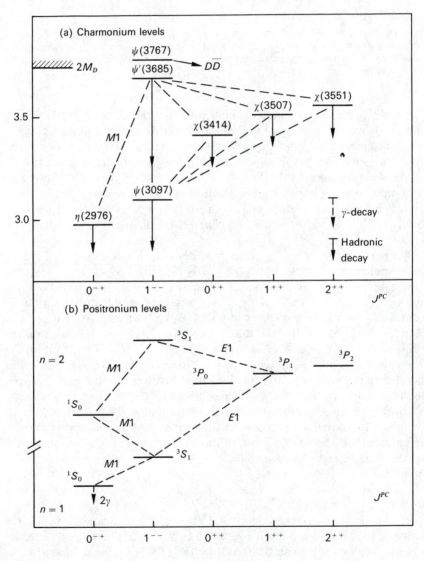

Fig. 5.15

direct and the one-photon channel (Fig. 5.12) is observed. From the radiative decays $\psi' \rightarrow \gamma\chi$, it is established that the χ-states have charge-conjugation parity $C = +1$, and from their decay into even numbers of pions, $G = +1$, and thus $I = 0$ as for ψ and ψ'. The pseudoscalar state $\eta_c(2960)$ is identified with the first singlet S-state $1\,^1S_0$.

The comparison with the energy levels in positronium, Fig. 5.15(b), hardly needs comment. The similarity in the levels in the two cases — although they differ in energy scale by a factor 10^8 — is the strongest possible reason for believing that the ψ, χ states result from the binding of a heavy quark-antiquark pair to form a hadron (see Problem 5.6).

Still heavier charmonium states are observed, but these are all above $D\bar{D}$ threshold; successive states rapidly become broader because the decays to $D\bar{D}$ and other charmed mesons are favored by the Zweig rule. Charmed pseudo-scalar mesons D^+ ($= c\bar{d}$), D^0 ($= c\bar{u}$), and their antiparticles; charmed vector mesons D^* decaying in the mode $D^* \rightarrow \pi D$; and mesons F ($= c\bar{s}$ etc.) with both charm and strangeness, have all been catalogued. The pseudoscalars decay by $\Delta C = 1$ weak interactions into noncharmed states, with the decay into kaons ($D^0 \rightarrow K^-\pi^+$ etc.) favored by the Cabibbo suppression factor (see Section 6.9). Figure 2.13 shows a nice example of production and decay of a vector meson D^* in a neutrino reaction in the BEBC hydrogen chamber at CERN.

5.11.2 Upsilon States (Υ)

The discovery of the narrow charmonium ($\psi = c\bar{c}$) states in 1974 was followed in 1977 by the observation of similar narrow resonances in the mass region 9.5–10.5 GeV, attributed to bound states of still heavier "bottom" quarks with charge $\frac{1}{3}$, and generically named $\Upsilon = b\bar{b}$ — see Table 5.8.

Figure 5.16 shows the results on the mass spectrum of muon pairs produced in 400-GeV proton-nucleus collisions

$$p + \text{Be, Cu, Pt} \rightarrow \mu^+ + \mu^- + \text{anything},$$

as observed in a two-arm spectrometer by Herb *et al.* (1977) and Innes *et al.* (1977) in an experiment at FNAL. A broad peak centered around 10 GeV is apparent against the falling continuum background. Since the total width ($\simeq 1.2$ GeV) was greater than that arising from the apparatus resolution (0.5 GeV), it was deduced that two or three resonances were present, with masses of 9.4, 10.01, and possibly 10.4 GeV — named Υ, Υ', and Υ'' respectively.

As in the case of charmonium, the states Υ, Υ' were also observed (one year later) in e^+e^- experiments at the DORIS storage ring in Hamburg, where they could be clearly resolved, and at CESR, Cornell, where the narrow state Υ'' and a fourth state Υ''' were identified. As for charmonium, the apparent widths of the three Υ-states are determined by the beam energy resolution. Their masses and leptonic widths are given in Table 5.8.

In the nonrelativistic quark model, the leptonic decay width of a vector meson coupling to a single virtual photon is expected to be proportional to the

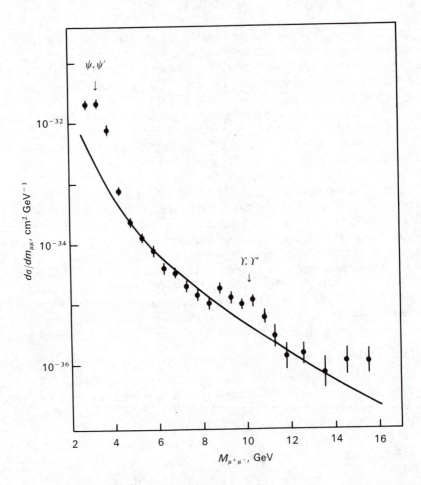

Fig. 5.16 First evidence for the upsilon resonances Υ, Υ'', obtained by Herb *et al.* (1977) from the spectrum of muon pairs observed in 400-GeV proton-nucleus collisions at Fermilab, near Chicago. The enhancement due to these resonances stands out against the rapidly falling continuum background.

TABLE 5.8 Upsilon states

	$\Upsilon(1\,^3S)$	$\Upsilon'(2\,^3S)$	$\Upsilon''(3\,^3S)$	$\Upsilon'''(4\,^3S)$
Mass, GeV	9.46 ± 0.01	10.02 ± 0.01	10.35	10.57
$\Gamma_{e^+e^-}$, keV	$1.2 \;\pm 0.2$	0.50 ± 0.14	0.40 ± 0.10	0.30 ± 0.10
$\Gamma_{(tot)}$, MeV		Less than a few		19 ± 3

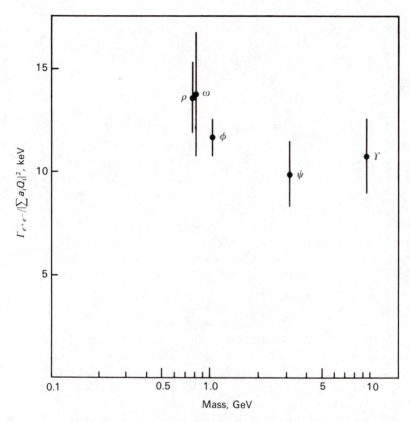

Fig. 5.17 The ratio $\Gamma_{e^+e^-}/|\sum a_i Q_i|^2$ of the leptonic width to the square of the mean quark charge, for the vector mesons $\rho = (u\bar{u} - d\bar{d})/\sqrt{2}$, $\omega = (u\bar{u} + d\bar{d})/\sqrt{2}$, $\phi = s\bar{s}$, and $\psi = c\bar{c}$. The value for $\Upsilon = b\bar{b}$ is obtained by assuming $\frac{1}{3}$ charge for the b-quark.

square of the quark charges (Rutherford scattering). Empirically, as in Fig. 5.17, the ratio $\Gamma_{e^+e^-}/Q^2$ for the ρ, ω, ϕ, and ψ are closely similar, and the leptonic width of Υ agrees with the rest if one assumes it is built from b, $\bar{b}$ quarks with $Q = \frac{1}{3}$. This is also clear from the value of

$$R = \sigma(e^+e^- \rightarrow \text{hadrons})/\sigma(e^+e^- \rightarrow \mu^+\mu^-)$$

above $b\bar{b}$ threshold (see Section 7.8).

Figure 5.18 shows the energy-level diagram of the ^3S states of the ψ and Υ families. The mass differences corresponding to the $2\,^3S$–$1\,^3S$ levels

$$\Delta m(\psi' - \psi) = 589 \pm 1\,\text{MeV}$$

and

$$\Delta m(\Upsilon' - \Upsilon) = 560 \pm 4\,\text{MeV} \tag{5.48}$$

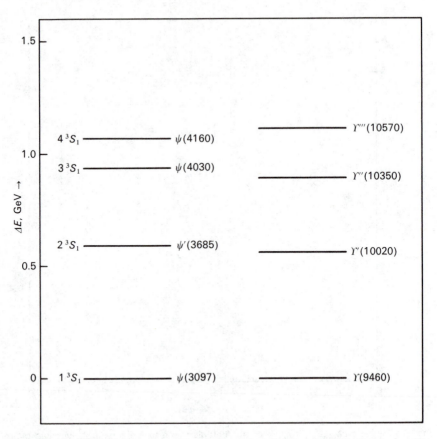

Fig. 5.18 The triplet S-states of the charmonium and upsilon systems. Note the near equality of the $2\,^3S_1$–$1\,^3S_1$ level separations in the two cases.

are closely similar. The expected level separations depend on the form of the interquark potential, and a combination of linear plus Coulomb potential, as in (5.47), is in agreement with the near equality (5.48). The terms in the potential (5.47) in fact depend differently on the quark mass μ. The Coulombic term k_1/r gives a binding energy, and hence a 2S-1S level separation, proportional to μ, whereas the $k_2 r$ term gives separations proportionate to $\mu^{-1/3}$. The values of k_1 and k_2 just happen to result in a mass-independent 2S-1S separation, for masses in the range $\mu \simeq 2\text{–}5$ GeV. If much heavier (top) quarks exist, which are more sensitive to the short-distance (Coulombic) term, this equality might not be expected to hold.

It will be noted from Table 5.8 that the $4\,^3S$ level Υ'''' has a much larger total width than the lower S-states, indicating that it is above threshold for production of pairs of pseudoscalar B-mesons ($\bar{b}u, b\bar{u}, b\bar{d}, \bar{b}d$) analogous to the D-mesons containing c-quarks. The B-meson mass is expected at 5.25 GeV.

5.12   THE SEARCH FOR FREE QUARKS

Ever since the quark hypothesis of hadron substructure was formulated in 1964 by Gell-Mann and by Zweig, intensive searches for the existence of free quarks have been carried out. The vast majority of these experiments have failed to find any such evidence, and this has led to the postulate of confinement, that the interaction between quarks is such that they cannot be separated as free particles once they are trapped inside a hadron.

The experimental evidence for or against free quarks has been reviewed extensively, and only a summary will be given here. The experiments have been of two broad types:

(a) Production *in situ* of quarks in high-energy collisions at accelerators, in hadron-hadron, lepton-hadron, and e^+e^- reactions, and among cosmic-ray secondaries.

(b) Search for quarks already existing in matter, either on earth or elsewhere.

5.12.1   Accelerator Experiments

The searches for quarks produced as secondaries in hadron-hadron (usually proton-nucleus) collisions have used a combination of Čerenkov counters, pulse-height and time-of-flight techniques, and magnetic spectrometers to identify particles of charge $|2e/3|$ and $|e/3|$. They provide limits of $< 10^{-11}$ for the ratio of quarks to pions, and are sensitive to free-quark masses up to about 10 GeV. Generally, such hadronic interactions consist of rather "soft" collisions in which the momentum transfers are small. Lepton-nucleon collisions (see Chapter 7) involve larger momentum transfers and might be more efficient at "kicking out" quarks. The limits obtained are $< 10^{-4}$ quarks per interaction. In e^+e^- experiments the ratio $\sigma(e^+e^- \rightarrow \text{quark} + \text{anything})/\sigma(e^+e^- \rightarrow \mu^+\mu^-)$ $< 10^{-2}$. The lepton experiments so far are sensitive only to quark masses up to 15 GeV.

Analyses of secondary cosmic-ray particles have, on more than one occasion, reported a possible quark signal, but the evidence was never convincing, and the only safe statement to be made is that the sea-level quark flux is less than 10^{-10} of the flux of primary nucleons incident on the atmosphere. The level of sophistication and redundancy built into the cosmic-ray experiments is far inferior to that of those at accelerators.

5.12.2   Search for Preexisting Quarks

If quarks exist in ordinary matter, they may be located in "quark atoms" with peculiar physical and chemical properties and with, of course, a net fractional charge. Some experiments have therefore sought to obtain enriched samples by application of electric fields. In this way, the free-quark density in sea water is found to be $< 10^{-24}$ quarks per nucleon. If quarks were present in cosmic rays, for example, and brought to rest in the oceans, this limit implies a cosmic-ray quark flux below 10^{-8} of the primary nucleon flux. Searches have also been made in moon rocks, meteorites, deep ocean sediments, and even oyster shells. The existence of quarks was investigated by optical or mass spectrometry using the same in an arc or ion source. All these searches have proved fruitless; and provided limits on quark density not far inferior to that from sea water. Of course, it can be argued that because of the peculiar physical and chemical properties of quark atoms, they are concentrated only in specific materials or even in regions of space outside the solar system.

So far, only one experiment has reported a positive result and actually claimed the existence of free particles of fractional charge (LaRue, Fairbank, and Hebard 1977). The principle of the electrometer method employed is similar to that of Millikan's oil-drop device. It is to levitate small (100-μg) diamagnetic spheres of niobium in a suitably shaped magnetic field, in which they oscillate vertically with a frequency $f \sim 1$ Hz. The spheres are located between two horizontal capacitor plates, 1 cm apart, to which an oscillatory electric field is applied at the frequency f, but 90° out of phase with the free oscillations. The rate of change of the oscillation amplitude is proportional to the electric charge on the sphere. The net charge on a sphere can be changed by $\pm 1e$ by bringing close a radioactive electron or positron source, and this provides the charge calibration. The niobium spheres examined were heat-treated in niobium or tungsten substrates to improve the Q-value of the natural oscillations. In their most recent paper (LaRue *et al.* 1981) a total of 39 measurements were reported on 13 spheres with radii 100–140 μm. Five of the spheres exhibited fractional charges on different occasions. Of these, 5 measurements gave a negative charge of mean value $- 0.343 \pm 0.011$, and 9 gave a positive charge of mean value $+ 0.328 \pm 0.007$. The remaining 25 measurements corresponded to integral charges (0.001 ± 0.003). The number of fractional charges observed corresponds to about 10^{-20} quarks per nucleon of niobium if we include the total number of nucleons in the spheres. This is comparable to the upper limits from other quark searches.

A second experiment, by Gallinaro, Marinelli, and Morpurgo (1977), employed (ferromagnetic) iron cylinders (200 μg) suspended in a magnetic field, but found only integral charges (i.e. $\Delta Q < 0.1e$). Their limit on the free-quark concentration in iron is $< 3 \times 10^{-21}$ quarks/nucleon.

In view of the above results, no conclusion can be reached about the existence of free quarks in terrestrial conditions, except that, if they do exist at all, it is at the level of $\leqslant 10^{-20}$ of bound quarks. Quark confinement in hadrons may even be absolute, and any existing particles of fractional charge be "left-overs" from creation in the early universe.

PROBLEMS

5.1 (a) Verify the expressions for the magnetic moments of baryons in Table 5.5. The magnetic moments of the proton and neutron, as well as appropriate combinations of those of the hyperons, depend only on the magnetic moments of the u- and d-quarks.

Assuming that the u- and d-quarks each have one-third of the mass of a nucleon, calculate the baryon moments and compare them with the experimental values.

(b) The anomalous magnetic moments of the neutron and proton are nearly equal in magnitude but opposite in sign. Show how this result can also be obtained by considering a nucleon to consist part of the time as a Dirac (pointlike) nucleon, and part of the time as a pointlike core with a charged circulating pion in a P-state, contributing to the overall moment as a circulating current.

5.2 Discuss the possible decay modes of the Ω^--hyperon allowed by the conservation laws, and show that weak decay is the only possibility.

5.3 Evaluate $|\psi(0)|^2$ in (5.30) from the typical size of a hadron, and hence estimate the absolute leptonic widths of the vector mesons ρ, ω, ϕ.

5.4 Show that the additive quark model predicts the following cross-section relations:

$$\sigma(\Lambda p) = \sigma(pp) + \sigma(K^- n) - \sigma(\pi^+ p),$$

$$\sigma(\Sigma^- p) = \sigma(pp) + \sigma(K^- p) - \sigma(\pi^- p) + 2[\sigma(K^+ n) - \sigma(K^+ p)],$$

$$\sigma(\Sigma^- n) = \sigma(pp) + \sigma(K^- p) - \sigma(\pi^- p).$$

5.5 Show that the decay mode

$$D^+ \to K^- \pi^+ \pi^+$$

is allowed, but that

$$D^+ \to K^+ \pi^0$$

$$\to K^+ \pi^+ \pi^-$$

are strongly suppressed.

5.6 For two particles of equal mass m interacting via an attractive $1/r$ potential, refer to a book on atomic physics to convince yourself that the energy level separations vary as $\alpha^2 m$ where α is the coupling constant of the $1/r$ interaction. In Fig. 5.15, the $2^3 S - 1^3 S$ separation is ~ 600 MeV for charmonium and ~ 5 eV for positronium. Justify this factor 10^8 in energy scale in terms of constituent masses and couplings in the two cases and show that $\alpha_s \sim 1$ is needed for the strong color coupling.

BIBLIOGRAPHY

Close, F. E., *An Introduction to Quarks and Partons*, Academic Press, 1979.

Feldman, G. J., and M. L. Perl, "Recent results in $e^+ e^-$ annihilation above 2 GeV", *Phys. Rep.* **33**, 285 (1977).

Gell-Mann, M., and Y. Ne'eman, *The Eightfold Way*, Benjamin, New York, 1964.

Glashow, S. L., "Quarks with color and flavour", *Sci. Am.* **233**, 38 (Oct. 1975).

Greenberg, O. W., "Quarks", *Ann. Rev. Nucl. Science* **28**, 327 (1978).

Jones, L. W., "A review of quark search experiments", *Rev. Mod. Phys.* **49**, 717 (1977).

Kim, Y. S., "The search for quarks in terrestrial matter", *Contemp. Phys.* **14**, 289 (1973).

Kokkedee, J. J., *The Quark Model*, Benjamin, New York, 1969.

Lederman, L., "The upsilon particle", *Sci. Am.* **239**, 60 (Oct. 1978).

Lipkin, H., *Lie Groups for Pedestrians*, North-Holland, Amsterdam, 1966.

Nambu, Y., "Confinement of quarks", *Sci. Am.* **235**, 48 (Nov. 1976).

Schwitters, R. F. "Fundamental particles with charm", *Sci. Am* **238**, 56 (Oct. 1977).

Chapter 6

Weak Interactions

6.1 CLASSIFICATION OF WEAK INTERACTIONS

Weak interactions were first observed in the slow process of nuclear β-decay. This takes place, and can be observed, in circumstances where the much faster strong or electromagnetic decays are forbidden by the conservation laws. Otherwise, although all hadrons and leptons take part in weak interactions, the effects are usually swamped by the stronger electromagnetic or strong couplings.

Weak interactions frequently involve leptons among the interaction products. In particular, the neutral leptons – neutrinos – are unique in that they can only take part in the weak interactions. Experiments with beams of neutrinos produced by the decay in flight of pions and kaons, carried out 20 years ago, together with recent e^+e^- experiments, showed the charged and neutral leptons (see Table 1.2) appear in doublets,

$$
\begin{array}{cccc}
Q & L_e = 1 & L_\mu = 1 & L_\tau = 1 \\
\hline
0 & \begin{pmatrix} v_e \\ e^- \end{pmatrix} & \begin{pmatrix} v_\mu \\ \mu^- \end{pmatrix} & \begin{pmatrix} v_\tau \\ \tau^- \end{pmatrix} \\
-1 & & &
\end{array}
\tag{6.1}
$$

to each of which one has to assign a conserved lepton number (L_e, L_μ, L_τ). Antileptons have opposite charge and lepton number to leptons. Thus muon neutrinos v_μ, produced by the decay $\pi^+ \to \mu^+ v_\mu$ in flight, were shown to result in muons in their subsequent interactions (e.g. $v_\mu n \to p\mu^-$) and never electrons. By the same token, the decay $\mu^+ \to e^+\gamma$ is forbidden.

Apart from conservation of lepton number (if leptons are involved), weak reactions may or may not conserve isospin I and strangeness, S. Some typical weak processes – with the reasons why electromagnetic or strong transitions are forbidden – are given below:

Process	Mean lifetime (sec)		
Neutron decay: $n \rightarrow pe^+\nu_e$	10^3	E.m. decay forbidden by charge conservation	(6.2)
Inverse neutron decay: $\bar{\nu}_e p \rightarrow ne^+$		Neutrinos have only weak interactions	(6.3)
Lambda decay: $\Lambda \rightarrow p\pi^-$	10^{-10}	$\Delta S = 1$: strong/e.m. decay forbidden	(6.4)
Pion decay: $\pi^+ \rightarrow \mu^+\nu_\mu$	10^{-8}	Leptons are the only lighter particles	(6.5)

The lifetimes for weak decays depend on phase-space factors as well as the weak coupling constant G, but are long compared with typical lifetimes for electromagnetic decays ($\sim 10^{-19}$ sec) or strong decays ($\sim 10^{-23}$ sec). Weak cross-sections are correspondingly small; for example, the cross-section for $\nu_\mu + N \rightarrow N + \pi + \mu^-$ at 1 GeV is 10^{-38} cm^2, a factor 10^{12} times smaller than for $\pi + N \rightarrow \pi + N$ at the same energy.

The weak interactions are sometimes classified as to whether they involve leptons only, leptons and hadrons, or hadrons only, as in Table 6.1.

On a cosmic scale, weak interactions are of great importance. They control the thermonuclear reaction rate in the main sequence stars. The first stage of energy production in the hydrogen of the solar core is believed to be the weak interaction

$$pp \rightarrow de^+\nu_e. \qquad (6.6)$$

The smallness of the cross-section is evidenced by the fact that, even at the enormous density in the core (~ 100 times that of water) any one proton survives on average for $\sim 10^{10}$ years before undergoing this reaction. Hence the long life of the solar system, and the impossibility of observing this collision process directly in the laboratory. Weak effects between nucleons can, however,

TABLE 6.1 Classification and examples of weak interactions

Leptonic	$\mu^+ \rightarrow e^+\nu_e\bar{\nu}_\mu$	$\nu_e e^- \rightarrow e^-\nu_e$
Semileptonic	$\Delta S = 0$: $n \rightarrow pe^-\bar{\nu}_e,$ $\bar{\nu}_e p \rightarrow ne^+$	$\Delta S = 1$: $K^+ \rightarrow \pi^0 + e^+ + \nu_e,$ $K^+ \rightarrow \mu^+\nu_\mu$
Nonleptonic	$N+N \rightarrow N+N,$ parity violation in nuclei	$\Lambda \rightarrow \pi^-p,$ $K^+ \rightarrow \pi^+\pi^0,$ $\rightarrow \pi^+\pi^-\pi^+$

be detected in suitable circumstances; they give rise to parity-violating effects in nuclear transitions (see Section 3.7).

6.2 NUCLEAR β-DECAY: FERMI THEORY

As described in Chapter 1, the prototype weak interactions associated historically with β-decay of atomic nuclei are represented by the transformation of neutrons to protons (or vice versa) with the spontaneous emission or absorption of electrons (or positrons) and neutrinos (or antineutrinos):

$$n \to pe^{-}\bar{v}_e \qquad \text{(electron emission),} \qquad (6.7a)$$

$$p \to ne^{+}v_e \qquad \text{(positron emission),} \qquad (6.7b)$$

$$e^{-}p \to nv_e \qquad \text{(K-capture),} \qquad (6.7c)$$

$$\bar{v}_e p \to ne^{+} \qquad \text{(antineutrino absorption).} \qquad (6.7d)$$

The reaction (6.7a) occurs for free neutrons, and (6.7b) in nuclei where the difference in binding energy of parent and daughter nuclei exceeds the negative Q-value of the reaction ($Q = -1.8$ MeV, representing the mass energy required to create the positron, plus the neutron-proton mass difference). The reaction (6.7c) occurs in heavy nuclei, where the K-electrons have an appreciable probability ($\propto Z^4$) of being inside the nuclear volume, and is an alternative to positron emission, over which it is favored energetically.

Both electron and neutrino in (6.7a) and (6.7b) have spin $\frac{1}{2}$, so their total angular momentum $\Delta \mathbf{J}$ must be integral. If no orbital angular momentum is carried away, it follows that $\Delta \mathbf{J} = 1$ or 0, and this must be the change in the angular momentum of the nucleus. The simplest β-transitions are the so-called *allowed transitions*, in which $\Delta \mathbf{J} = 1$ or 0 and there is no change in the spatial configuration of the nucleons, so that parent and daughter nuclei have the same parity. A proton changes into a neutron, or vice versa, with ($\Delta \mathbf{J} = 1$) or without ($\Delta \mathbf{J} = 0$) spin-flip. $\Delta \mathbf{J} = 0$ transitions are called Fermi transitions, $\Delta \mathbf{J} = 1$ Gamow-Teller transitions:

Fermi	Gamow-Teller	
$n \to p + e^{-} + \bar{v}_e$	$n \to p + e^{-} + \bar{v}_e$	
$\uparrow \quad \uparrow \quad \downarrow \quad \uparrow$	$\uparrow \quad \downarrow \quad \uparrow \quad \uparrow$	(6.8)
$\Delta \mathbf{J} = 0$	$\Delta \mathbf{J} = 1 \ (\lvert \Delta J \rvert = 0, \pm 1)$	
$J_i = 0 \to J_f = 0$ allowed	$J_i = 0 \to J_f = 0$ forbidden	

where spin directions are indicated by vertical arrows.

There are then three types of allowed β-transitions: pure Fermi, pure Gamow-Teller, and mixed transitions. Examples are:

$$\begin{array}{ll} {}^{6}\text{He} \to {}^{6}\text{Li} + e^{-} + \bar{\nu}, & \Delta J = 1 & \text{(pure Gamow-Teller)}, \\ J^{P} \quad 0^{+} \quad 1^{+} & & \end{array}$$

$$\begin{array}{ll} {}^{14}\text{O} \to {}^{14}\text{N*} + e^{+} + \nu, & J = 0 \to J = 0 & \text{(pure Fermi)}, \quad (6.9) \\ J^{P} \quad 0^{+} \quad\quad 0^{+} & & \end{array}$$

$$\begin{array}{ll} n \quad \to p + e^{-} + \bar{\nu}, & J \to J \quad (J \neq 0) & \text{(mixed)}. \\ J^{P} \quad \frac{1}{2}^{+} \quad \frac{1}{2}^{+} & & \end{array}$$

Note that in all cases, the parity of the initial and final nuclear states is the same. The effect of the decay is to flip the isospin of one nucleon ($n \to p$ or vice versa) and, in the Gamow-Teller case, also to flip the nucleon spin, but otherwise to leave the nuclear configuration unaltered.

Nuclear β-decay transitions with very long lifetimes, called forbidden transitions, are also observed. These are associated with values of nuclear angular-momentum change $\Delta J > 1$ (and sometimes as large as 4). Such processes necessarily involve change in the orbital quantum number of the affected nucleon in the nuclear shell-model structure, with or without a change in nuclear parity.

The shape of the spectrum of β-decay electrons emitted in allowed transitions was first deduced by Fermi. Since the interaction is known to be weak, the transition probabilities for β-decays can be calculated from first-order perturbation theory. The transition probability per unit time is then [Eq. (1.27)]

$$W = \frac{2\pi}{\hbar} G^{2} |M|^{2} \frac{dN}{dE_{0}}, \quad (6.10)$$

where E_{0} is the available energy of the final state, dN/dE_{0} is the density of final states per unit energy, and G^{2} is a universal constant typical of the β-decay coupling. $|M|^{2}$ is the square of the matrix element for the transition, which involves an integral over the nuclear (interaction) volume of the four fermion wave functions. For the present we shall regard $|M|^{2}$ as some constant and calculate the phase-space factor dN/dE_{0}. This is determined by the number of

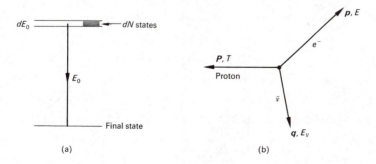

Fig. 6.1 (a) An initial state with a spread in energy dE_{0} decays to a final (stable) state of unique energy, with energy release E_{0}. (b) The momentum vectors in neutron β-decay.

ways it is possible to share out the available energy $E_0 \to E_0 + dE_0$ between, for example, p, e^-, and v^- in the neutron decay (6.8) above. The quantity dE_0 arises because of the spread in energy of the final state corresponding to the finite lifetime of the initial state. In Fig. 6.1, **p**, **q**, and **P** are the momenta of the electron, neutrino, and proton; E, E_v, and T are their kinetic energies.

Then in the rest frame of the initial state (neutron),

$$\mathbf{P} + \mathbf{q} + \mathbf{p} = 0,$$

$$T + E_v + E = E_0.$$

Assume $m_v = 0$, so $E_v = qc$. In order of magnitude, $E_0 \sim 1$ MeV, so $Pc \sim 1$ MeV. Thus, if the recoiling nucleon mass is M, its kinetic energy $T = P^2/2M \sim 10^{-3}$ MeV only, and can be neglected. The nucleon serves to conserve momentum, but we can regard E_0 as shared entirely between electron and neutrino. Thus $qc = E_0 - E$. The number of states in phase space available to an electron, confined to a volume V with momentum $p \to p + dp$ defined inside the element of solid angle $d\Omega$, is

$$\frac{V \, d\Omega}{h^3} p^2 \, dp.$$

If the fermion wave functions are normalized to unit volume $(V = 1)$ and we integrate over all space angles, the electron phase-space factor is

$$\frac{4\pi p^2 \, dp}{h^3}.$$

Similarly, for the neutrino it is

$$\frac{4\pi q^2 \, dq}{h^3}.$$

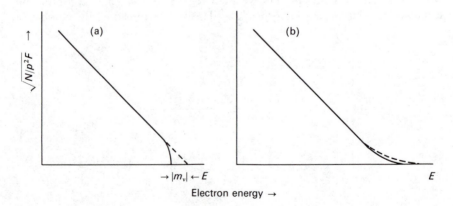

Fig. 6.2 Kurie plot of allowed transition for zero and finite neutrino mass: (a) perfect resolution, (b) finite resolution.

We disregard any possible correlation in angle between **p** and **q**, and treat these two factors as independent, since the proton will take up the resultant momentum. [There is no phase-space factor for the proton, since its momentum

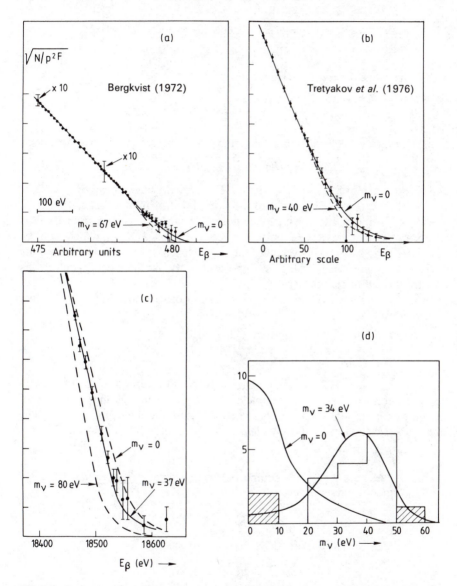

Fig. 6.3 Kurie plots from recent measurements of tritium β-decay. (a) Bergkvist (1972); (b) Tretyakov *et al.* (1976); (c) and (d) Lyubimov *et al.* (1980). (c) shows the average of several runs, and (d) the "best" mass estimate for m_v from each of 16 runs. The expected distributions for $m_v = 0$ and $m_v = 35$ eV/c^2 are indicated.

is now fixed: $\mathbf{P} = -(\mathbf{p} + \mathbf{q})$.] So the number of final states is

$$d^2 N = \frac{16\pi^2}{h^6} p^2 q^2 \, dp \, dq.$$

Also, for given values of p and E the neutrino momentum is fixed,

$$q = (E_0 - E)/c,$$

within the range $dq = dE_0/c$. Hence

$$\frac{dN}{dE_0} = \frac{16\pi^2}{h^6 c^3} p^2 (E_0 - E)^2 \, dp.$$

If $|M|^2$ is regarded as a constant, this last expression gives the electron spectrum

$$N(p) \, dp \propto p^2 (E_0 - E)^2 \, dp, \tag{6.11}$$

and thus if we plot $[N(p)/p^2]^{1/2}$ against E, a straight line cutting the x-axis at $E = E_0$ should result. This is called a Kurie plot. For many β-transitions, the Kurie plot is linear, as shown for the decay $^3\text{H} \rightarrow {}^3\text{He} + e^- + \bar{\nu}$ (mixed). It is necessary to include a Coulomb correction factor $F(Z, p)$, which can be calculated exactly, to take account of the energy lost (e^-) or gained (e^+) from the nuclear Coulomb field. This is important only for low-energy electrons (positrons) and nuclei of large Z.

For a nonzero neutrino mass m_ν, it is straightforward to show that the effect is to modify (6.11) to

$$N(p) \, dp \, F(Z, p) \propto p^2 (E_0 - E)^2 \sqrt{1 - \left(\frac{m_\nu c^2}{E_0 - E}\right)^2} \, dp. \tag{6.12}$$

It is left as an exercise to show that the Kurie plot turns over to cut the axis vertically at $E = E_0 - m_\nu c^2$, as depicted in Fig. 6.2. Thus the shape of the plot near the endpoint allows one in principal to determine the neutrino mass. The present best estimate (Lyubimov *et al.* 1980) gives $m_\nu \simeq 30 \text{ eV}/c^2$ – see Fig. 6.3.

TABLE 6.2 Some neutrino (ν_e) mass estimates from Kurie plot of tritium β-decay

	eV/c^2
Langer and Moffatt (1952)	< 10,000 [a]
Bergkvist (1972)	< 65 [a]
Tretyakov *et al.* (1976)	< 35 [a]
Lyubimov *et al.* (1980)	~ 30 [b]

[a] 90% confidence limit.
[b] $16 < m_\nu < 46 \text{ eV}/c^2$.

These and the other results in Table 6.2 were obtained from study of tritium (^{3}H) β-decay, which has a very small endpoint energy ($E_0 = 18.6$ keV). The present value of the neutrino mass is comparable with the energy resolution ($\simeq 20$ eV) of the spectrometers employed; this has to be folded into the analysis, as in Fig. 6.2(b). Complications also arise from the fact that the daughter ^{3}He ion may be left in an excited state.

The value of the neutrino mass is of great importance for astrophysics and cosmology. In the "big-bang" theory of the origin of the universe, neutrino-antineutrino pairs are presumed to have been created with intensity similar to that of photons (which now constitute the 3°K background radiation). A neutrino mass of order 40 eV/c^2 would increase the gravitational potential energy associated with galactic neutrino "halos" sufficiently as to make the universe "closed" or even possibly to oscillate with repeated "big bangs" (see Section 8.8.3).

Returning to earth, the total decay rate is obtained by integrating (6.11) over the electron spectrum. This can be done exactly, but as a crude approximation we can consider the electrons as extreme relativistic, $E \simeq pc$ (not for tritium!), whence we obtain

$$N \simeq \int_0^{E_0} E^2(E_0 - E)^2 \, dE = \frac{E_0^5}{30}. \qquad (6.13)$$

Under these conditions, the disintegration constant (i.e. the decay rate) varies as the fifth power of the disintegration energy – the *Sargent rule*.

More generally, and including the Coulomb term, the transition rate will be

$$W = |M|^2 G^2 \frac{4\pi^2}{h} \frac{16\pi^2}{h^6 c^3} \int_0^{P_{max}} F(Z,p)N(p) \, dp.$$

If p, E are expressed in units of the electron mass, $p = p' \cdot mc$ and $E = E' \cdot mc^2$, the integral becomes

$$m^5 c^7 \int_0^{p'_{max}} F(Z,p')N(p') \, dp' = m^5 c^7 f, \qquad (6.14)$$

where f is a dimensionless number depending on p'_{max} and the daughter nuclear charge Z. Then

$$W = |M|^2 G^2 \frac{64\pi^4 m^5 c^4}{h^7} f \text{ transitions/sec.}$$

If t denotes the half-life, then $W = (\ln_e 2)/t$, so that

$$ft = \frac{K}{|M|^2} \quad \text{where} \quad K = \frac{h^7 \ln_e 2}{64 G^2 \pi^4 m^5 c^4}. \qquad (6.15)$$

The quantity ft therefore gives information about the matrix element M, which depends on the overlap of the initial and final nuclear wavefunctions. For allowed transitions, where the overlap is essentially complete, $|M|^2 = 1$.

From (6.15) we may calculate G for a pure Fermi transition by taking the observed ft value of ^{14}O-decay, 3100 ± 20 sec, and assuming $|M|^2 = 1$. (This value of ft is a mean of recent values and incorporates some corrections.) Actually, we need to double the value of ft, since ^{14}O is a ^{12}C core plus *two* protons, either of which can decay. Inserting the numerical constants, one gets

$$G = \left(\frac{h^7 \ln_e 2}{64\pi^4 m^5 c^4 ft} \right)^{1/2} = 1.4 \times 10^{-49} \, \text{erg cm}^3$$

$$= 1.0 \times 10^{-5} \, \hbar c \left(\frac{\hbar}{M_p c} \right)^2$$

$$= \frac{10^{-5}}{M_p^2} \, \text{in units } \hbar = c = 1. \tag{6.16}$$

We have purposely quoted G here to only one decimal place: in the more refined theory (Section 6.9 below) a correction factor $\sec^2 \theta_c = 1.08$ is required to these numbers, giving finally

$$G = \frac{1.02 \times 10^{-5}}{M_p^2}. \tag{6.17}$$

We may also compare the ^{14}O pure Fermi transition rate with that of the neutron, which is mixed. Denote by C_F and C_{GT} the coefficients corresponding to the Fermi and Gamow-Teller couplings. The neutron decay rate is then proportional to $C_F^2 + 3C_{GT}^2$. The factor 3 enters because there are $2J + 1 = 3$ possible orientations for the total angular momentum $J = 1$ of the leptons in a Gamow-Teller transition. If we consider only the total event rate, the two contributions add in quadrature. For the ^{14}O rate, for the reason stated above, we write $2C_F^2$. Then

$$\frac{(ft)_{^{14}O}}{(ft)_n} = \frac{3100 \pm 20 \, \text{sec}}{1080 \pm 16 \, \text{sec}} = \frac{C_F^2 + 3C_{GT}^2}{2C_F^2},$$

which yields

$$\lambda = \left| \frac{C_{GT}}{C_F} \right| = 1.25 \pm 0.02. \tag{6.18}$$

Thus if we define the constant G for Fermi transitions, then for the Gamow-Teller transitions it has the slightly larger value λG. Note that the sign of λ is not determined here (actually it is negative).

6.3 INTERACTION OF FREE NEUTRINOS: INVERSE β-DECAY

The cross-section for the inverse reaction (6.3) of free antineutrinos on protons can be calculated from (6.10). In this case, there are only two particles in the final state, so that using (1.29) we obtain (in units $\hbar = c = 1$)

$$\sigma(\bar{v}_e p \rightarrow n e^+) = \frac{W}{v_i} = \frac{G^2}{\pi}|M|^2 \frac{p^2}{v_i v_f}, \tag{6.19}$$

where v_i, v_f are the relative velocities of the particles in the initial and final states ($v_i = v_f \simeq c$) and p is the numerical value of the CMS momentum of the neutron and positron. We are dealing with a mixed transition, with $M_F^2 = 1$ for the Fermi contribution ($\Delta J = 0$) and $M_{GT}^2 \simeq 3$ for the spin multiplicity factor for the Gamow-Teller contribution ($\Delta J = 1$). Thus

$$\sigma = \frac{M_F^2 + M_{GT}^2}{\pi} G^2 p^2 \simeq \frac{4G^2 p^2}{\pi}. \tag{6.20}$$

For neutrinos in the MeV energy range, incident on a fixed nucleon target, the CMS momentum and laboratory neutrino energy above threshold ($Q = 1.8$ MeV) are related by $p \simeq (E_v - Q)/c$. For $pc \simeq 1$ MeV and G from (6.17) one obtains therefore

$$\sigma = \frac{4}{\pi} \times 10^{-10} \left(\frac{\hbar}{M_p c}\right)^2 \left(\frac{p}{M_p c}\right)^2 \simeq 10^{-43} \text{ cm}^2. \tag{6.21}$$

This corresponds to a mean free path for antineutrino absorption in water of 10^{20} cm or 100 light years. The first observation of such interactions was made by Reines and Cowan in 1959. They employed a reactor as the source. The uranium fission fragments are neutron-rich and undergo β-decay, emitting electrons and antineutrinos (on average, 6 per fission with a spectrum centered at 1 MeV). For a 1000-MW reactor, the useful flux of antineutrinos is of order 10^{13} cm^{-2} sec^{-1}. The reaction

$$\bar{v}_e p \rightarrow n e^+$$

was observed, using a target of cadmium chloride ($CdCl_2$) and water. The positron produced in this reaction rapidly comes to rest by ionization loss and forms positronium, which annihilates to γ-rays, in turn producing fast electrons by the Compton effect. The electrons are recorded in a liquid scintillation counter. The time scale for this process is of order 10^{-9} sec, so that the positron gives a so-called "prompt" pulse. The function of the cadmium is to capture the neutron after it has been moderated (i.e. reduced to thermal energy by successive elastic collisions with protons) in water—a process which delays by several microseconds the γ-rays coming from eventual radiative capture of the neutron in cadmium. Thus the signature of an event consists of two pulses microseconds apart. Figure 6.4 shows schematically the experimental arrangement employed.

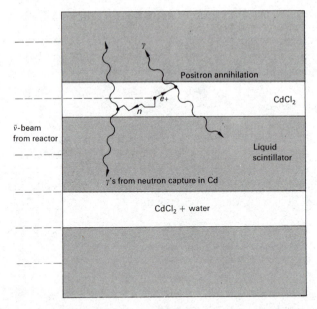

Fig. 6.4 Schematic diagram of the experiment by Reines and Cowan (1959), detecting the interactions of free antineutrinos from a reactor.

6.4 PARITY NONCONSERVATION IN β-DECAY

In 1956, following a critical review of the experimental data then available, Lee and Yang came to the conclusion that the weak interactions were not invariant under spatial inversion, i.e. did not conserve parity — largely on the basis of the fact that the K^+ could decay in the two decay modes $K^+ \to 2\pi$, $K^+ \to 3\pi$, in which the final states have opposite parities (so-called "τ-θ paradox").

To test parity conservation, an experiment was carried out by Wu *et al.* (1957), who employed a sample of ^{60}Co at $0.01°$K inside a solenoid. At this temperature a high proportion of ^{60}Co nuclei are aligned. ^{60}Co $(J = 5)$ decays to ^{60}Ni* $(J = 4)$ — a pure Gamow-Teller transition. The relative electron intensities along and against the field direction were measured (see Fig. 6.5). The degree of ^{60}Co alignment could be determined from observations of the angular distribution of γ-rays from ^{60}Ni*. The results for electron intensity were consistent with a distribution of the form

$$I(\theta) = 1 + \alpha \left(\frac{\boldsymbol{\sigma} \cdot \mathbf{p}}{E} \right)$$

$$= 1 + \alpha \frac{v}{c} \cos \theta, \qquad (6.22)$$

where $\alpha = -1$; $\boldsymbol{\sigma}$ is a unit spin vector in the direction $\mathbf{J}$; $\mathbf{p}$ and E are the

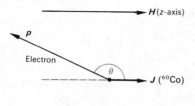

Fig. 6.5

electron momentum and total energy, and θ is the angle of emission of the electron with respect to **J**. The variation with electron velocity was checked over the range $0.4 < v/c < 0.8$.

The fore-aft asymmetry of the intensity in (6.22) implies that the interaction as a whole violates parity conservation; for imagine the whole system reflected in a mirror normal to the z-axis. The first term (unity) does not change sign under reflection – it is *scalar* (even parity). σ, being an axial vector, does not change sign, while the polar vector $\mathbf{p} \to -\mathbf{p}$. Thus the product $\sigma \cdot \mathbf{p} \to -\sigma \cdot \mathbf{p}$ under reflection – it is called a *pseudoscalar* (odd parity). The existence of two terms of opposite parity means that parity is not a well-defined quantum number in weak interactions.

As already stated, the nuclear parity itself does not change in an allowed transition, so that parity non-conservation for the whole system implies that the leptons must be emitted in a mixture of s $(L = 0)$ and p $(L = 1)$ orbital states. In the case $\theta = \pi$, where $I(\theta)$ is a maximum, $L_z = 0$ necessarily and both electron and neutrino spins must be aligned along the z-axis, in other words the *leptons must be longitudinally spin-polarized*.

This longitudinal polarization of electrons and positrons is found to be the same form for Fermi, Gamow-Teller, and mixed transitions. If σ is now the electron spin vector (normalized so that $\sigma_z = \pm 1$), the intensity is

$$I = 1 + \alpha \frac{\sigma \cdot \mathbf{p}}{E}. \tag{6.23}$$

The polarization or *helicity* H is defined as

$$H = \frac{I_+ - I_-}{I_+ + I_-} = \alpha \frac{v}{c},$$

where I_+ and I_- represent the intensities for σ parallel and antiparallel to **p**. Experimentally,

$$\alpha = \begin{cases} +1 & \text{for } e^+, H = +v/c, \\ -1 & \text{for } e^-, H = -v/c. \end{cases} \tag{6.24}$$

The property of helicity is not limited to leptons. We have already seen that massless photons possess two possible spin states, $J_z = \pm 1$ (where the z-axis defines the propagation vector). Right-handed $(J_z = +1)$ and left-handed $(J_z = -1)$ photons, have helicities $+1$ and -1 respectively. Parity is conserved in electromagnetic processes involving photons, because the two types of photon are always emitted with equal amplitude, and one does not therefore observe a net circular polarization. On the contrary, in weak interactions, β-processes consist of the emission of *either* electrons, with a net left-handed spin polarization, *or* positrons, which are predominantly right-handed.

6.5 LEPTON POLARIZATION IN β-DECAY

6.5.1 Measurement of Electron Polarization

Experimentally, the longitudinal polarization of electrons has been observed by a variety of methods. Three of these are sketched diagrammatically in Fig. 6.6. In (a), the longitudinal polarization is turned into a transverse polarization by electrostatic bending through 90°. Provided the electrons are nonrelativistic $(v^2 \ll c^2)$, the spin direction σ is virtually unaltered by the field (as $v \to c$, σ tends to follow the trajectory). A second technique, (b), employs crossed electric and magnetic fields. When $E/H = v/c$, the beam is undeflected, but σ precesses about **H**. The value and extent of the fields is adjusted to give a 90° precession. A third method, (c), makes use of the fact that Coulomb scattering in a light element (in this case a 0.5-mm-thick aluminium foil) does not affect the spin direction. The

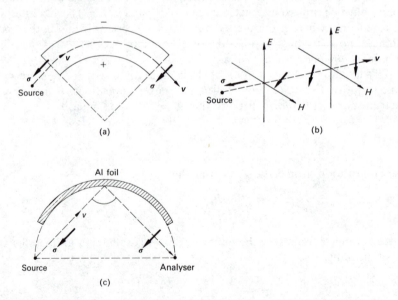

Fig. 6.6 Methods employed to transform longitudinal polarization of electrons from β-decay into transverse polarization. (a) Frauenfelder *et al.* (1957), (b) Cavanagh *et al.* (1957), (c) De Shalit *et al.* (1957).

arrangement employes a semicircular strip of foil, with source and analyser at each end of a diameter, resulting in a 90° scattering angle.

In each case the sense of polarization may be analysed by observing the right-left asymmetry in the scattering by a foil of heavy element. The effect can be seen by a classical argument (Fig. 6.7). Let l be the orbital angular momentum of the electron relative to the scattering nucleus Z. In the coordinate frame of the electron, the approaching nucleus appears as a positive current, clockwise or anticlockwise according to whether the electron passes to left or right. The magnetic field **H** associated with this current is therefore in the direction of l. Let **σ** represent the spin vector of the electron; the spin magnetic moment **μ** then points in the direction − **σ**. It follows that the magnetic energy **μ** · **H** is positive (i.e. one gets a repulsive force) when l and **σ** are parallel, and negative (attractive force) when l and **σ** are antiparallel. Thus the magnetic interaction arising from the spin-orbit coupling adds to the electrical force when l · **σ** is negative, and more electrons are scattered to right than to left. From the right-left asymmetry and a suitable calibration, the degree of transverse polarization can be determined.

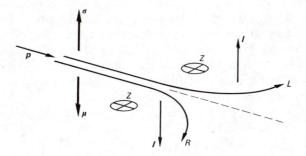

Fig. 6.7 Schematic diagram of scattering of a transversely polarized electron by a nucleus Z.

The longitudinal polarization of electrons can also be found directly, without spin-twisting, by using a magnetized iron sheet and observing the electron-electron (Møller) scattering. Both electrons have to be observed in coincidence to distinguish the effect from the more intense nuclear scattering. The electron beam from the β-decay source falls obliquely on an iron foil magnetized in its plane, so that **p** and **H** are parallel. The scattering is greatest when the spins of incident and target electrons are parallel (i.e. **σ** · **H**/H = + 1). On reversing the field **H**, the scattering ratio determines the longitudinal polarization. Another method is to observe the circular polarization of γ-rays from high-energy bremsstrahlung when the electrons are stopped in an absorber. The sense of polarization of the forward γ-rays is the same as that initially possessed by the electrons, and is in turn determined by scattering in magnetized iron, as described below. Observations on the polarization of β-rays were carried out in numerous laboratories during 1957–1958 and, within the experimental errors, clearly verified the helicity assignments (6.24).

6.5.2 Helicity of the Neutrino

The result (6.23), if applied to a neutrino ($m = 0$), implies that such a particle must be fully polarized, $H = +1$ or -1. In order to determine the type of operators occurring in the matrix element, the sign of the neutrino helicity turns out to be crucial. The neutrino is here defined as the neutral particle emitted together with the positron in β^+-decay, or following K-capture. The anti-neutrino then accompanies negative electrons in β^--decay. The neutrino helicity was determined in a classic and beautiful experiment by Goldhaber *et al.* in 1958. The steps in this experiment are indicated in Fig. 6.8:

(i) ^{152}Eu undergoes K-capture to an excited state of ^{152}Sm with $J = 1$ [Fig. 6.8(a)]. To conserve angular momentum, **J** must be parallel to the spin of the electron but opposite to that of the neutrino, so the recoiling ^{152}Sm* has the same polarization sense as the neutrino [Fig. 6.8(b)].

(ii) Now, in the transition ^{152}Sm* $\rightarrow$ ^{152}Sm $+ \gamma$, γ-rays emitted in the forward (backward) direction with respect to the line of flight of ^{152}Sm* will be polarized in the same (opposite) sense to the neutrino, as in Fig. 6.8(c). Thus the polarization of the "forward" γ-rays is the same as that of the neutrino.

(iii) The next step is to observe resonance scattering of the γ-rays in a ^{152}Sm target. Resonance scattering is possible with γ-rays of just the right frequency to "hit" the excited state:

$$\gamma + {}^{152}\text{Sm} \rightarrow {}^{152}\text{Sm}^* \rightarrow \gamma + {}^{152}\text{Sm}. \tag{6.25}$$

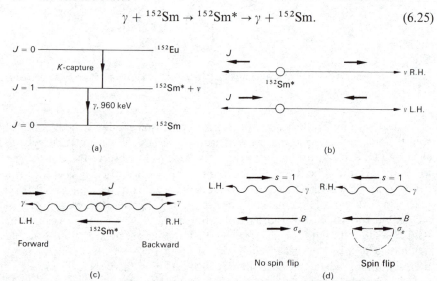

Fig. 6.8 Principal steps in the experiment of Goldhaber *et al.* to determine the neutrino helicity.

To produce resonance scattering, the γ-ray energy must slightly exceed the 960 keV to allow for the nuclear recoil. It is precisely the "forward" γ-rays, carrying with them a part of the neutrino-recoil momentum, which are able to do this, and which are therefore automatically selected by the resonance scattering.

(iv) The last step is to determine the polarization sense of the γ-rays. To do this, they were made to pass through magnetized iron before impinging on the ^{152}Sm absorber. An electron in the iron with spin σ_e opposite to that of the photon can absorb the unit of angular momentum by spin-flip; if the spin is parallel it cannot. This is indicated in Fig. 6.8(d). If the γ-ray beam is in the same direction as the field **B**, the transmission of the iron is greater for left-handed γ-rays then for right-handed.

A schematic diagram of the apparatus is shown in Fig. 6.9. By reversing **B** the sense of polarization could be determined from the change in counting rate.

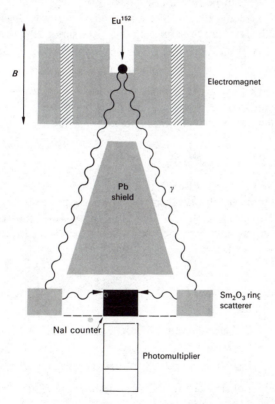

Fig. 6.9 Schematic diagram of apparatus used by Goldhaber *et al.* in which γ-rays from the decay of ^{152}Sm*, produced following *K*-capture in ^{152}Eu, undergo resonance scattering in Sm_2O_3, and are recorded by a sodium iodide scintillator and photomultiplier. The transmission of photons through the iron surrounding the ^{152}Eu source depends on their helicity and the direction of the magnetic field **B**.

When allowance was made for various depolarizing effects, it was concluded that neutrinos were left-handedly spin polarized.

In conclusion, the helicity assignments for the leptons emitted in nuclear β-decay are therefore as follows:

Particle	e^+	e^-	ν	$\bar{\nu}$	
Helicity	$+v/c$	$-v/c$	-1	$+1$	(6.26)

6.6 THE V-A INTERACTION

The formulae (6.22), (6.24), (6.26) have been presented from a purely empirical viewpoint. Their actual form follows from the so-called V-A theory of weak interactions. This involves the Dirac description of relativistic fermions; a brief account of this theory, the definition of the γ-matrices and the derivation of the results given below can be found in Appendix F.

Fermi developed his theory of β-decay in analogy with electromagnetic interactions. Consider first the process of electromagnetic scattering of an electron and a proton:

$$e^- + p \to e^- + p. \qquad (6.27)$$

Since electrons and baryons are conserved in this process, we may describe it as the interaction of two *currents*, via single virtual-photon exchange (see Fig. 6.18(c)), and the matrix element is proportional to

$$M \propto \frac{e^2}{q^2} J_{\text{baryon}} \cdot J_{\text{lepton}}, \qquad (6.28)$$

where q is the momentum transfer.

Relativistic fermions in the Dirac theory are described by 4-component, or spinor, wavefunctions, and they are operated on by 4×4 matrix operators O. Electromagnetic currents involve the *vector* operator $O_{\text{e.m.}} = \gamma_4 \gamma_\mu$ (where $\mu = 1, \ldots, 4$ indicates space-time components) and the currents have the form

$$J_{\text{lepton}} = \psi_e^* \gamma_4 \gamma_\mu \psi_e \equiv \bar{\psi}_e \gamma_\mu \psi_e,$$

$$J_{\text{baryon}} = \bar{\psi}_p \gamma_\mu \psi_p, \qquad (6.29)$$

where $\bar{\psi} = \psi^* \gamma_4$.

By analogy, Fermi assumed that for the weak process of neutron decay one could write

$$M = G J_{\text{baryon}}^{\text{weak}} \cdot J_{\text{lepton}}^{\text{weak}} = G(\bar{\psi}_p O \psi_n)(\bar{\psi}_e O \psi_\nu), \qquad (6.30)$$

where the grouping of the wavefunctions is made more plausible if we write the β-decay process

$$n \to p + e^- + \bar{\nu}$$

in the equivalent form

$$\nu + n \to p + e^-.$$

Fermi assumed that the matrix operator O in (6.30) would again be a vector operator as in (6.29). The only differences are that in β-decay one has a constant G instead of e^2, that the weak currents are assumed to interact at a point [i.e., there is no $1/q^2$ factor as in (6.28)], and that the electric charges of the lepton and baryon change by one unit in the interaction. Hence these β-decay reactions are referred to as charge-changing or simply charged-current weak interactions.

The vector interaction was satisfactory (prior to the discovery of parity violation in 1956) in describing Fermi transitions. It did not account for Gamow-Teller transitions, since it could not produce a flipover of the nucleon spin. Within certain requirements of relativistic invariance, one can in fact have five possible independent forms for the operator O in (6.30). They are called scalar (S), vector (V), tensor (T), axial vector (A), and pseudoscalar (P). These names are associated with the transformation properties of the weak currents under space inversions. The S- and V-interactions produce Fermi transitions, while T and A produce Gamow-Teller transitions. The pseudoscalar interaction P is unimportant in β-decay, since it couples spinor components proportional to particle velocity and thus introduces a factor in M^2 of v^2/c^2, where v is the nucleon velocity ($v^2/c^2 \sim 10^{-6}$ in β-decay).

The various operators O_i lead to different predictions for the angular correlations of the leptons in β-decay, and these are shown in Fig. 6.10. In these diagrams, v refers to the velocity of the positron.

We note that, since in a Fermi transition the total angular momentum of the leptons is zero, the S-interaction must produce leptons of the same helicity, and the V-interaction, leptons of opposite helicity. In Gamow-Teller transitions, the total lepton angular momentum is $J = 1$, so that the T- and A-interactions produce leptons of the same and opposite helicities respectively. The foregoing experiments on lepton polarization, summarized in (6.26), show therefore that only the V- and A-interactions can produce the helicities actually observed. A similar conclusion is arrived at by studying the shape of the momentum spectra of the nuclear recoils in Fermi and Gamow-Teller transitions. We may also remark that the factor $\frac{1}{3}$ in the $\cos \theta$ term for A- and T-couplings arises because the total angular momentum ($J = 1$) of the leptons possesses three possible orientations in space, and this reduces the angular correlation. Thus we can write in place of (6.30) for a general β-interaction

$$M = G \sum_{i=V,A} C_i(\bar{\psi}_p O_i \psi_n)(\bar{\psi}_e O_i \psi_\nu), \tag{6.31}$$

where C_V and C_A are appropriate coefficients. For example, for a pure Fermi transition, $C_A = 0$. The matrix element (6.31) is, however, a scalar quantity. It implies that parity is conserved, and that for each diagram in Fig. 6.10 one should add a second diagram with lepton spins reversed, so that there is no net lepton polarization. This is clearly wrong. We need to add another term to (6.31) to give us a pseudoscalar quantity, so that the matrix element contains both scalars and pseudoscalars, and thus has no well-defined parity. We can do

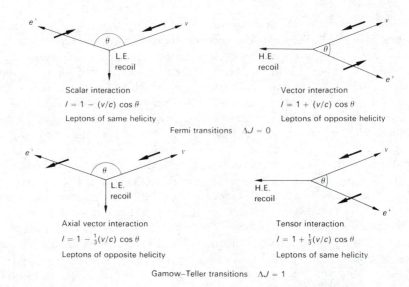

Fig. 6.10 Angular distributions in β-decay predicted by various types of interaction. The thick arrows denote lepton spin directions which result from angular-momentum conservation, assuming the neutrino is always left-handed. L.E. and H.E. represent low- and high-energy recoils respectively.

this by adding a similar expression, with one bracket multiplied by a matrix γ_5 ($\gamma_5 = \gamma_1\gamma_2\gamma_3\gamma_4$). Thus, inserting a $1/\sqrt{2}$ factor to retain the definition of G

$$M = \frac{G}{\sqrt{2}} \sum_{i=V,A} [\bar{\psi}_p O_i \psi_n][\bar{\psi}_e O_i (C_i + C'_i \gamma_5)\psi_\nu]. \qquad (6.32)$$

If, as all the evidence suggests, the interaction is invariant under time reversal, it can be shown that C_i and C'_i must be real coefficients; furthermore, if a neutrino (or antineutrino) is to be completely spin-polarized ($H = \pm 1$), one must have $C'_i = \pm C_i$. Since scalar and pseudoscalar terms then occur with equal magnitude, this is called the principle of *maximum parity violation*. For this case, the operator $(1 \pm \gamma_5)$ acting on the neutrino wave function in (6.32) projects out one sign of helicity:

$$(1 + \gamma_5) \to \text{L.H. neutrino state (R.H. antineutrino)},$$

$$(1 - \gamma_5) \to \text{R.H. neutrino state (L.H. antineutrino)}.$$

Since the experiment in Section 6.5.2 shows the neutrino to be left-handed, we need to take the first expression, corresponding to $C'_i = C_i$. This type of weak interaction is called the *two-component neutrino theory*, and was proposed independently by Lee and Yang, by Landau, and by Salam. It also correctly predicts the helicities $\pm v/c$ for the charged leptons in (6.26). Hence

(6.32) becomes

$$M = G/\sqrt{2} \sum_{i=V,A} C_i[\bar{\psi}_p O_i \psi_n][\bar{\psi}_e O_i(1 + \gamma_5)\psi_v].$$

Inserting the Dirac matrix expressions for the operators $O_V = \gamma_\mu$ and $O_A = i\gamma_\mu\gamma_5$, and setting the ratio $C_A/C_V = -\lambda$, we obtain

$$M = [G/\sqrt{2}]C_V[\bar{\psi}_p\gamma_\mu(1 + \lambda\gamma_5)\psi_n][\bar{\psi}_e\gamma_\mu(1 + \gamma_5)\psi_v]. \qquad (6.33)$$

The value and sign of λ have been determined by observation of the angular correlation between electron and proton in the decay of polarized neutrons, relative to each other and the neutron spin vector. These results give

$$-\frac{C_A}{C_V} = \lambda = +1.26 \pm 0.02, \qquad (6.34)$$

agreeing in magnitude with the independent result (6.18). We note that, had the weak nucleon current been pure V-A (with $\lambda = 1$), the operators in nucleon and lepton brackets would have been identical, so that *all* fermions in β-decay would have helicity $H = -v/c$, all antifermions $H = +v/c$; and that the result would differ from the pure vector interaction of Fermi (6.29) only by the inclusion of the projection operator $(1 + \gamma_5)$ in both terms.

6.7 PARITY VIOLATION IN Λ-DECAY

Parity nonconservation is a general property of weak interactions not solely associated with leptonic processes. The "τ-θ paradox", associated with the nonleptonic decay of kaons into pions, has already been mentioned. Another nonleptonic decay is that of the Λ-hyperon, for which the dominant decay modes are

$$\Lambda \to \pi^- p, \qquad \Lambda \to \pi^0 n. \qquad (6.35)$$

The Λ-hyperon does in fact also undergo leptonic β-decay, $\Lambda \to p + e^- + \bar{v}$, with a small branching ratio.

Parity violation can be demonstrated by considering the decay of Λ-hyperons produced in the associated production process

$$\pi^- p \to \Lambda K^0. \qquad (6.36)$$

In this process, the Λ can be (and in general is) spin-polarized. Parity conservation in such a strong interaction implies that the Λ must be polarized with spin $\boldsymbol{\sigma}$ *transverse* to the production plane, i.e. of the form $\boldsymbol{\sigma} \propto (\mathbf{p}_\Lambda \times \mathbf{p}_K)$, which does not change sign under inversion. Note that spin polarization *in* the production plane in general does change sign, and is not allowed.

In practice the mean transverse polarization

$$P_\Lambda = (N\uparrow - N\downarrow)/(N\uparrow + N\downarrow) \simeq 0.7$$

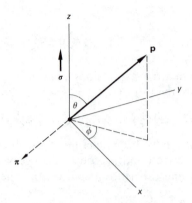

Fig. 6.11   Definition of axes and directions in Λ-decay.

for an incident pion momentum just above 1 GeV/c. In the decay process (6.35) let us define the direction of $\boldsymbol{\sigma}$ as the z-axis of the Λ rest frame (Fig. 6.11). In this frame, the distribution in angle of emission (θ, ϕ) of the pion or proton will depend on their orbital angular momentum l.

Since $J_\Lambda = \frac{1}{2}$, $J_z = \pm\frac{1}{2}$, we can have either $l = 0$ (proton and Λ-spins parallel) or $l = 1$ (spins antiparallel). Thus, we generally expect a combination of s- and p-waves. Denote by m_1 the z-component of the proton spin vector, and by m_2 the z-component of $\boldsymbol{l}$. In the s-wave case, $m_2 = 0$ and the angular-momentum wavefunction is $Y_m^l = Y_0^0$. Thus, for $J_z = +\frac{1}{2}$, the total wavefunction is the product

$$\psi_s = a_s Y_0^0 \chi^+, \tag{6.37}$$

where a_s denotes the s-wave amplitude and χ^+ the proton spin-up state of $m_1 = +\frac{1}{2}$. For the p-wave, $m_1 + m_2 = J_z = \frac{1}{2}$, with either $m_1 = +\frac{1}{2}$ and $m_2 = 0$, or $m_1 = -\frac{1}{2}$ and $m_2 = +1$.

Referring to Table III in the Appendix for the Clebsch-Gordan coefficients for adding $J = 1$ and $J = \frac{1}{2}$, the entry for $J_{\text{total}} = m = +\frac{1}{2}$ gives

$$\psi_p = a_p[\sqrt{\tfrac{2}{3}}\, Y_1^1 \chi^- - \sqrt{\tfrac{1}{3}}\, Y_1^0 \chi^+]. \tag{6.38}$$

Here, a_s and a_p are, in general, complex amplitudes. Thus if *both* s- and p-waves are present, the total amplitude will be

$$\psi = \psi_s + \psi_p = [a_s Y_0^0 - a_p\sqrt{\tfrac{1}{3}}\, Y_1^0]\chi^+ + [a_p\sqrt{\tfrac{2}{3}}\, Y_1^1]\chi^-.$$

Recalling the orthogonality of the spin states χ^+ and χ^-, the angular distribution becomes

$$\psi\psi^* = (a_s Y_0^0 - a_p\sqrt{\tfrac{1}{3}}\, Y_1^0)(a_s Y_0^0 - a_p^*\sqrt{\tfrac{1}{3}}\, Y_1^0) + a_p^2(\sqrt{\tfrac{2}{3}}\, Y_1^1)^2,$$

where one of the phases must be arbitrary and we take a_s to be real. Also, $Y_0^0 = 1$, $\sqrt{\frac{1}{3}}Y_1^0 = \cos\theta$, $\sqrt{\frac{2}{3}}Y_1^1 = -\sin\theta$, so that

$$\psi\psi^* = |a_s|^2 + |a_p|^2\cos^2\theta + |a_p|^2\sin^2\theta - a_s\cos\theta[a_p + a_p^*]$$

$$= |a_s|^2 + |a_p|^2 - 2a_s\,\mathrm{Re}\,a_p^*\cos\theta. \tag{6.39}$$

Setting

$$\alpha = \frac{2a_s\,\mathrm{Re}\,a_p^*}{|a_s|^2 + |a_p|^2},$$

the angular distribution has the form

$$I(\theta) = 1 - \alpha\cos\theta. \tag{6.40}$$

The angle θ is defined relative to $\boldsymbol{\sigma}$; physically, one can only measure relative to the normal to the production plane, so that if we redefine θ in this way, the above result becomes

$$I(\theta) = 1 - \alpha P\cos\theta,$$

where P is the average polarization; experiment shows that $\alpha P \simeq -0.7$. Thus, the parity violation in the Λ-decay is manifested as an up-down asymmetry of the decay pion (or proton) relative to the production plane.

Note that (6.40) has the same form as (6.23), that the parity violation parameter α is finite only if *both* a_s and a_p are finite, and thus that the parity violation arises from interference of the s (even) and p (odd) waves.

6.8 PION AND MUON DECAY

The lepton helicities first observed in 1957 in nuclear β-decay were detected simultaneously in the decay of pions and muons. We recall that the pion and muon decay schemes are

$$\pi^+ \to \mu^+ v, \tag{6.41}$$

$$\mu^+ \to e^+ v\bar{v}. \tag{6.42}$$

Since the pion has spin zero, the neutrino and muon must have antiparallel spin vectors, as shown in Fig. 6.12. If the neutrino has helicity $H = -1$, as in β-decay, the μ^+ must have negative helicity. In the subsequent muon decay, the positron spectrum is peaked in the region of the maximum energy, so the most

Pion rest – frame

Muon rest – frame

Fig. 6.12 Sketches indicating sense of spin polarization in pion and muon decay.

likely configuration is that shown – the positron having positive helicity. In fact, the positron spectrum has the shape indicated in Fig. 6.14. In the experiments, positive pions decayed in flight and those decay muons projected in the forward direction – thus with negative helicity – were selected. These μ^+ were stopped in a carbon absorber, and the e^+ angular distribution relative to the original muon momentum $\mathbf{p}_\mu$ was observed. The muon spin $\boldsymbol{\sigma}$ should be opposite to $\mathbf{p}_\mu$, if there is no depolarization of the muons in coming to rest (true in carbon). The angular distribution observed was of the form

$$\frac{dN}{d\Omega} = 1 - \frac{\alpha}{3}\cos\theta, \tag{6.43}$$

where θ is the angle between $\mathbf{p}_\mu$, the initial muon momentum vector, and $\mathbf{p}_e$, the electron momentum vector, and $\alpha = 1$ within the errors of measurement. The same value of α was found for μ^+ and μ^-. Equation (6.43) is exactly the form predicted by the V-A theory. The helicity of the electrons (positrons) was also measured and shown to be $\mp v/c$.

6.8.1 The $\pi \rightarrow \mu$ and $\pi \rightarrow e$ Branching Ratios

To represent pion decay in terms of a four-fermion interaction, we treat the pion as a quark-antiquark pair (Fig. 6.13). We shall try to calculate the decay branching ratio

$$\frac{\pi \rightarrow e v}{\pi \rightarrow \mu v}, \tag{6.44}$$

which constitutes a test of the V-A theory. Since the pion has $J^P = 0^-$, the transition

$$\pi^+ = u\bar{d} \rightarrow e^+ v$$

may be thought of as the process

$$u \rightarrow d e^+ v,$$

which is the quark description of the nuclear β-decay

$$p \rightarrow n e^+ v, \quad \Delta\mathbf{J} = 0, \quad \text{parity change.}$$

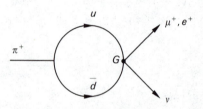

Fig. 6.13

In the terminology of β-decay, the last reaction is a forbidden transition, in which the overall change in nuclear angular momentum is zero, resulting from spin flip and the emission of the final nucleon in an orbital state $l = 1$ – hence odd parity. Such a transition can only be produced by the axial-vector (A) and pseudoscalar (P) couplings. There is no evidence for P-coupling in β-decay, but this might be ascribed to the very difficulty of generating a nucleon of $l = 1$ at the low energies available in such processes. Therefore, we must consider it as a possibility in the higher-energy process of pion decay. As we have seen (Fig. 6.10), the A-coupling tends to produce leptons of opposite helicity. The P-coupling (not shown in Fig. 6.10) produces leptons of the same helicity. On the other hand, conservation of angular momentum compels the leptons to have the same helicity (Fig. 6.12). Thus, for A-coupling, we expect for the matrix element $M^2 \propto 1 + (-v/c)$ – there being no factor $\frac{1}{3}$ here (as in Fig. 6.10), because we are considering a $\Delta\mathbf{J} = 0$ transition instead of $\Delta\mathbf{J} = 1$, with three projections of the lepton angular momentum in an arbitrary direction. For P-coupling, one finds instead $M^2 \propto 1 + v/c$.

Prediction of the branching ratio (6.44) then follows if one includes the phase-space factor. This is

$$\frac{dN}{dE_0} = \text{const}\, p^2 \frac{dp}{dE_0}.$$

Here, $\mathbf{p}$ is the momentum of the charged lepton in the pion rest frame, v its velocity, and m its rest mass. The neutrino momentum is then $-\mathbf{p}$:

In units $c = 1$, the total energy is

$$E_0 = m_\pi = p + \sqrt{p^2 + m^2}.$$

Hence

$$p^2 \frac{dp}{dE_0} = \frac{(m_\pi^2 + m^2)(m_\pi^2 - m^2)^2}{4m_\pi^4},$$

$$1 + \frac{v}{c} = \frac{2m_\pi^2}{m_\pi^2 + m^2},$$

$$1 - \frac{v}{c} = \frac{2m^2}{m_\pi^2 + m^2}.$$

Thus, for A-coupling the decay rate is given by

$$p^2 \frac{dp}{dE_0}\left(1 - \frac{v}{c}\right) = \frac{m^2}{2}\left(1 - \frac{m^2}{m_\pi^2}\right)^2,$$

and for P-coupling it is

$$p^2 \frac{dp}{dE_0}\left(1 + \frac{v}{c}\right) = \frac{m_\pi^2}{2}\left(1 - \frac{m^2}{m_\pi^2}\right)^2.$$

The predicted branching ratios become, with the approximation $m_e^2/m_\pi^2 \ll 1$,

$$A\text{-coupling:} \quad R = \frac{\pi \to e\nu}{\pi \to \mu\nu} = \frac{m_e^2}{m_\mu^2} \frac{1}{(1 - m_\mu^2/m_\pi^2)^2} = 1.275 \times 10^{-4}, \qquad (6.45)$$

$$P\text{-coupling:} \quad R = \frac{\pi \to e\nu}{\pi \to \mu\nu} = \frac{1}{(1 - m_\mu^2/m_\pi^2)^2} = 5.5. \qquad (6.46)$$

The dramatic difference in the branching ratio for the two types of coupling just stems from the fact that angular momentum conservation compels the electron or muon to have the "wrong" helicity for A-coupling. The phase-space factor for the electron decay is greater than for muon decay, but the factor $(1 - v/c)$ strongly inhibits decay to the lighter lepton.

The current measured value for the ratio is

$$R_{\text{exp}} = (1.267 \pm 0.023) \times 10^{-4}. \qquad (6.47)$$

This result was a major triumph for the $V\text{-}A$ theory, and proves that the pseudoscalar coupling is zero or extremely small. Figure 6.14 shows a typical

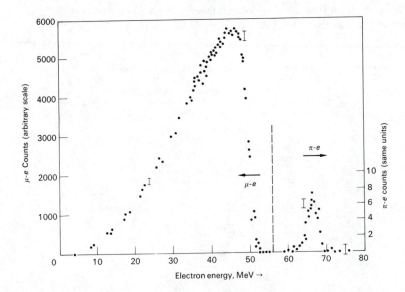

Fig. 6.14 Electron spectrum from stopping positive pions, as measured in the experiment of Anderson *et al.* (1960). The broad distribution extending up to 53 MeV is from $\mu^+ \to e^+ + \nu + \bar{\nu}$. The narrow peak around 70 MeV is from $\pi^+ \to e^+ + \nu$ decay. Note the change in vertical scale for these very rare events.

positron spectrum observed from positive pions stopping in an absorber in one experiment. The rare $\pi \to ev$ process yields positrons of unique energy, about 70 MeV. They are accompanied by the much more numerous positrons from the decay sequences $\pi \to \mu v$, $\mu \to ev\bar{v}$. The spectrum from muon decay extends to 50 MeV. Rejection of electrons from $\pi \to \mu \to e$ decays is based on momentum, timing (the mean life of the pion is 25 ns, that of the muon 2200 ns) as well as the absence of a muon pulse in the counters.

The above formulae may be applied also to the kaon-decay branching ratio, for which the predicted value for A-coupling is

$$R_K = \frac{K \to ev}{K \to \mu v} = 2.5 \times 10^{-5},$$

compared with an experimentally measured value of $(2.43 \pm 0.14) \times 10^{-5}$. Here the electron is even more relativistic, and therefore this decay is more strongly suppressed than in pion decay.

In this analysis, it will be noted that in Fig. 6.13 we have tacitly assumed the same coupling G for $\pi \to \mu v$ and $\pi \to ev$, and the data support this. This hypothesis of universality in the coupling of electrons and muons is discussed in Section 6.9.

6.9 WEAK DECAYS OF STRANGE PARTICLES: CABIBBO THEORY

The nonleptonic decays of strange particles appear to be characterized by the selection rule $\Delta S = 1$, $\Delta I = \frac{1}{2}$, as we might expect if one replaces a strange quark s $(S = -1, I = 0)$ by a nonstrange quark d $(S = 0, I = \frac{1}{2})$, and the effects of the other quarks may be neglected. As evidence we cite the branching ratios for Λ decay:

Decay	Quark description	
$\Lambda \to p\pi^-$	$uds \to udd \to uud + \bar{u}d$	(6.48)
	ᴡɪ ꜱɪ	
$\to n\pi^0$	$\to udd + \bar{u}u$	

Since $I = 0$, the $\Delta I = \frac{1}{2}$ rule states that the nucleon and pion must be in a state of $I = \frac{1}{2}$. Referring to the table of Clebsch-Gordan coefficients (Table III, Appendix), the rule predicts

$$\frac{\Gamma(\Lambda \to n\pi^0)}{\Gamma(\Lambda \to n\pi^0) + \Gamma(\Lambda \to p\pi^-)} = \frac{1}{3} \times 1.036 = 0.345, \qquad (6.49)$$

where the correction factor accounts for the slightly different phase space in $n\pi^0$ and $p\pi^-$ states. The observed value is 0.358 ± 0.005, in agreement with this prediction, when account is taken of small electromagnetic radiative corrections.

In semileptonic decays of strange particles, the isospin of the final state cannot be specified, but they obey the selection rule $\Delta Q = \Delta S$, where ΔQ and ΔS

are the changes in charge and strangeness of the hadrons. $\Delta Q = \Delta S = 1$ is expected if $\Delta I_3 = \frac{1}{2}$, from the relation

$$Q = I_3 + \tfrac{1}{2}(B + S).$$

As examples of the rule, the decay

Decay	Quark description
$\Sigma^- \rightarrow n + e^- + \bar{v}_e$	$dds \rightarrow ddu + e^- + \bar{v}_e \,(\Delta S = \Delta Q = 1)$

(6.50)

is observed, with branching ratio 1.08×10^{-3}, but not

Decay	Quark description
$\Sigma^+ \rightarrow n + e^+ + v$	$uus \rightarrow udd + e^+ + v_e \,(\Delta S = -\Delta Q = 1)$

(6.51)

for which the branching ratio is less than 5×10^{-6}.

Deviations from the $\Delta I = \frac{1}{2}$ rule are certainly observed in cases where the final-state hadrons can have both $I = \frac{1}{2}$ and $I = \frac{3}{2}$. For example, the branching ratio

$$\frac{\Gamma(\Omega^- \rightarrow \Xi^0 \pi^-)}{\Gamma(\Omega^- \rightarrow \Xi^- \pi^0)} = 2.94 \pm 0.35,$$

while the $\Delta I = \frac{1}{2}$ rule predicts 2.00. Thus the $I = \frac{3}{2}$ amplitude is also present, although heavily suppressed.

Of more interest are the absolute decay rates for $\Delta S = 1$ transitions, as compared with $\Delta S = 0$ transitions. $\Delta S = 1$ transitions are suppressed by a factor of about 20 relative to ones with $\Delta S = 0$. The reduced decay rate for $\Delta S = 1$ processes is accounted for by the Cabibbo theory (1963). In this model, the d and s quark states participating in the weak interactions are "rotated" by a mixing angle θ_C, called the Cabibbo angle. In analogy with the lepton doublets involved in charge-changing weak lepton currents, the u, d, s quarks are organized in a doublet:

$$\overset{\text{Leptons}}{\begin{pmatrix} \mu \\ v_\mu \end{pmatrix}}, \quad \begin{pmatrix} e \\ v_e \end{pmatrix}, \quad \overset{\text{Quarks}}{\begin{pmatrix} u \\ d_C \end{pmatrix}} = \begin{pmatrix} u \\ d\cos\theta_C + s\sin\theta_C \end{pmatrix}$$

(6.52)

For either of these sets of doublets, the weak coupling is specified by the Fermi constant G. For a $\Delta S = 0$ semileptonic decay, involving u- and d-quarks, the coupling will then be $G\cos\theta_C$, and for a $\Delta S = 1$ decay, involving s-quarks, it will be $G\sin\theta_C$. We illustrate this in Table 6.3 by writing out muon decay, ^{14}O β-decay, pion β-decay ($\pi^- \rightarrow \pi^0 e^- \bar{v}_e$), and leptonic K-decay ($K^- \rightarrow \pi^0 e^- \bar{v}_e$) in terms of the quarks involved.

We note that (6.53) and (6.54) are allowed Fermi transitions ($J^P = 0^+ \rightarrow 0^+$ or $0^- \rightarrow 0^-$) and should have the same ft-value — as indeed they do (see

TABLE 6.3

	Hadronic spin-parity	Quark description	Rate	Eq.
$p \to ne^+ v_e$ (^{14}O)	$0^+ \to 0^+$	$u \to de^+ v_e$	$G^2 \cos^2 \theta_C$	(6.53)
$\pi^- \to \pi^0 e^- \bar{v}_e$	$0^- \to 0^-$	$d \to ue^- \bar{v}_e$	$G^2 \cos^2 \theta_C$	(6.54)
$K^- \to \pi^0 e^- \bar{v}_e$	$0^- \to 0^-$	$s \to ue^- \bar{v}_e$	$G^2 \sin^2 \theta_C$	(6.55)
$\mu^+ \to e^+ v_e \bar{v}_\mu$	—	—	G^2	(6.56)

Problem 6.1). From the various processes of semileptonic decay of hadrons we can obtain estimates of θ_C via the relations

$$\Delta S = 0, \qquad G_\pi^2 = G_n^2 = G_\mu^2 \cos^2 \theta_C,$$

$$\Delta S = 1, \qquad G_K^2 = G_\mu^2 \sin^2 \theta_C. \tag{6.57}$$

Thus, comparison of the ^{14}O decay rate with that for μ-decay [(6.53) and (6.56)] gives $\cos^2 \theta_C$ and hence

$$\theta_C(\text{vector}) = 0.210 \pm 0.025, \tag{6.58}$$

while comparison of reactions (6.55) and (6.54) for pure vector $\Delta S = 1$ and $\Delta S = 0$ transitions gives

$$\theta_C(\text{vector}) = 0.25 \pm 0.01. \tag{6.59}$$

An estimate of θ_C can also be obtained by comparing the pure axial-vector transition rates

$$\frac{K^+ \to \mu^+ v}{\pi^+ \to \mu^+ v} \to \theta_C(\text{axial}) = 0.269 \pm 0.001. \tag{6.60}$$

The results (6.58), (6.59), and (6.60) do not give precisely equal values, but the reasons for the small discrepancies can be understood. This would take us on to subjects like nonconservation of the axial-vector currents and quark mass effects. It is sufficient to state that when these are taken into consideration, the relative rates of strangeness-changing and -nonchanging decays are consistent with a unique value of the Cabibbo angle. However, this theory was formulated at a time when only three flavors of quark (u, d, s) were known. Modifications to be expected in a six-flavor model are discussed in Section 6.12.

6.10 WEAK NEUTRAL CURRENTS

The production at accelerators of intense beams of high-energy neutrinos and antineutrinos, from the early 1960s onwards, led to dramatic developments in our understanding of the weak interactions. The layout of a neutrino beam

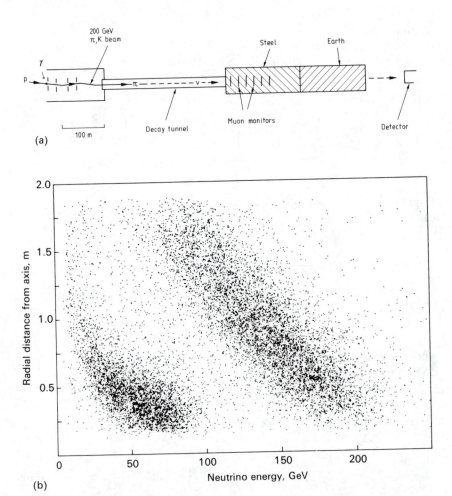

Fig. 6.15 (a) Layout of CERN 200-GeV narrow-band neutrino beam. (b) There is a kinematic relation between the energy and direction of neutrinos from decay of a monochromatic beam of pions or kaons (see Problem 6.2). Hence the energies of neutrino events are correlated with the distance from the beam axis. The high-energy band is due to neutrinos from kaon decay, the low-energy events to those from pion decay. Data from the calorimeter detector of Fig. 2.21 (de Groot *et al.* 1979).

is sketched out in Fig. 6.15. The basic principle is to first produce secondary pions and kaons from high-energy proton collisions in a target T. Secondaries of one sign of charge and in a small band of momentum can be selected by means of dipole bending magnets, quadrupole focusing magnets, and collimating slits. The pions and kaons enter a decay tunnel, where a fraction decay to muons and neutrinos ($\pi^+ \to \mu^+ \nu_\mu$, $\pi^- \to \mu^- \bar{\nu}_\mu$). Muons are ranged out by a thick iron shield, and the interactions of the neutrinos observed in the detector.

Figures 2.21 and 2.12 show examples of large electronic and bubble-chamber neutrino detectors.

As mentioned before, such experiments gave the first evidence, in 1961, for the separate identity of electron and muon neutrinos, v_e and v_μ (see Figs. 6.16 and 6.17 for examples of events). In 1973, the existence of v_μ interactions without a charged lepton in the final state were first demonstrated in a bubble-chamber experiment at CERN (Hasert *et al.* 1973, 1974). These were termed *neutral-current* events and ascribed to reactions of the form

$$v_\mu N \to v_\mu X,$$
$$\bar{v}_\mu N \to \bar{v}_\mu X, \tag{6.61}$$

where X is any hadronic state. They occurred at a fraction of the rate of the more usual *charged-current* events

$$v_\mu N \to \mu^- X,$$
$$\bar{v}_\mu N \to \mu^+ X. \tag{6.62}$$

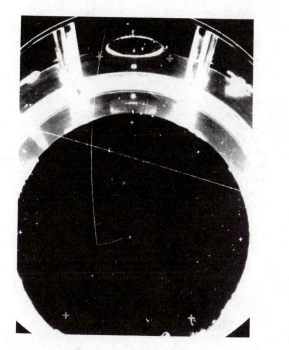

Fig. 6.16 Event attributed to an "elastic" neutrino interaction $v_\mu + n \to p + \mu^-$ in the CERN heavy-liquid bubble chamber, filled with freon (CF_3Br). The negatively charged particle passes out of the chamber without interaction; it was identified as a muon because, in many events of this type, the observed interaction length of the negative particles was very much larger than for strongly interacting particles. The original interaction takes place in a heavy nucleus, and one observes a proton plus a short nuclear fragment.

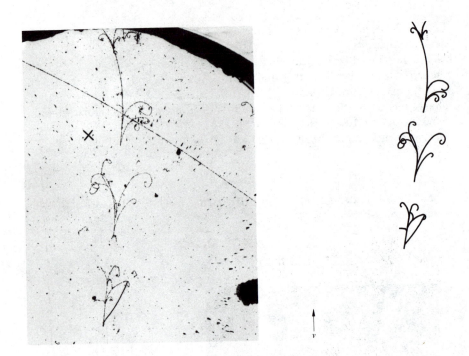

Fig. 6.17 Event produced by interaction of an electron-neutrino v_e: $v_e + n \rightarrow p + e^-$. The incident beam consists mostly of muon-neutrinos v_μ, with a very small admixture of v_e ($\sim \frac{1}{2}\%$) from the 3-body decays in flight, $K^+ \rightarrow \pi^0 + e^+ + v_e$. The high-energy electron secondary is recognized by the characteristic shower it produces by the processes of bremsstrahlung and pair production. The chamber diameter is 1.1 m, and the radiation length in CF_3Br is 0.11 m. The relative numbers of events giving electron and muon secondaries are consistent with the calculated fluxes of v_e and v_μ in the beam, and thus confirms conservation of muon number. (Courtesy CERN.)

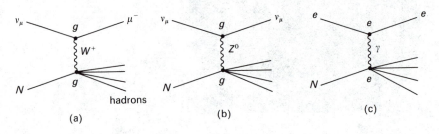

Fig. 6.18

A better if clumsier nomenclature would perhaps be non-charge-changing and charge-changing currents respectively. The current-current interaction introduced in Section 6.6 can be visualised in terms of exchange of charged and neutral bosons, $W^\pm$ and Z^0, as in Fig. 6.18(a) and (b), in analogy with photon exchange in electromagnetic (neutral-current) reactions. The introduction of a mediating boson Z^0 or $W^\pm$ means that the matrix element must be given, in terms of the momentum transfer q, by

$$M \sim \frac{g_W^2}{q^2 + M_W^2},$$

where g_W is the coupling of $W^\pm$ to lepton or hadron and the denominator is the propagator term. If $M_W^2 \gg q^2$, the current-current interaction is effectively pointlike and we can identify the four-fermion coupling G as

$$\frac{G^2}{2} = \lim_{q^2 \to 0} \frac{g_W^4}{(q^2 + M_W^2)^2}, \tag{6.63}$$

or

$$g_W^2 = M_W^2 G / \sqrt{2}.$$

Predictions exist for the masses of the bosons, as discussed in Section 8.5: they are $M_W \simeq 80$ GeV and $M_{Z^0} \simeq 90$ GeV. The present experimental limit

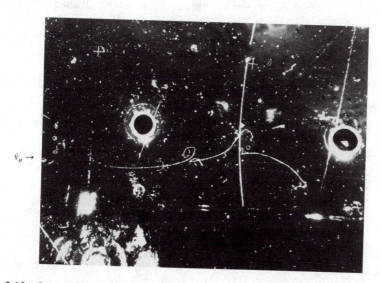

$\bar{\nu}_\mu \rightarrow$

Fig. 6.19 One of the historic pictures of high-energy physics. The event was found in the bubble chamber Gargamelle, filled with freon (CF_3Br) and exposed to an antineutrino ($\bar{\nu}_\mu$) beam at the CERN PS (Hasert *et al.* 1973). It consists of a single electron of energy 400 MeV projected at an angle of $1.5 \pm 1.5°$ to the beam (see Problem 6.13). It is identified by the characteristic bremsstrahlung energy loss and pair production along the track. The event constituted the first evidence for a weak neutral current in the process $\bar{\nu}_\mu + e^- \rightarrow e^- + \bar{\nu}_\mu$. A total of three such events were observed in 1.4 million pictures (with $\sim 10^9$ antineutrinos per pulse) over a two-year period. Even today (1980) only about 100 events exist from six experiments.

(1981), based on absence of measurable propagator effects in neutrino scattering, is $M_W > 30$ GeV. Thus the weak reaction is effectively pointlike (range $R \sim 1/M_W$) for the values of momentum transfer q considered in experiments to date.

The first experiments gave for the ratio of the cross-sections of neutral and charged current reactions (6.61) and (6.62)

$$\frac{\sigma(v_\mu N \to v_\mu X)}{\sigma(v_\mu N \to \mu^- X)} \simeq 0.25,$$

$$\frac{\sigma(\bar{v}_\mu N \to \bar{v}_\mu X)}{\sigma(\bar{v}_\mu N \to \mu^+ X)} \simeq 0.45, \tag{6.64}$$

so that, on the basis of a hundred or so events, the rates for neutral and charged current reactions were comparable.

In addition to these semileptonic reactions, purely leptonic neutral current events of the type

$$\bar{v}_\mu + e^- \to e^- + \bar{v}_\mu \tag{6.65}$$

were also observed in the same experiment; indeed, the first such event, shown in Fig. 6.19, was obtained before the reactions (6.61) were established. The results of the more recent data on semileptonic and leptonic neutral current reactions are given in Section 8.6. The main point to be made here is that neutral-current couplings exist, with strength comparable to that of charged currents.

6.11 ABSENCE OF $\Delta S = 1$ NEUTRAL CURRENTS: THE GIM MODEL AND CHARM

All neutral current processes observed are characterized by the selection rule $\Delta S = 0$. Indeed, one of the reasons that the early theories of weak interactions, incorporating neutral currents, were discounted was that they had never been observed in decay processes. For example, the ratio of neutral- to charged-current rates in kaon decay is

$$\frac{K^+ \to \pi^+ v \bar{v}}{K^+ \to \pi^0 \mu^+ v_\mu} < 10^{-5} \qquad (\Delta S = 1). \tag{6.66}$$

In Chapter 8, we shall discuss interpretation of the particles W^+, W^-, Z^0 introduced in Fig. 6.18 as members of a "weak isospin" triplet of bosons interacting with quarks (and leptons). Leaving aside the matrix operators of (6.33), the charged (or charge-raising) weak current, for example in neutron decay, will be $J^+ = \bar{u}d \cos \theta_C$. We then expect from (6.52) that the neutral current coupling (Fig. 6.20) will be of the form

$$\underbrace{u\bar{u} + (d\bar{d} \cos^2 \theta_C + s\bar{s} \sin^2 \theta_C)}_{\Delta S = 0} + \underbrace{(s\bar{d} + \bar{s}d) \sin \theta_C \cos \theta_C,}_{\Delta S = 1} \tag{6.67}$$

so that $\Delta S = 1$ neutral currents should be possible, since $\sin \theta_C \neq 0$. In a classic paper in 1970, Glashow, Iliopoulos and Maiani (GIM) proposed the introduction of a new quark with flavor labelled c for "charm", with charge $+\frac{2}{3}$. They proposed for the quark states in weak interactions, a further doublet, consisting of the c-quark and the s, d combination orthogonal to d_c in (6.52). Thus the two quark doublets were

$$\begin{pmatrix} u \\ d_c \end{pmatrix} = \begin{pmatrix} u \\ d\cos\theta_C + s\sin\theta_C \end{pmatrix}, \quad \begin{pmatrix} c \\ s_c \end{pmatrix} = \begin{pmatrix} c \\ s\cos\theta_C - d\sin\theta_C \end{pmatrix}. \quad (6.68)$$

In this way, extra terms have to be added to Fig. 6.20, as shown in Fig. 6.21, and when these are included we obtain for the weak-interaction neutral-current matrix element

$$\underbrace{u\bar{u} + c\bar{c} + (d\bar{d} + s\bar{s})\cos^2\theta_C + (s\bar{s} + d\bar{d})\sin^2\theta_C}_{\Delta S = 0}$$

$$\underbrace{+ (s\bar{d} + \bar{s}d - \bar{s}d - s\bar{d})\sin\theta_C\cos\theta_C.}_{\Delta S = 1} \quad (6.69)$$

Hence, at the price of a new quark and a second quark doublet, the unwanted $\Delta S = 1$ neutral currents have been canceled.

It was therefore a tremendous triumph for theory when the heavy c-quark states ($\psi = c\bar{c}$, $D = c\bar{u}$, etc.) were observed in 1974–1976, as described in Section 5.11. We note that the assignment (6.68) also predicts that in the decays ($\Delta C = 1$) of charmed mesons to noncharmed mesons, we expect $c \to s$

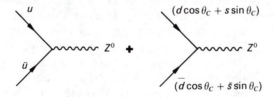

Fig. 6.20

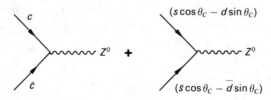

Fig. 6.21

transitions (coupling $\cos^2 \theta_C$) to dominate $c \to d$ transitions (coupling $\sin^2 \theta_C$). Thus the $D^0(1893)$ meson decays predominantly in the mode $D^0 \to K^- + \pi's$, and $D^0 \to \pi's$ is strongly suppressed (see Fig. 2.13 for an example of such a decay).

By the same mechanism that cancels $\Delta S = 1$ neutral currents, we also expect $\Delta C = 1$ neutral currents to be absent. Limits on these rates can be set from neutrino experiments, looking for (single) charmed-particle production in neutral-current reactions. They are currently known to be less than 3% of the $\Delta C = 0$ neutral current cross-sections.

6.12   WEAK MIXING ANGLES WITH SIX QUARKS

In Sections 6.9 and 6.11 we have restricted ourselves to u, d, s, and c quarks, and the Cabibbo-GIM formalism for the mixing matrix which specifies the quark states involved in the weak interactions. The weak charged current is then expressed in matrix form by

$$J^+_{\text{weak}} = (\bar{u}, \bar{c}) \begin{pmatrix} \cos \theta_C & \sin \theta_C \\ -\sin \theta_C & \cos \theta_C \end{pmatrix} \begin{pmatrix} d \\ s \end{pmatrix}, \tag{6.70}$$

where the coupling constant and the space-time structure operator $\gamma_\mu(1 + \gamma_5)$ have been omitted. The mixing is specified here by a single parameter, the Cabibbo angle θ_C. All data on weak decays involving u, d, s, c quarks seem to be consistent with a unique value of θ_C, as described previously.

The treatment must be extended if we intend to incorporate further quarks – the b-quark of $Q = -\frac{1}{3}$, which had to be postulated following the discovery of the Υ-resonances (Section 5.11), and its proposed partner, the t-quark of $Q = +\frac{2}{3}$, whose existence has not yet been established (1981). The existence of three pairs of quarks, in parallel with three pairs of leptons (see Section 6.14), is esthetically attractive, and even receives some justification on the basis of the absence of certain "triangle anomalies". We do not discuss these here, but remark that the condition for canceling such anomalies is that the net charge of all fermions should be zero. The three lepton doublets $(e, \nu_e; \mu, \nu_\mu; \tau, \nu_\tau)$ contribute $- 3|e|$; and the three quarks (u, c, t) of $Q = +\frac{2}{3}$, plus three (d, s, b) of charge $Q = -\frac{1}{3}$, will contribute $+ 3|e|$, if allowance is made for the factor 3 for color (Section 5.2).

With six quark flavors, the weak currents will be described by unitary transformations among three quark doublets, characterized by three Euler angles and six phases (Kobayashi and Maskawa 1972). Of the latter, five are arbitrary and unobservable, leaving one nontrivial phase, δ. Then

$$J^+_{\text{weak}} = (\bar{u}, \bar{c}, \bar{t}) M \begin{pmatrix} d \\ s \\ b \end{pmatrix}, \tag{6.71a}$$

where the Kobayashi-Maskawa 3×3 matrix M is given by

$$M = \begin{vmatrix} c_1 & c_3 s_1 & s_1 s_3 \\ -c_2 s_1 & c_1 c_2 c_3 - s_2 s_3 e^{i\delta} & c_1 c_2 s_3 + c_3 s_2 e^{i\delta} \\ s_1 s_2 & -c_1 c_3 s_2 - c_2 s_3 e^{i\delta} & -c_1 s_2 s_3 + c_2 c_3 e^{i\delta} \end{vmatrix} \qquad (6.71b)$$

with $c_i = \cos\theta_i$ and $s_i = \sin\theta_i$. Here $\theta_1, \theta_2, \theta_3$ are the three mixing angles, replacing the single angle θ_C of the 2×2 matrix (6.70). Since under time reversal $e^{i\delta} \to e^{-i\delta}$, the phase angle δ introduces the possibility of a T- or CP-violating amplitude (see Section 6.13.3).

Comparing (6.70) and (6.71), it will be seen that the $u \to d$ mixing, specified by a coefficient $\cos\theta_C$ in the Cabibbo theory, is given by $\cos\theta_1$ in the more general model. This coefficient, it will be recalled, is equal to the ratio of coupling G in μ-decay and $G\cos\theta_C$ in nuclear β-decay. The semileptonic decays of hyperons and kaons, previously used to determine $\sin\theta_C$ [Eqs. (6.59) and (6.60)] now determine the product $\sin\theta_1 \cos\theta_3$. Since θ_C from β-decay and from kaon decay are equal within errors, $\cos\theta_3 \simeq 1$ and thus $\sin\theta_3$ must be small.

Information on θ_2 can be obtained from observations on single charmed-particle production in high-energy neutrino collisions; written in terms of quarks, this reaction is expressed by

$$\nu_\mu + d \to \mu^- + c$$
$$\hookrightarrow s + \mu^+ + \nu_\mu. \qquad (6.72)$$

In this equation, the charmed quark produced in the $\Delta C = 1$ weak reaction is indicated as decaying semileptonically into an s-quark and a lepton pair. Such events are very characteristic — they contain a pair of opposite-sign muons $(\mu^+ \mu^-)$ of high energy. The rate of the $d \to c$ transitions, compared with the more common $\Delta C = 0$ $d \to u$ transitions, can be obtained from the observed dilepton event rate and the known branching ratios for semileptonic decay of charmed mesons (D^+, D^0 etc.). According to (6.71), it determines the product $\sin\theta_1 \cos\theta_2$; experiments indicate that $\cos\theta_2 \simeq 1$ and hence $\sin\theta_2$ is small.

Although the precision of available data is not very high, it seems therefore that s_1, s_2, and s_3 are all small (< 0.4). If this is so, the model makes predictions about the weak decay sequence of mesons containing b-quarks. The decay chain

$$b \to c \to s \to u,$$

amplitude: $\qquad c_1 c_2 s_3 + c_3 s_2 e^{i\delta},$

should be favored over

$$b \to u,$$

amplitude: $\qquad s_1 s_3,$

and hence, in the decay of such bottom states, strange particles (kaons) should be frequent among the hadron secondaries. Initial indications from the CESR $e^+ e^-$ collider suggest that this is the case. We also note that such decays may also be *CP*-violating, although the effects are likely to be at the 1% level or less and thus extremely difficult to detect.

6.13 K^0 DECAY

The Gell-Mann–Nishijima formula

$$\frac{Q}{e} = I_3 + \frac{B + S}{2}$$

indicates, in addition to the charged kaons $K^{\pm}$, of $S = \pm 1$, two neutral kaons, K^0 and $\bar{K}^0$ to complete $I = \frac{1}{2}$ doublets:

S	I_3	
	$+\frac{1}{2}$	$-\frac{1}{2}$
$+1$	K^+	K^0
-1	$\bar{K}^0$	K^-

$$(6.73)$$

The K^0 can be produced by nonstrange particles in association with a hyperon:

$$\pi^- + p \rightarrow \Lambda + K^0.$$ (6.74)

$S \quad\;\; 0 \quad\; 0 \quad -1 \quad +1$

However, a $\bar{K}^0$ can be produced only in association with a kaon or antihyperon, of $S = +1$:

$$\pi^+ + p \rightarrow K^+ + \bar{K}^0 + p,$$ (6.75)

$S \quad\;\; 0 \quad\; 0 \quad +1 \quad -1 \quad\; 0$

$$\pi^- + p \rightarrow \bar{\Lambda} + \bar{K}^0 + n + n.$$ (6.76)

$S \quad\;\; 0 \quad\; 0 \quad +1 \quad -1 \quad 0 \quad 0$

The threshold pion energy for (6.74) is 0.91 GeV, while for (6.75) and (6.76) it is much higher: 1.50 and 6.0 GeV respectively. Thus it is possible to produce a pure K^0 beam by choosing incident pions of suitable energy.

K^0 and $\bar{K}^0$ are particle and antiparticle, and are connected by the process of charge conjugation, which involves a reversal of the value of I_3 and a change of strangeness, $\Delta S = 2$. Strong interactions conserve I_3 and S, so that as far as *production* is concerned, the separate neutral-kaon eigenstates are K^0 and $\bar{K}^0$.

Now suppose K^0 and $\bar{K}^0$ particles propagate through empty space. Since both are neutral, both can decay to pions by the weak interaction, with $|\Delta S| = 1$. Thus, *mixing* can occur via (virtual) intermediate pion states:

$$K^0 \underset{3\pi}{\overset{2\pi}{\rightleftarrows}} \bar{K}^0$$

These transitions are $\Delta S = 2$ and thus second-order weak interactions. Although extremely weak therefore, this implies that, if one has a pure K^0 state at $t = 0$, at any later time t one will have a superposition of both K^0 and $\bar{K}^0$, so that the state can be written

$$|K(t)\rangle = \alpha(t)|K^0\rangle + \beta(t)|\bar{K}^0\rangle. \tag{6.77}$$

We know that objects which decay by weak interactions are eigenstates of CP, not of strangeness S. The operation of CP on the states K^0 and $\bar{K}^0$ gives

$$CP|K^0\rangle \to \eta|\bar{K}^0\rangle,$$

$$CP|\bar{K}^0\rangle \to \eta'|K^0\rangle, \tag{6.78}$$

where η, η' are arbitrary phase factors, which we can define as $\eta = \eta' = 1$. Clearly $|K^0\rangle$ and $|\bar{K}^0\rangle$ are not CP-eigenstates, but we can form the linear combinations

$$|K_1\rangle = \frac{1}{\sqrt{2}}(|K^0\rangle + |\bar{K}^0\rangle), \qquad CP = +1,$$

$$|K_2\rangle = \frac{1}{\sqrt{2}}(|K^0\rangle - |\bar{K}^0\rangle), \qquad CP = -1, \tag{6.79}$$

where

$$CP|K_1\rangle \to |K_1\rangle,$$

$$CP|K_2\rangle \to -|K_2\rangle. \tag{6.80}$$

Unlike K^0 and $\bar{K}^0$, distinguished by their mode of *production*, K_1 and K_2 are distinguished by their mode of *decay*. Consider 2π and 3π decay modes:

(a) $\pi^0\pi^0, \pi^+\pi^-$. By Bose symmetry, the total wavefunction in either case must be symmetric under interchange of the two particles. Since no spin is involved, this is equivalent to the operation C followed by P, so that $CP = +1$.

(b) $\pi^+\pi^-\pi^0, 3\pi^0$. The small Q-value (70 MeV) of the decay suggests $l = 0$, that is, the three pions are in a relative S-state. By the previous argument, the CP parity of $\pi^+\pi^-$ is $+1$. The π^0 has $C = +1$ (since it decays to two γ's) and $P = -1$, and therefore $CP = -1$. So, combining the π^0 with the $\pi^+\pi^-$ system, we obtain one of $CP = -1$. For $l > 0$, both positive and negative CP eigenvalues can result, but such decays are strongly suppressed by angular-momentum barrier effects.

TABLE 6.4 Neutral-kaon eigenstates

Production	Decay	Lifetime, sec
$|K^0\rangle$ $(S = +1)$	$|K_1\rangle = (1/\sqrt{2})(|K^0\rangle + |\bar{K}^0\rangle) \to 2\pi$ $(CP = +1)$	$\tau_1 = 0.9 \times 10^{-10}$
$|\bar{K}_0\rangle$ $(S = -1)$	$|K_2\rangle = (1/\sqrt{2})(|K^0\rangle - |\bar{K}^0\rangle) \to 3\pi$ $(CP = -1)$	$\tau_2 = 0.5 \times 10^{-7}$

To summarize, the 2π state must have $CP = +1$, while the 3π state can have $CP = +1$ or -1, with $CP = -1$ heavily favored kinematically. The two- and three-pion decay modes have different Q-values, and thus different phase-space factors and disintegration rates, the two-pion decay being much faster. In summary, the production and decay states of neutral kaons are listed in Table 6.4.

6.13.1   Strangeness Oscillations

So far, we have discussed K_1- and K_2-amplitudes without regard to their precise time dependence. To be more specific we should include phase factors with the amplitudes. The relative phase of K_1- and K_2-states of a given momentum will only be constant with time if these particles have identical masses. K_1 and K_2 are not charge-conjugate states, having quite different decay modes and lifetimes, so that, in the same sense that the mass difference between neutron and proton can be attributed to differences in their electromagnetic coupling, a $K_1 - K_2$ mass difference — but a very much smaller one — is to be expected because of their different weak couplings.

The amplitude of the state K_1 at time t can be written as

$$a_1(t) = a_1(0)e^{-(iE_1/h)t}e^{-\Gamma_1 t/2h}, \tag{6.81}$$

where E_1 is the total energy of the particle, so that E_1/h is the circular frequency ω_1, and $\Gamma_1 = h/\tau_1$ is the width of the state, τ_1 being the mean lifetime in the frame in which the energy E_1 is defined. The second term must have the form shown, in accord with the law of radioactive decay of the intensity:

$$I(t) = a_1(t)a_1^*(t) = a_1(0)a_1^*(0)e^{-\Gamma_1 t/h}$$

$$= I(0)e^{-t/\tau_1}.$$

Set $h = c = 1$ and measure all times in the rest frame, so that τ_1 is the proper lifetime and $E_1 = m_1$, the particle rest mass. Then the K_1-amplitude is

$$a_1(t) = a_1(0)e^{-(\Gamma_1/2 + im_1)t}. \tag{6.82}$$

Similarly, for K_2,

$$a_2(t) = a_2(0)e^{-(\Gamma_2/2 + im_2)t}. \tag{6.83}$$

Now suppose that at $t = 0$, a beam of unit intensity consists of pure K^0. Then from (6.79) $a_1(0) = a_2(0) = 1/\sqrt{2}$. After a time t for free decay *in vacuo*, the K^0-intensity will be

$$I(K^0) = \frac{(a_1(t) + a_2(t))}{\sqrt{2}} \frac{(a_1^*(t) + a_2^*(t))}{\sqrt{2}}$$

$$= \tfrac{1}{4}[e^{-\Gamma_1 t} + e^{-\Gamma_2 t} + 2e^{-[(\Gamma_1 + \Gamma_2)/2]t}\cos \Delta m\, t], \tag{6.84}$$

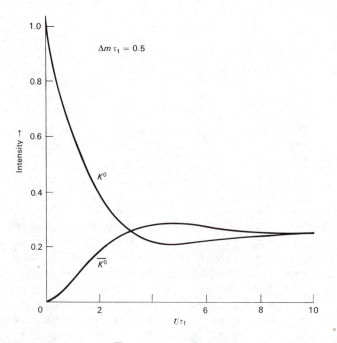

Fig. 6.22 Oscillations of K^0- and $\bar{K}^0$-intensities, for an initially pure K^0-beam, as calculated from Eqs. (6.84) and (6.85). A value $\Delta m \tau_1 = 0.5$ has been assumed.

where $\Delta m = |m_2 - m_1|$. Similarly, the $\bar{K}^0$-intensity will be, writing the amplitude as $[a_1(t) - a_2(t)]/\sqrt{2}$,

$$I(\bar{K}^0) = \tfrac{1}{4}[e^{-\Gamma_1 t} + e^{-\Gamma_2 t} - 2e^{-[(\Gamma_1 + \Gamma_2)/2]t} \cos \Delta m \, t]. \tag{6.85}$$

Thus, the K^0- and $\bar{K}^0$-intensities *oscillate* with the frequency Δm. Figure 6.22 shows the variation to be expected for $\Delta m = 0.5/\tau_1$. If one measures the number of $\bar{K}^0$ interaction events (i.e. the hyperon yield) as a function of position from the K^0 source, one can therefore deduce $|\Delta m|$. The present value is given by

$$\Delta m \, \tau_1 = 0.477 \pm 0.002, \tag{6.86}$$

where $\Delta m = m_2 - m_1$, and in separate regeneration experiments, it is found that $m_2 > m_1$.

The actual value of the mass difference is tiny:

$$\Delta m = 3.52 \times 10^{-6} \text{ eV},$$

or a fractional mass difference

$$\frac{\Delta m}{m} = 0.7 \times 10^{-14}. \tag{6.87}$$

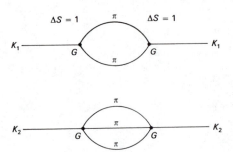

Fig. 6.23

The magnitude of the self-energy (i.e. contribution to the mass) of K_1 and K_2 can be estimated crudely by inspection of Fig. 6.23. In these diagrams the "weak" mass generated is clearly proportional to G^2 (all higher-order terms being negligible in comparison), and thus to an effect of the second order in the weak coupling constant – the only one so far observed experimentally. Since $G = 10^{-5}/M_p^2$, we must introduce a mass in order to get the dimensions right. It is reasonable to choose this as the kaon mass. Since we have two strangeness-changing transitions, we should in fact replace G by $G \sin \theta_C$, as in Eq. (6.57). Hence we may estimate

$$\Delta m(\text{for } \Delta S = 1 \text{ coupling}) \sim G^2 m_K^5 \sin^2 \theta_C \sim 10^{-4} \text{ eV}, \qquad (6.88)$$

within an order of magnitude of the observed value (6.87). By a similar argument, one can set very stringent limits on the strength of a possible direct $\Delta S = 2$ transition between K^0- and $\bar{K}^0$-states. If fG denotes the coupling strength, then

$$\Delta m_K(\text{for } \Delta S = 2 \text{ coupling}) \simeq fGm_K^3 = 10^3 f \text{ eV}, \qquad (6.89)$$

giving $f < 10^{-8}$.

6.13.2 The K^0 Regeneration Phenomenon

The K_2 decay mode was not observed until some time after its prediction by Gell-Mann and Pais. However, Pais and Piccioni (1955) had observed that the $K_1 K_2$ phenomenon would lead to the process of *regeneration*. Suppose we start with a pure K^0-beam, and let it coast *in vacuo* for the order of 100 K_1 mean lives, so that all the K_1-component has decayed and we are left with K_2 only. Now let the K_2-beam traverse a slab of material and interact. Immediately, the strong interactions will pick out the strangeness $+1$ and -1 components of the beam, i.e.

$$|K_2\rangle = \frac{1}{\sqrt{2}}(|K^0\rangle - |\bar{K}^0\rangle). \qquad (6.90)$$

Thus, of the original K^0-beam intensity, 50% has disappeared by K_1-decay. The remainder, called K_2, upon traversing a slab where its nuclear interactions can be observed, should consist of 50% K^0 and 50% $\bar{K}^0$. The existence of $\bar{K}^0$ ($S = -1$) a long way from the source of an originally pure K^0-beam ($S = +1$) was confirmed by the observation of production of hyperons in 1956. Thus $\bar{K}^0 + p \to \Lambda + \pi^+$, for example.

The K^0- and $\bar{K}^0$-components in (6.90) must be absorbed differently; K^0-particles can only undergo elastic and charge exchange scattering, while $\bar{K}^0$-particles can also undergo strangeness exchange giving hyperons. With more strong channels open, the $\bar{K}^0$ is therefore absorbed more strongly than K^0. After emerging from the slab, we shall therefore have a K^0-amplitude $f|K^0\rangle$ and a $\bar{K}^0$-amplitude $\bar{f}|\bar{K}^0\rangle$ say, where $\bar{f} < f < 1$. If we then ask what are the characteristics of the emergent beam with respect to decay, we must write, in place of (6.90):

$$\frac{1}{\sqrt{2}}(f|K^0\rangle - \bar{f}|\bar{K}^0\rangle) = \frac{f+\bar{f}}{2\sqrt{2}}(|K^0\rangle - |\bar{K}^0\rangle) + \frac{f-\bar{f}}{2\sqrt{2}}(|K^0\rangle + |\bar{K}^0\rangle)$$

$$= \tfrac{1}{2}(f+\bar{f})|K_2\rangle + \tfrac{1}{2}(f-\bar{f})|K_1\rangle. \tag{6.91}$$

Since $f \neq \bar{f}$, it follows that some of the K_1-state has been regenerated (Fig. 6.24). This regeneration of short-lived K_1's in a long-lived K_2-beam was confirmed by experiment.

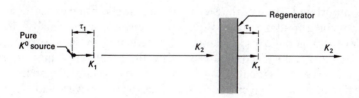

Fig. 6.24 Regeneration of short-lived K_1-mesons when a pure K_2-beam traverses a regenerator.

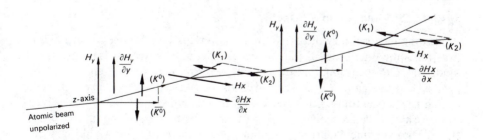

Fig. 6.25 Analogy between deflection of atomic beam in two orthogonal planes, by means of inhomogeneous magnetic fields, and the K^0-decay and regeneration phenomena.

The K^0 regeneration phenomenon is simply a consequence of the concepts of superposition and quantization in quantum mechanics. The behavior of an atomic beam in an inhomogeneous magnetic field – the classical Stern-Gerlach experiment – provides a fairly close analogy. Referring to Fig. 6.25, suppose we have an initially unpolarized atomic beam of spin $\frac{1}{2}$ moving along the z-axis, which then traverses an inhomogeneous field H_y directed along the y-axis (representing the strong interaction in the K^0 case). The atoms are then quantized in two spin eigenstates, $\sigma_y = +\frac{1}{2}$ and $\sigma_y = -\frac{1}{2}$, particles in the first state being deflected upward and those in the second downward. (These are analogous to the K^0 and $\bar{K}^0$ S-eigenstates.) We select the part of the beam deflected upward, and pass it through an inhomogeneous field along the x-axis, H_x (the analogue of the weak interactions). The beam again splits into two components, 50% having $\sigma_x = -\frac{1}{2}$ and being deflected to the left, and 50% having $\sigma_x = +\frac{1}{2}$, deflected to the right. Note that *all* information about quantization along the y-axis has necessarily been lost. In turn, if we take the $\sigma_x = \frac{1}{2}$ component (corresponding to K_2), we may again pass it through a field along the y-axis, and recover the components $\sigma_y = \pm\frac{1}{2}$, and lose all knowledge of σ_x (analogous to regeneration of K^0 and $\bar{K}^0$ in an absorber). Finally, passage through a field H_x yields as eigenvalues $\sigma_x = \pm\frac{1}{2}$ (corresponding to reappearance of K_1). To make the analogy more exact, i.e. account for the different decay and nuclear absorption probabilities in the K^0 problem, one could arrange to absorb components of the beam deflected in two of the four directions.

The important feature of the Stern-Gerlach atomic-beam experiment is that it is impossible simultaneously to quantize the spin components of the beam along two orthogonal axes. This may also be stated by saying that the spin operators σ_x and σ_y do not commute, or, of the three matrix operators σ_x, σ_y, and σ_z, only one may be diagonalized at a time (i.e. possess only diagonal elements, and have real, nonzero eigenvalues). Since an operator with real eigenvalues must commute with the Hamiltonian (energy) operator, the axis singled out in space is necessarily that defined by the magnetic field. In a similar fashion, the CP- and S-operators do not commute, so that one can have states which are eigenstates of CP or S, but not both.

6.13.3 CP-Violation in K^0 Decay

Following the discovery of parity violation in weak decay processes in 1957, it was for some time believed that the weak interactions were at least invariant under the CP-operation, and the ensuing description of the CP-eigenstates of neutral kaons, K_1 and K_2, has been given above. In 1964 however, an experiment by Christenson, Cronin, Fitch and Turlay first demonstrated that the long-lived state we have called K_2 could also decay to $\pi^+\pi^-$ with a branching ratio of order 10^{-3}. The experimental arrangement of Christenson *et al.* is shown in Fig. 6.26. The nomenclature K_1 (for a state of $CP = +1$) and K_2 ($CP = -1$) has therefore been superseded by K_S (shortlived component) and K_L (long-lived component). The arguments quoted above on regeneration and mass difference,

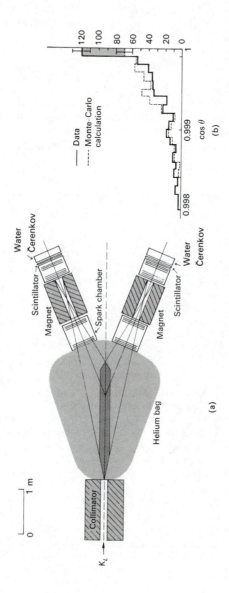

Fig. 6.26 (a) Experimental arrangement of Christenson *et al.* (1964) demonstrating the decay $K_L \to \pi^+ \pi^-$. The K^0-beam is incident from the left, and consists of K_L only, the K_S-component having died out. K_L-decays are observed in a helium bag, the charged products being analyzed by two spectrometers, consisting of bending magnets and spark chambers triggered by scintillators. The rare two-pion decays are distinguished from the common three-pion and leptonic decays on the basis of the invariant mass of the pair and the direction θ of their resultant momentum vector relative to the incident beam. (b) $\cos \theta$ distribution of events of $490 < M_{\pi\pi} < 510$ MeV. The distribution is that expected for three-body decays (dashed line), but with some fifty events (shaded) exactly collinear with the beam due to the $\pi^+\pi^-$ decay mode.

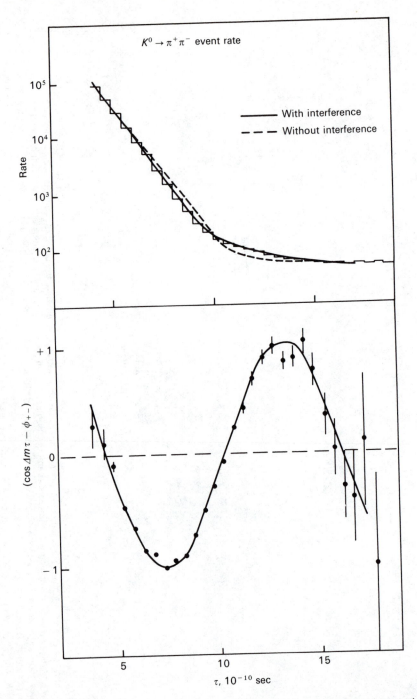

Fig. 6.27 (a) Event rate for $\pi^+\pi^-$ decays from a neutral kaon beam as a function of proper time, demonstrating that the best fit needs the existence of interference between K_L and K_S amplitudes. (b) The interference term extracted from the results in (a). From the fit one can obtain the K_L-K_S mass difference Δm and the phase angle ϕ_{+-} between the two amplitudes. (After Geweniger *et al.* 1974.)

which follow from the superposition principle, remain essentially unchanged, although the formulae (6.79) need to be modified slightly if they are to describe K_S and K_L. The measure of the degree of CP-violation is usually quoted as the amplitude ratio

$$|\eta_{+-}| = \frac{\text{Ampl}(K_L \to \pi^+\pi^-)}{\text{Ampl}(K_S \to \pi^+\pi^-)} = (2.27 \pm 0.02) \times 10^{-3}. \qquad (6.92)$$

The state K_S consists principally of a $CP = +1$ amplitude, but with a little $CP = -1$, and K_L vice versa. Since both K_L and K_S decay to two pions, interference effects in the $\pi^+\pi^-$ signal are expected as a function of the time development of an initially pure K^0 beam. From (6.84) we expect the intensity to vary as

$$I_{2\pi}(t) = I_{2\pi}(0)[e^{-\Gamma_S t} + |\eta_{+-}|^2 e^{-\Gamma_L t}$$
$$+ 2|\eta_{+-}|e^{-[(\Gamma_L + \Gamma_S)/2]t}\cos(\Delta m\, t + \phi_{+-})], \qquad (6.93)$$

where t measures proper time and ϕ_{+-} is an appropriate phase angle between the $K_S \to 2\pi$ and $K_L \to 2\pi$ amplitudes. Such interference effects are indeed observed (see Fig. 6.27). Similar CP-violating effects are found for the decay $K_L \to 2\pi^0$. The quantities η_{+-} and η_{00} depend on the amplitudes A_0 and A_2 for finding the final-state pions in a state of isospin $I = 0$ or 2 ($I = 1$ is forbidden by Bose symmetry), and one obtains the expressions

$$\eta_{+-} = \varepsilon + \varepsilon',$$
$$\eta_{00} = \varepsilon - 2\varepsilon', \qquad (6.94)$$

where

$$\varepsilon' = \frac{i}{\sqrt{2}} \frac{\text{Im } A_2}{A_0} e^{i(\delta_2 - \delta_0)}.$$

In general, η_{+-} and η_{00} will be complex numbers and can be written as

$$\eta_{+-} = |\eta_{+-}|e^{i\phi_{+-}},$$
$$\eta_{00} = |\eta_{00}|e^{i\phi_{00}}, \qquad (6.95)$$

where ϕ_{+-}, ϕ_{00} are related to ε' in (6.94).

If the $\Delta I = \frac{1}{2}$ rule holds, we expect $A_2 = 0$ and hence $|\eta_{+-}| = |\eta_{00}|$, $\phi_{+-} = \phi_{00}$. The experimental values are

$$\eta_{+-} = (2.274 \pm 0.022) \times 10^{-3}, \qquad \phi_{+-} = 44.6 \pm 1.2^0,$$
$$\eta_{00} = (2.33 \pm 0.08) \times 10^{-3}, \qquad \phi_{00} = 54 \pm 5^0, \qquad (6.96)$$

and are thus compatible with the rule. The upper limit $\varepsilon'/\varepsilon \leqslant 0.02$ results from the above definitions and numbers.

CP-noninvariance is also demonstrated in the leptonic decay modes of K_L. These modes are

$$K_L \rightarrow e^+ \nu_e \pi^-, \tag{6.97a}$$

$$K_L \rightarrow e^- \bar{\nu}_e \pi^+, \tag{6.97b}$$

with similar ones for muons replacing electrons.

The decays (6.97a) and (6.97b) transform one into the other under the *CP*-operation, and if *CP*-invariance is violated, a small charge asymmetry is expected. The asymmetry is

$$\Delta = \frac{\text{rate}(K_L \rightarrow e^+ \nu_e \pi^-) - \text{rate}(K_L \rightarrow e^- \bar{\nu}_e \pi^+)}{\text{rate}(K_L \rightarrow e^+ \nu_e \pi^-) + \text{rate}(K_L \rightarrow e^- \bar{\nu}_e \pi^+)}$$

$$= (0.330 \pm 0.012) \times 10^{-2}. \tag{6.98}$$

All data available on *CP*-violation in K^0 decay are consistent with the "superweak model" of Wolfenstein (1964), in which the effects are those due to a specific, new type of interaction, much smaller than the usual weak coupling and visible only in the neutral K^0 system. Among the predictions of this model are that $\varepsilon' = 0$ and thus $|\eta_{+-}| = |\eta_{00}|$, $\phi_{+-} = \phi_{00}$, and that, if one neglects terms of order δ^2 compared with 1, one obtains the relation

$$\Delta = 2\,\text{Re}\,\eta = 2|\eta|\cos\phi = 0.322 \times 10^{-2} \tag{6.99}$$

from (6.96), and in agreement with (6.98). There are, however, other models of *CP*-violation which are for practical purposes almost indistinguishable from the superweak theory. As discussed in Section 6.12, models involving six or more quark flavors and generalized Cabibbo mixing do include a possible finite *CP*-violating phase angle δ [Eq. (6.71)]. The value of the *CP*-violating parameter in K^0 decay in the six-flavor model is of order $\varepsilon \sim s_1 s_3 \sin \delta$. A finite value for $\varepsilon'/\varepsilon \sim 10^{-2}$ is also expected, which may be just about measurable in future experiments. The important point however is that, as distinct from the superweak theory, where *CP*-violation is detectable only in the narrow cul-de-sac of the neutral kaon system, other theories predict effects which are in principle detectable in other systems, for example in bottom meson decays (Section 6.12), or in the form of a nonzero value for the electric dipole moment of the neutron (Section 3.16). On a much grander scale, *CP*-violation and baryon instability have been postulated in order to account for the mismatch between baryons and antibaryons and the ratio of baryons to photons in the universe (see Section 8.11.3).

A general remark may be made in conclusion. Until the advent of *CP*-violation, there was no unambiguous way of defining left-handed and right-handed coordinate systems, or of differentiating matter from antimatter, on a cosmic scale. Thus, aligned ^{60}Co nuclei emit negative electrons with forward-backward asymmetry and left-handed polarization (as we define left-handed). This is insufficient information to describe what we mean by a left-handed system, with the aid of light signals, to an intelligent being in a distant part of the universe, unless one can also uniquely define negative and positive charge, or equivalently ^{60}Co and anti-^{60}Co. *CP*-violation, however, provides an unambiguous definition. Positive charge is now defined as that of the lepton associated with the more abundant leptonic decay mode of the long-lived K_L-particle, $(K_L \to \pi^- e^+ v)/(K_L \to \pi^+ e^- \bar{v}) > 1$. This lepton has the same (or opposite) charge as the local atomic nuclei of matter (or antimatter).

6.14 LEPTON FAMILIES, NEUTRINO MASSES, AND NEUTRINO OSCILLATIONS

Our discussion of leptons thus far has been largely in terms of the electron $(e^-, v_e; e^+, \bar{v}_e)$ and muon $(\mu^-, v_\mu; \mu^+, \bar{v}_\mu)$ leptons. In 1975, experiments by Perl and his coworkers at Stanford produced evidence for the existence of a new heavy lepton, named the tau and associated presumably with a new family $(\tau^-, v_\tau; \tau^+, \bar{v}_\tau)$. The discovery was based on the observation, in $e^+ e^-$ annihilation, of 24 $e\mu$ events consisting of a secondary high-energy electron and a high-energy muon and nothing else, i.e. attributable to

$$e^+ + e^- \to \tau^+ + \tau^-$$
$$\hookrightarrow \mu^- + \bar{v}_\mu + v_\tau$$
$$\hookrightarrow e^+ + v_e + \bar{v}_\tau. \tag{6.100}$$

The mass of the τ-lepton is 1.78 GeV, so that the threshold for (6.100), 3.56 GeV, is rather close to that for charmed-meson production, 3.74 GeV, and the disentangling of this and other hadronic background effects constituted a superb piece of scientific detective work. There is not space here to describe the analysis of this and subsequent confirmatory experiments. Figure 6.28 shows just one example of an analysis which indicates spin $\frac{1}{2}$ for the τ. The lepton momentum spectrum indicates a *V-A*-type coupling, as in the leptonic weak interactions: the relative leptonic branching ratios (i.e. to electrons and muons) are consistent with e/μ universality and the same (Fermi) coupling as describes μ-decay; and the semileptonic branching ratios are in agreement with the theory. For example, $\tau \to \pi v_\tau$ can be related to $\pi \to \mu v_\mu$, and $\tau \to \rho v_\tau$ to $e^+ e^- \to \rho$.

The mass value for the τ, and an upper limit to that for v_τ, are included in Table 6.5.

Our discussion of leptons thus far has been conducted almost exclusively under the assumption that neutrino masses are identically zero, so that the distinction between neutrinos and antineutrinos—apart from their mode of production, for example $\pi^+ \to \mu^+ v_\mu$, $\pi^- \to \mu^- \bar{v}_\mu$—can be based on their helicity: neutrinos are left-handed ($H = -1$) and antineutrinos right-handed ($H = +1$), as in (6.26). This is a relativistically invariant description, just as for right and left circularly polarized light. If the neutrino mass were finite, this would not be true: one could always find a reference frame traveling faster than the neutrino, from which the sign of the helicity would be reversed.

The assumption that neutrinos are all massless and occur in several flavors is perhaps somewhat surprising, and is actively questioned at the present time.

It has been necessary to invoke a quantum number (L_e, L_μ, L_τ) or internal degree of freedom to distinguish the three types, and such internal structure might be expected to manifest itself in a difference in masses, as it does for the charged leptons. One experiment (Table 6.1) does in fact suggest a finite mass for v_e. There are, in addition, hints from astrophysics that massless neutrinos might

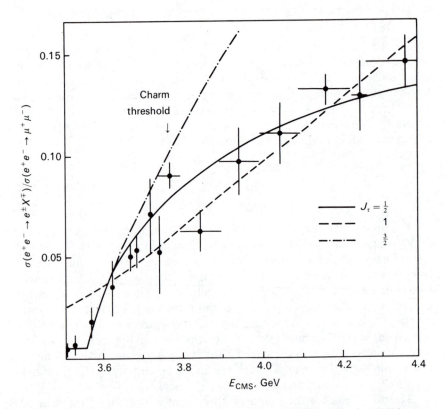

Fig. 6.28 Analysis of τ-lepton production by DELCO collaboration at SLAC e^+e^- collider SPEAR (Bacino *et al.* 1978). Pair production of τ-leptons is recognized as events consisting of one electron track, and one other track due to a muon or hadron, of opposite sign of charge. The data clearly favor spin $\frac{1}{2}$ for the τ.

TABLE 6.5   Lepton masses

	Electron family (e, v_e)	Muon family (μ, v_μ)	Tau family (τ, v_τ)
Charged lepton (e, μ, τ)	0.511 MeV	105.66 MeV	1784 ± 4 MeV
Neutral lepton (v_e, v_μ, v_τ)	$\leqslant 30$ eV	< 0.5 MeV	< 250 MeV

lead to difficulties in our understanding of the expanding universe. Basically, endowing neutrinos with mass leads to dramatic differences in the gravitational potential energy, and would correct the apparent mismatch between the motional energy of the galaxies and their gravitational energy, the former being an order of magnitude larger than the latter. If one or more of the masses are in fact of the order of electron volts, as the astrophysicists suggest, then laboratory experiments may settle this matter in the near future, at least indirectly. There seems little hope of dramatically improving the present mass limits on v_μ and v_τ by direct measurement.

If lepton number is not absolutely conserved (and there are no compelling reasons for believing it should be) and neutrinos have finite masses, then mixing may occur between the different types of neutrino (v_e, v_μ, v_τ). This possibility was first envisaged many years ago (Maki *et al.* 1962, Pontecorvo 1968). The weak-interaction eigenstates v_e, v_μ, v_τ are expressed as combinations of mass eigen-states, say v_1, v_2, v_3, which propagate with slightly different frequencies due to their mass differences. As a result, if one were to start off with a pure v_e-beam, for example, oscillations would occur and at subsequent times one would have admixtures of v_e with v_μ, v_τ. In order to simplify the treatment we shall consider the case of two types of neutrino, say v_e and v_μ. Each will be a linear combination of the two mass eigenstates, say v_1 and v_2, as given by the unitary transformation involving an arbitrary mixing angle θ:

$$\begin{pmatrix} v_\mu \\ v_e \end{pmatrix} = \begin{pmatrix} \cos\theta & \sin\theta \\ -\sin\theta & \cos\theta \end{pmatrix} \begin{pmatrix} v_1 \\ v_2 \end{pmatrix}, \tag{6.101}$$

so that the wavefunctions

$$v_\mu = v_1 \cos\theta + v_2 \sin\theta \quad \text{and} \quad v_e = -v_1 \sin\theta + v_2 \cos\theta$$

are orthonormal states. The states v_μ and v_e are those produced in a weak decay process, for example $\pi \to \mu + v_\mu$. However, propagation in space-time is determined by the characteristic frequencies of the mass eigenstates,

$$v_1(t) = v_1(0)e^{-iE_1 t},$$

$$v_2(t) = v_2(0)e^{-iE_2 t}, \tag{6.102}$$

where we set $\hbar = c = 1$. Since we have to consider spatially coherent states, $v_1(t)$ and $v_2(t)$ must have the same momentum p. Then, if the mass $m_i \ll E_i$ ($i = 1, 2$),

$$E_i = p + \frac{m_i^2}{2p}. \tag{6.103}$$

Suppose we start at $t = 0$ with muon-type neutrinos, so $v_\mu(0) = 1$ and $v_e(0) = 0$. Then from (6.101) we find

$$v_2(0) = v_\mu(0) \sin\theta,$$

$$v_1(0) = v_\mu(0) \cos\theta, \tag{6.104}$$

and

$$v_\mu(t) = \cos\theta\, v_1(t) + \sin\theta\, v_2(t).$$

Using (6.102) and (6.104), one then obtains

$$\frac{v_\mu(t)}{v_\mu(0)} = \cos^2\theta\, e^{-iE_1 t} + \sin^2\theta\, e^{-iE_2 t},$$

and the intensity is

$$\frac{I_\mu(t)}{I_\mu(0)} = \left|\frac{v_\mu(t)}{v_\mu(0)}\right|^2 = \cos^4\theta + \sin^4\theta + \sin^2\theta\cos^2\theta\,[e^{i(E_2-E_1)t} + e^{-i(E_2-E_1)t}]$$

$$= 1 - \sin^2 2\theta \sin^2\left[\frac{(E_2-E_1)t}{2}\right].$$

Writing $\Delta m^2 = m_2^2 - m_1^2$ and with the help of (6.103), this takes the following form for the probability of finding v_μ or v_e after time t:

$$P(v_\mu \to v_\mu) = 1 - \sin^2 2\theta \sin^2\left(\frac{1.27\,\Delta m^2\, L}{E}\right),$$

$$P(v_\mu \to v_e) = 1 - P(v_\mu \to v_\mu). \qquad (6.105)$$

The numerical constant, 1.27, applies if Δm^2 is expressed in $(eV/c^2)^2$, while L, the distance from the source, is expressed in metres, and E, the beam energy, is in MeV. Equation (6.105) shows that the intensities of v_μ and v_e oscillate as a function of distance from the source. For example, at a reactor source (of antineutrinos), $E \sim 1$ MeV, and for $\Delta m \sim 1\ eV/c^2$ the oscillation length will be a few metres.

The subject of neutrino oscillations acquired considerable impetus as a result of the so-called solar neutrino problem: The rate of detection of solar neutrinos in the reaction $v_e + {}^{37}\mathrm{Cl} \to {}^{37}\mathrm{Ar} + e^-$ is found to be about a factor 3 smaller than expected (Davis *et al.* 1968). However, there are uncertainties in the solar model calculation and the discrepancy is hardly evidence for neutrino oscillations (although mixing v_e with v_μ and v_τ would clearly give just such a factor). There is however at present (1981) no convincing evidence for oscillation phenomena from laboratory experiments using reactors or accelerators as the neutrino sources.

PROBLEMS

6.1 The ground states of ${}^{34}_{17}\mathrm{Cl}$ and ${}^{34}_{16}\mathrm{S}$ have $J^P = 0^+$ and belong to an $I = 0$ multiplet. The decay ${}^{34}_{17}\mathrm{Cl} \to {}^{34}_{16}\mathrm{S} + \beta^+ + v$ has a mean lifetime $\tau = 2.3$ sec and a maximum positron energy of 4.5 MeV. The lifetime of the pion is 26 ns. Estimate the branching ratio for the decay $\pi^+ \to \pi^0 e^+ v$.

6.2 A "narrow-band" neutrino beam (see Fig. 6.15) is produced by bombarding a Be target with 400-GeV protons, and forming a pencil secondary beam with a small spread in momentum centered at 200 GeV/c. This beam contains charged pions and kaons of one sign of charge and traverses an evacuated decay tunnel of length 300 m, where a fraction of them decay to muons and neutrinos. This is followed by an absorber of 200 m steel and 150 m rock. A cylindrical detector of radius 2 m is placed 400 m beyond the end of the decay tunnel, and aligned with the beam axis.

(a) Find a relation between the laboratory energy of a neutrino and its angle relative to the beam axis, for neutrinos from pion and from kaon decays.

(b) What are the maximum and minimum energies of neutrinos produced in the two cases?

(c) Above what neutrino energy do all neutrinos from kaon decay pass through the detector?

(d) If 10^{10} pions per burst enter the decay tunnel, how many neutrinos from pion decay traverse the detector? (The divergence of the pion beam may be neglected.)

(e) If the detector has mass 100 tons, how many neutrinos from pion decay interact in it per burst, if the cross-section per nucleon at energy E is $0.6E \times 10^{-38}$ cm^2 with E in GeV?

(f) Why is 350 m of absorber necessary? Refer to Problem 2.1 for estimates of straggling in range.

6.3 The neutron has a mean lifetime $\tau_n = 930$ sec, and the muon $\tau_\mu = 2.2 \times 10^{-6}$ sec. Show that the couplings involved in the two cases are of the same order of magnitude, when account is taken of the phase-space factors.

6.4 Obtain an estimate of the branching ratio $(\Sigma^- \to \Lambda e^- \bar{\nu})/(\Sigma^- \to n\pi^-)$, assuming that the matrix element for the electron decay is the same as that of the neutron, and that baryon recoil may be neglected. (See Table IV of Appendix for neutron and Σ-decay data.)

6.5 Strangeness-changing decays do not conserve isospin but appear to obey the $\Delta I = \frac{1}{2}$ rule. Use this rule to compute the ratios

(a) $K_S \to \pi^+\pi^-/K_S \to \pi^0\pi^0$,

(b) $\Xi^- \to \Lambda\pi^-/\Xi^0 \to \Lambda\pi^0$.

Compare your answers with the experimental values in Table IV. [*Hint*: the $\Delta I = \frac{1}{2}$ rule may be applied by postulating that a hypothetical particle of $I = \frac{1}{2}$ — called a "spurion" — is added to the left-hand side in a weak decay process, and then treating the decay as an isospin-conserving reaction.]

6.6 What is the expected ratio $K_L \to 2\pi^0/K_L \to \pi^+\pi^-$ if the pions are in (a) an $I = 0$ state ($\Delta I = \frac{1}{2}$ rule), or (b) an $I = 2$ state ($\Delta I = \frac{3}{2}$ or $\frac{5}{2}$)?

6.7 Using the $\Delta I = \frac{1}{2}$ rule, find a relation between the amplitudes for

$$\Sigma^+ \to n\pi^+, \qquad \text{say } a_+,$$

$$\Sigma^- \to n\pi^-, \qquad \text{say } a_-,$$

$$\Sigma^+ \to p\pi^0, \qquad \text{say } a_0.$$

You will need to introduce $I = \frac{3}{2}$ and $I = \frac{1}{2}$ amplitudes. The triangle relation you should obtain is

$$a_+ + \sqrt{2}a_0 = a_-.$$

Experimentally, $|a_+| \simeq |a_0| \simeq |a_-|$, so that a right-angled triangle results.

6.8 Given that the width of the $\Delta(1234)$ πp resonance is 150 MeV, estimate the branching ratio for β-decay,

$$\frac{\Delta^{++} \to pe^+ v}{\Delta^{++} \to p\pi^+}.$$

6.9 How can one produce experimentally a beam of pure, monoenergetic K^0-particles? If a short pulse of such particles travels through a vacuum, compute the intensity of K^0 and $\bar{K}^0$ as a function of proper time, assuming the mass difference Δm equal to (a) $0.5/\tau_1$, (b) $2/\tau_1$. Display the phenomena on a graph.

6.10 An experiment in a gold mine in South Dakota has been carried out to detect solar neutrinos, using the reaction

$$v + {}^{37}\text{Cl} \to {}^{37}\text{Ar} + e^-.$$

The detector contained approximately 4×10^5 liters of tetrachlorethylene (C_2Cl_4). Estimate how many atoms of ^{37}Ar per day would be produced, making the following assumptions:

(a) solar constant $= 2 \text{ cal cm}^{-2} \text{min}^{-1}$;
(b) 10% of thermonuclear energy of sun appears in neutrinos, of mean energy 1 MeV;
(c) 1% of all neutrinos are energetic enough to induce the above reaction;
(d) cross-section per ^{37}Cl nucleus for "active" neutrinos is 10^{-45} cm^2;
(e) ^{37}Cl isotopic abundance is 25%;
(f) density of C_2Cl_4 is 1.5 g ml^{-1}.

Do you expect any difference between day rate and night rate? [For description of the experiment, see R. Davis, D. S. Harmer, and K. C. Hoffman, *Phys. Rev. Lett.* **20**, 1205 (1968).]

6.11 Energetic neutrinos may produce single pions in the following reactions on protons and neutrons:

$$v + p \to \pi^+ + p + \mu^-, \tag{i}$$

$$v + n \to \begin{Bmatrix} \pi^0 + p \\ \pi^+ + n \end{Bmatrix} + \mu^-. \tag{ii}$$

Assume the process is dominated by the first pion-nucleon resonance $\Delta(1234)$, so that the π-nucleon system has $I = \frac{3}{2}$ only. As in weak decay processes of $\Delta S = 0$, the isospin of the hadronic state then changes by one unit ($\Delta I = 1$ rule). Show that this rule predicts a rate for (i) 3 times that of (ii). Also show that, on the contrary, for a $\Delta I = 2$ transition (for which there is no present evidence), the rate for (ii) would be 3 times that for (i).

6.12 Use the $\Delta I = \frac{1}{2}$ rule to demonstrate the following relations between the rate of three-pion decays of charged and neutral kaons:

$$\Gamma(K_L \to 3\pi^0) = \tfrac{3}{2}\Gamma(K_L \to \pi^+ \pi^- \pi^0),$$

$$\Gamma(K^+ \to \pi^+ \pi^+ \pi^-) = 4\Gamma(K^+ \to \pi^+ \pi^0 \pi^0),$$

$$\Gamma(K_L \to \pi^+ \pi^- \pi^0) = 2\Gamma(K^+ \to \pi^+ \pi^0 \pi^0).$$

[*Comment*: These results depend on the (reasonable) assumption that the three pions are in a relative S-state. Then, any pair of pions must be in a symmetric isospin state, i.e., $I = 0$ and/or 2. The third pion ($I = 1$) must then be combined with the pair to yield a three-pion state of $I = 1, I_3 = 1$ for K^+, or $I = 1, I_3 = 0$ for K^0 (from the $\Delta I = \frac{1}{2}$ rule). It is necessary to write the three-pion wave function in a manner which is, as required for identical bosons, completely symmetric under pion label interchange. Thus, the $\pi^+\pi^+\pi^-$ state must be written

$$(+ + -) = \frac{1}{\sqrt{6}}(\pi_1^+\pi_2^+\pi_3^- + \pi_2^+\pi_1^+\pi_3^- + \pi_3^-\pi_2^+\pi_1^+ + \pi_3^-\pi_1^+\pi_2^+ + \pi_2^+\pi_3^-\pi_1^+ + \pi_1^+\pi_3^-\pi_2^+).$$

Each term in this expression will be the product of an isospin state of $I = 0$ or 2 for the first two pions, and one of $I = 1$ for the third pion, with appropriate Clebsch-Gordan coefficients.]

6.13 Show why observation of the process $\bar{\nu}_\mu + e^- \to e^- + \bar{\nu}_\mu$ (Fig. 6.19) constitutes unique evidence for neutral currents, whereas $\bar{\nu}_e + e^- \to e^- + \bar{\nu}_e$ does not. Prove that the maximum angle of emission of the recoil electron relative to the beam is $\sqrt{2m/E}$ where m, E are the electron mass and energy.

BIBLIOGRAPHY

Commins, E. D., *Weak Interactions*, McGraw-Hill, New York, 1973.

Feinberg, G., and L. Lederman, "Physics of muons and muon neutrinos", *Ann. Rev. Nucl. Science* **13**, 431 (1963).

Gasiorowicz, S., *Elementary Particle Physics*, John Wiley, New York, 1966.

Hung, P. Q., and C. Quigg, "Intermediate bosons; weak interaction carriers", *Science* **210**, 1205 (1980).

Kleinknecht, K., "*CP* violation and K^0 decays", *Ann. Rev. Nucl. Science* **26**, 1 (1976).

Konopinski, E. J., "Experimental clarification of the laws of β-radioactivity", *Ann. Rev. Nucl. Science* **9**, 99 (1959).

Lederman, L., "Neutrino physics", in Burhop (ed.), *Pure and Applied Physics*, Academic Press, New York, 1967, Vol. 25-II.

Lee, T. D., and C. S. Wu, "Weak interactions", *Ann. Rev. Nucl. Science* **15**, 381 (1965).

Okun, L. B., *Weak Interactions of Elementary Particles*, Pergamon, London, 1965.

Perl, M. L., and W. T. Kirk, "Heavy leptons", *Sci. Am.* **238**, 50 (Mar. 1978).

Reines, F., "Neutrino interactions", *Ann. Rev. Nucl. Science* **10**, 1 (1960).

Chapter 7

Lepton-Quark Interactions: The Parton Model of Hadrons

In our discussions thus far, the evidence for quarks has been largely circumstantial: the quark hypothesis is the simplest one to account for the observed regularities among the hadron multiplets such as the mass relations, magnetic moments, and other static properties of hadrons. It has been outstandingly successful in accounting for the energy levels of the ψ and Υ series of heavy meson states. But the evidence for quarks is not completely overwhelming: one might still argue that they are mathematical fictions, a convenient shorthand to describe the symmetries among the hadrons. Indeed, in the early 1960s, the emphasis was placed on such unitary symmetry, rather than the actual physical quark constituents.

A dynamical (rather than static) understanding of quark substructure had its origins in 1968, when new evidence for quarks started to come from experiments on deep-inelastic lepton-nucleon scattering. These showed firstly that the complicated process of leptoproduction of many hadrons in such a collision could be simply interpreted as (quasi) elastic scattering of the lepton by a pointlike, or nearly pointlike, constituent or *parton*, later to be identified with the quark; secondly that the analysis of precise and detailed scattering experiments could give valuable information on the quark-quark interactions. This evidence was strongly reinforced from the results of studies of e^+e^- annihilation to hadrons at high energy, and of the production of lepton pairs in hadron-hadron collisions.

7.1 NEUTRINO-NUCLEON TOTAL CROSS-SECTIONS: THE EVIDENCE FOR PARTONS

Perhaps the most dramatic demonstration of the constituent nature of hadrons is provided by the total cross-section, as a function of energy, for neutrino-nucleon scattering. Figure 7.1 shows an example of the interaction of a 200-GeV neutrino with a proton or neutron in the liquid neon-hydrogen mixture in a large bubble chamber,

$$\nu_\mu + N \to \mu^- + \text{hadrons}, \tag{7.1}$$

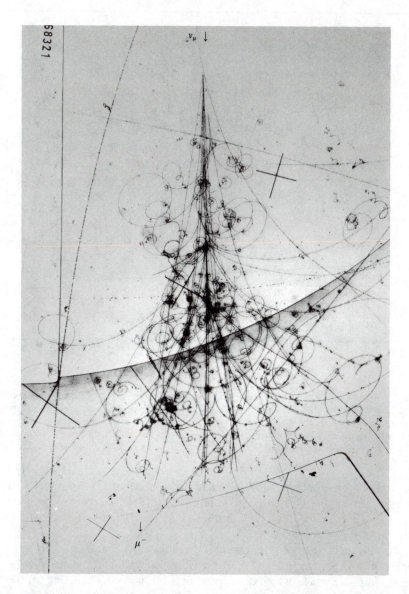

Fig. 7.1 Neutrino-nucleon collision in the BEBC chamber filled with Ne-H$_2$ mixture, and exposed to the 200-GeV narrow-band neutrino beam from the CERN SPS (see Fig. 6.15). An incident muon-neutrino (ν_μ) transforms to a 100-GeV negative muon, appearing as a very straight track at an angle to the general "jet" of charged and neutral hadrons. The muon is identified by its penetration through iron and recorded in external proportional chambers (see Fig. 2.12). The hadrons produced carry off the remaining (100 GeV) energy. γ-rays from neutral pion decay convert to pairs in the liquid, which point back to the main vertex. The development of electromagnetic cascades and secondary interactions of the hadrons are visible.

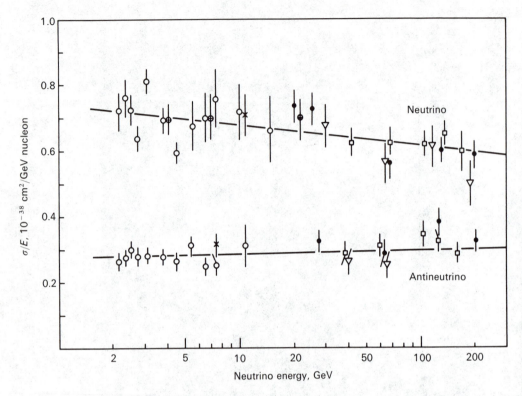

Fig. 7.2 Total neutrino and antineutrino cross-sections on nucleons as a function of energy, from experiments at CERN, Fermilab, and Serpukhov. The ratio of cross-section per nucleon to incident energy is practically constant over two orders of magnitude, and is a direct demonstration of pointlike constituents (partons) inside the nucleon. The gentle dependence of σ/E on energy is predicted in the field theory of interacting quarks (QCD) described in Chapter 8.

where about 10 secondary hadrons, in addition to the muon, are produced. The event looks very complicated, but if we just measure the total cross-section σ as a function of energy, the result is a simple one – an almost linear rise of σ with the neutrino energy (see Fig. 7.2).

This is exactly the result we expect if we replace the complicated process of hadron production, as in Fig. 7.1, by the *elastic* scattering of the neutrino by a *single pointlike particle*, which can depend only on the Fermi coupling constant G and the phase space. Indeed, if the neutrino energy E is large compared with any of the masses involved, we expect, as in (6.20)

$$\sigma \sim G^2 p^2 \sim G^2 E, \tag{7.2}$$

where p is the momentum of the two particles in the CMS, and a simple calculation gives $p^2 = mE/2$, where m is the mass of the pointlike target. So the linear dependence of σ on E just tells us that the two particles undergo a contact interaction with a cross-section rising like phase space. In this discussion, we have neglected any possible "spread" of the weak interaction due to effects of a propagator term associated with the $W^{\pm}$ boson, which would multiply the cross-section by $(q^2 + M_W^2)^{-2}$, where q ($\leqslant 2p$) is the momentum transfer in the collision. This is permissible because we believe that $M_W \sim 80$ GeV and $E^2 \ll M_W^2$ over the range in Fig. 7.2. Indeed, the linearity of the cross-section with energy up to $E = 200$ GeV can be used to set a lower limit $M_W > 30$ GeV.

This result can be qualitatively understood on the basis of the *parton model* of Feynman (1969), which was actually developed to describe hadron-hadron collisions. In this model, the complicated process of inelastic scattering is interpreted very simply as the *elastic* scattering of the lepton by a pointlike, quasifree constituent of the nucleon, or parton, described in detail below. The miraculous success of the parton description naturally leads to two questions: what is the nature of the partons (the answer to which turns out to be that some of them, but not all, are quarks); and why does such a simple model work so well? Surely it cannot be right to treat a nucleon as composed of a cluster of essentially free constituents, when we know they must be very tightly bound together, despite our most determined attempts to free the quarks? It is indeed this very success of the parton model, and more importantly the small deviations from it, which have given us essential clues on the nature of quark-quark interactions. To proceed further, on a quantitative basis, we must first introduce some of the formalism of lepton-nucleon scattering, starting off with the simplest example, that of elastic scattering of electrons by atomic nuclei.

7.2 ELECTRON-NUCLEUS SCATTERING

7.2.1 Elastic Scattering of Spinless Electrons by Nuclei

The Rutherford scattering of "spinless" electrons by nuclei can be derived from first-order perturbation theory (and also classically). We employ (1.27) for the transition probability:

$$W = \frac{2\pi}{h}|M_{if}|^2 \rho_f. \qquad (7.3)$$

For a perturbing central potential $V(r)$ provided by a stationary nucleus Ze, the matrix element becomes the volume integral

$$M_{if} = \int \psi_f^* V(r) \psi_i \, d\tau, \qquad (7.4)$$

where ψ_i, ψ_f are the initial- and final-state wave functions of the scattered electron. The Born approximation assumes the perturbation to be weak, so that only single scattering is considered (this can be shown to be equivalent to the requirement $Z < 137$). We can therefore represent ψ_i and ψ_f as plane waves, before and after scattering. Writing $\mathbf{k}_0$ and $\mathbf{k}$ for the initial and final propagation vectors, we obtain from (7.4)

$$M_{if} = \int e^{i(\mathbf{k}_0 - \mathbf{k})\cdot\mathbf{r}} V(\mathbf{r}) \, d^3\mathbf{r}. \tag{7.5}$$

As in Eq. (1.26), the differential scattering cross section is W/v, where v is the velocity of the incident beam relative to the scattering center. Setting the density-of-states factor

$$\rho_f = \frac{p^2 \, d\Omega}{h^3} \frac{dp}{dE_f},$$

where $\mathbf{p} = \hbar\mathbf{k}$ is the momentum of the scattered electron and E_f is the total energy in the final state, gives us

$$\frac{d\sigma}{d\Omega} = \frac{1}{(2\pi)^2\hbar^4} \frac{p^2}{v} \frac{dp}{dE_f} |M_{if}|^2. \tag{7.6}$$

So far, we have assumed the nucleus to be infinitely massive. In practice we have to consider the nuclear recoil. Let $\mathbf{p}'$, W, and M denote the momentum, total energy, and rest mass of the recoiling nucleus, and θ the angular deflection of the electron (see Fig. 7.4). Using units $\hbar = c = 1$, and assuming both incident and scattered electrons are extreme relativistic, we have

$$p_0 = k_0 = E_0, \qquad p = k = E, \qquad v \simeq 1.$$

All quantities refer to the laboratory system. Applying energy-momentum conservation

$$E_i = p_0 + M = E_f = p + W, \qquad \mathbf{p}_0 = \mathbf{p} + \mathbf{p}',$$

one finds

$$E_f = p + \sqrt{p'^2 + M^2} = p + \sqrt{p_0^2 + p^2 - 2pp_0 \cos\theta + M^2},$$

$$E_f - p_0 \cos\theta = Mp_0/p,$$

$$\frac{dp}{dE_f} = \frac{W}{E_f - p_0 \cos\theta} = \frac{W}{M} \frac{p}{p_0}, \qquad W \gg p_0$$

and

$$\frac{p}{p_0} = \frac{1}{1 + \dfrac{p_0}{M}(1 - \cos\theta)}. \tag{7.7}$$

Equation (7.6) then becomes

$$\frac{d\sigma}{d\Omega} = \frac{1}{4\pi^2} p^2 \frac{W}{M} \frac{p}{p_0} \left| \int e^{i\mathbf{q}\cdot\mathbf{r}} V(\mathbf{r}) \, d^3\mathbf{r} \right|^2, \tag{7.8}$$

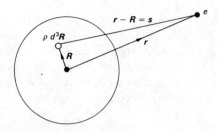

Fig. 7.3

where the momentum transfer

$$\mathbf{q} = \mathbf{p}_0 - \mathbf{p}.$$

Now let us represent the nucleus by a sphere of charge density $\rho(\mathbf{R})$ as shown in Fig. 7.3, normalized so that

$$\int_0^\infty \rho(\mathbf{R})\, d^3\mathbf{R} = 1.$$

Then

$$V(\mathbf{r}) = \left(Ze^2/4\pi\right) \int \frac{\rho(\mathbf{R})\, d^3\mathbf{R}}{|\mathbf{r}-\mathbf{R}|}, \tag{7.9}$$

and

$$M_{if} = \left(Ze^2/4\pi\right) \int\int \frac{\rho(\mathbf{R})e^{i\mathbf{q}\cdot\mathbf{R}}\, d^3\mathbf{R}\, e^{i\mathbf{q}\cdot(\mathbf{r}-\mathbf{R})}\, d^3\mathbf{r}}{|\mathbf{r}-\mathbf{R}|}$$

$$= \left(Ze^2/4\pi\right) \int \rho(\mathbf{R})e^{i\mathbf{q}\cdot\mathbf{R}}\, d^3\mathbf{R} \int \frac{e^{iqs\cos\alpha}2\pi s^2\, ds\, d(\cos\alpha)}{s}, \tag{7.10}$$

where $\mathbf{s} = \mathbf{r} - \mathbf{R}$ and α is the polar angle between $\mathbf{s}$ and $\mathbf{q}$, the momentum-transfer vector. We define the nuclear form factor by

$$F(q^2) = \int \rho(\mathbf{R})e^{i\mathbf{q}\cdot\mathbf{R}}\, d^3\mathbf{R}. \tag{7.11}$$

Then

$$M_{if} = \left(Ze^2/2\right) F(q^2) \int s\, ds \int e^{iqs\cos\alpha}\, d(\cos\alpha)$$

$$= \left(Ze^2/2\right) F(q^2) \int \frac{s\, ds\left(e^{iqs} - e^{-iqs}\right)}{iqs}. \tag{7.12}$$

This integral unfortunately diverges — as indeed it should, since the cross-section for scattering of two particles via an inverse-square-law field is infinite. In fact, the nucleus is of course *screened* at large distances by atomic electrons. The trick is therefore to modify $V(\mathbf{r})$ by a factor $e^{-r/a}$, where a represents a typical atomic radius; afterwards we let $a \to \infty$ if we wish. Since $a \gg R$ (by a factor $\sim 10^4$), we can set $e^{-r/a} = e^{-s/a}$. Then (7.12) becomes

$$
\begin{aligned}
M_{if} &= \frac{(Ze^2/2)F(q^2)}{iq}\left\{ \int e^{-s(1/a - iq)}\,ds - \int e^{-s(1/a + iq)}\,ds \right\} \\
&= \frac{(Ze^2/2)F(q^2)}{iq}\left\{ \frac{1}{(1/a) - iq} - \frac{1}{(1/a) + iq} \right\} \\
&= \frac{(Ze^2/2)F(q^2)}{q^2 + (1/a)^2}.
\end{aligned}
\tag{7.13}
$$

Note that if $q < 1/a$ (i.e. very small momentum transfer), $M^2 \propto d\sigma/d\Omega \to$ constant. Now $a \sim 10^{-8}$ cm and thus $1/a \sim 1$ keV only; for the region of interest, $q \gg 1/a$ and we obtain for the differential cross section

$$
\frac{d\sigma}{d\Omega} = \frac{4Z^2(e^2/4\pi)^2}{q^4}\, p^2\, \frac{W}{M}\, \frac{p}{p_0}\, \left[F(q^2)\right]^2.
\tag{7.14}
$$

For reasons that will appear later, $(W/M) - 1 = q^2/2M^2 \ll 1$, so we can set $W/M = 1$ in all practical cases — the recoiling nucleus is nonrelativistic. Under the further assumption that the nuclear recoil momentum, $p' = q \ll p_0$, we can set $p = p_0$ and

$$
q^2 = 2p_0^2 - 2p_0^2 \cos\theta = 4p_0^2 \sin^2 \tfrac{1}{2}\theta,
\tag{7.15}
$$

so that

$$
\frac{d\sigma}{d\Omega} = \frac{Z^2(e^2/4\pi)^2\left[F(q^2)\right]^2}{4p_0^2 \sin^4 \tfrac{1}{2}\theta}.
\tag{7.16}
$$

Note that for a *point* nucleus, $F(q^2) = 1$ for all q^2; (7.16) then gives the *Rutherford scattering formula*.

7.2.2   Four-Momentum Transfer: Mott Scattering

Equation (7.16) is expressed in terms of the 3-momentum transfer $\mathbf{q}$ in the laboratory system, between incident and target particle. In order to express the scattering in a form which is independent of the reference frame, it is better to consider q as the 4-*momentum transfer*. In Fig. 7.4, P_0, P, and Q represent the 4-momenta of the particles involved. Each 4-momentum has three space and one time component.

The square of the 4-momentum vector (of a real particle) is given by

$$
P^2 = \mathbf{p}^2 + (iE)^2 = p^2 - E^2 = -m^2,
$$

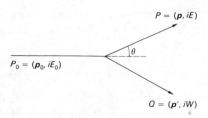

Fig. 7.4

and is thus invariant (see Appendix A). Similarly, the 4-momentum transfer squared, between the incident and emergent electron, is invariant, with a value

$$q^2 = (\mathbf{p}_0 - \mathbf{p})^2 - (E_0 - E)^2 = -2m^2 - 2pp_0 \cos\theta + 2EE_0$$
$$= 2pp_0(1 - \cos\theta) = 4pp_0 \sin^2\tfrac{1}{2}\theta, \tag{7.17}$$

if we neglect the electron mass ($m^2 \ll q^2$). Since the scattering angle is a real quantity ($-1 < \cos\theta < 1$), q^2 is positive. For exchange of a real particle, q^2 would be negative, i.e. with the same sign as the energy or time component $(iE)^2$. A value $q^2 > 0$ is sometimes referred to as a *spacelike* momentum transfer, to distinguish it from $q^2 < 0$, which is called *timelike*. All scattering processes necessarily refer to the spacelike region of q^2.

An alternative expression for q^2 is obtained by considering the transfer to the nucleus:

$$q^2 = (-\mathbf{p}')^2 - (M - W)^2 = -2M^2 + 2MW = 2MT, \tag{7.18}$$

where the kinetic energy acquired by the nucleus is $T = W - M$. Thus

$$\frac{W}{M} = 1 + \frac{q^2}{2M^2}. \tag{7.19}$$

If the nucleus is to recoil coherently, it turns out that $q^2 \ll 2M^2$, so that $W/M \simeq 1$ in a practical case, as assumed in (7.16).

Although the values of q^2 in (7.17) and (7.18) refer to quantities measured in the laboratory frame, its numerical value is the same in all reference frames. The result of evaluating $|M_{if}|^2$ is as before, except that now, q being a 4-momentum transfer, (7.11) is strictly an integral over space-time, with qR as a scalar product of 4-vectors. However, if $W/M \simeq 1$, the energy transfer (time component) is small, and we can still interpret (7.11) as the integral over a spatial charge distribution.

If we describe electrons using Dirac (4-component) wave functions, i.e. we incorporate spin, then an extra term $\cos^2\tfrac{1}{2}\theta$ appears in the cross-section. (This arises essentially because, as indicated by the Dirac theory, relativistic electrons are longitudinally aligned; a $180°$ scattering would then involve a flipover of the electron spin, which is forbidden by angular-momentum conservation along the beam axis.) Inserting the expression (7.7) for p/p_0, we obtain the Mott formula for the scattering of relativistic spin-$\frac{1}{2}$ electrons by spinless pointlike nuclei:

$$\left(\frac{d\sigma}{d\Omega}\right)_{\text{Mott}} = \frac{Z^2\left(e^2/4\pi\right)^2\cos^2\frac{1}{2}\theta}{4p_0^2\sin^4\frac{1}{2}\theta\left[1+(2p_0/M)\sin^2\frac{1}{2}\theta\right]}. \tag{7.20}$$

Note that for small scattering angles this is identical with the Rutherford formula. The final term in the denominator allows for the nuclear recoil. It is important only for high-energy electrons and large scattering angles.

7.2.3 Nuclear Form Factor

In order to simplify the physics picture we shall revert to the case where the 4-momentum transfer is given essentially by the 3-momentum transfer, i.e., the energy transfer is small. The effect of the finite nuclear size is to introduce the term

$$F(q^2) = \int \rho(\mathbf{R})e^{i\mathbf{q}\cdot\mathbf{R}}\,d^3\mathbf{R}, \tag{7.21}$$

so that

$$\frac{d\sigma}{d\Omega} = \left(\frac{d\sigma}{d\Omega}\right)_{\text{Mott}}|F(q^2)|^2. \tag{7.22}$$

Equation (7.21) is simply the Fourier transform of the nuclear charge-density distribution $\rho(\mathbf{R})$, and $F(q^2)$ is termed the nuclear form factor. By integrating (7.21) over angles the reader should prove that

$$F(q^2) = \int \rho(R)\frac{\sin qR}{qR}4\pi R^2\,dR. \tag{7.23}$$

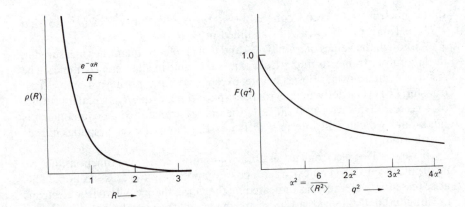

Fig. 7.5 Yukawa-type charge distribution, and corresponding form factor.

The interpretation of $F(q^2)$ as a (three-dimensional) Fourier transform of a charge distribution in space is strictly valid *only* when q is essentially equal to the 3-momentum transfer.

As an example, let us calculate $F(q^2)$ for a Yukawa-type charge distribution (Fig. 7.5):

$$\rho(R) = \rho_0 e^{-\alpha R}/R. \tag{7.24}$$

Then

$$F(q^2) = 4\pi\rho_0 \int \frac{e^{-\alpha R}}{2iqR^2}(e^{iqR} - e^{-iqR})R^2\, dR$$

$$= 4\pi\rho_0 \frac{1}{\alpha^2 + q^2}.$$

From normalization, one finds $\alpha^2 = 4\pi\rho_0$; hence

$$F(q^2) = \frac{1}{1 + \dfrac{q^2}{\alpha^2}}. \tag{7.25}$$

The rms radius of the charge distribution is given by

$$\langle R^2\rangle = \frac{\int \rho(R)R^2 \cdot R^2\, dR}{\int \rho(R)R^2\, dR} = \frac{6}{\alpha^2}.$$

Hence

$$F(q^2) = \frac{1}{1 + \dfrac{q^2\langle R^2\rangle}{6}}. \tag{7.26}$$

For small values of $q^2\langle R^2\rangle$, *all* form factors reduce to the expression

$$F(q^2) = 1 - \frac{q^2\langle R^2\rangle}{6} + \cdots. \tag{7.27}$$

This may be demonstrated by expanding the exponential in (7.21) or the sine in (7.23) and retaining the first two terms.

At this point, we may also note that, just as in (7.24), a Yukawa potential of the form (1.3)

$$U(R) = \frac{g}{R}e^{-\alpha R}$$

will result in a q^2-dependence for a scattering process of the form given in (1.5):

$$f(q^2) = \frac{1}{q^2 + m^2}, \tag{7.28}$$

where $m = \hbar\alpha/c$ is the mass of the boson associated with the short-range field.

7.2.4   Experiments on Nuclear Form Factors

Electron scattering experiments at 400 to 600 MeV have been employed to obtain fairly detailed charge distributions on many nuclei. From (7.23) $\rho(R)$ may be obtained from the inverse Fourier transform:

$$\rho(R) = \frac{1}{2\pi^2} \int F(q^2) \frac{\sin qR}{qR} q^2 \, dq. \tag{7.29}$$

The charge density observed in this way is found to vary slowly through the bulk of the nucleus and falls off sharply (by a factor 10) over an outer shell thickness of $\sim 2 \times 10^{-13}$ cm (see Fig. 7.6). The shape of the ρ-distribution in the outer regions (the so-called "nuclear stratosphere") is not well known. For details, the reader is referred to the review article by R. Hofstadter in the bibliography.

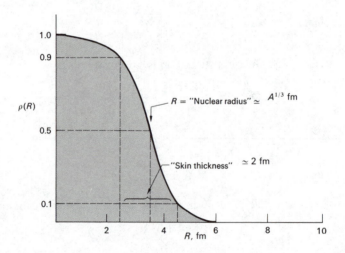

Fig. 7.6   Approximate radial variation of the charge density distribution in a medium-weight nucleus.

7.3   ELECTRON-NUCLEON ELASTIC SCATTERING

So far we have considered the scattering of electrons by spinless nuclei, in which the interaction was the electric (or Coulomb) interaction. If both projectile and target carry spin, there is in addition a spin-spin, or magnetic, interaction between the particles. This is the case in electron scattering by protons. The derivation of the cross-section formula involves the Dirac theory of relativistic electrons, and we quote only the result for an unpolarized target. Assume both electron and proton are pointlike, with magnetic moments $eh/2mc$ and $eh/2Mc$ respectively, where m, M are the electron and proton masses. Then

$$\left(\frac{d\sigma}{d\Omega}\right)_{\text{Dirac}} = \frac{\left(e^2/4\pi\right)^2}{\left(4p_0^2\sin^4\frac{1}{2}\theta\right)\left[1+(2p_0/M)\sin^2\frac{1}{2}\theta\right]}\left(\cos^2\frac{1}{2}\theta + \frac{q^2}{2M^2}\sin^2\frac{1}{2}\theta\right).$$

(7.30)

Here the factor in square brackets represents the recoil correction. This equation differs from the Mott formula in the inclusion of the second factor in parentheses, due to magnetic scattering. Large angles and high momentum transfers correspond to "close" collisions where magnetic scattering dominates (recall that the magnetic dipole potential varies as $1/r^2$, compared with $1/r$ for the electric potential).

Protons and neutrons are not pointlike, and their magnetic moments are anomalous, in the sense that they differ from the predictions for pointlike Dirac particles:

	μ (Dirac)	μ (observed)
Proton	$eh/2Mc = 1$ n.m.	$+ 2.79$ n.m.
Neutron	0	$- 1.91$ n.m.

The structure of protons and neutrons, like that of nuclei, is described empirically by suitable form factors: two form factors (one electric, one magnetic) are needed for each. In the way in which they are defined, the cross-section assumes the form

$$\frac{d\sigma}{d\Omega} = \left(\frac{d\sigma}{d\Omega}\right)_{\text{Mott}}\left\{\left(\frac{G_E^2 + \frac{q^2}{4M^2}G_M^2}{1 + \frac{q^2}{4M^2}}\right) + \frac{q^2}{4M^2}\cdot 2G_M^2\tan^2\frac{\theta}{2}\right\},$$

(7.31)

where $G_E = G_E(q^2)$, $G_M = G_M(q^2)$ with the normalization $G_E^P(0) = 1$, $G_E^N(0) = 0$, $G_M^P(0) = 2.79$, $G_M^N(0) = -1.91$. This is called the *Rosenbluth formula*. The essential feature is that

$$\left(\frac{d\sigma}{d\Omega}\right)\bigg/\left(\frac{d\sigma}{d\Omega}\right)_{\text{Mott}} = A(q^2) + B(q^2)\tan^2\frac{\theta}{2},$$

(7.32)

so that, if one plots the cross-section for different incident momenta and different scattering angles, such that q^2 remains fixed, a linear dependence on $\tan^2(\theta/2)$ should be obtained. The basic assumption on which (7.32) is derived is that of the Born approximation – that only a single collision, mediated by *single virtual photon exchange*, is involved, as depicted in Fig. 7.8. In principle, double

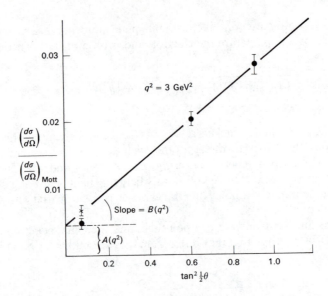

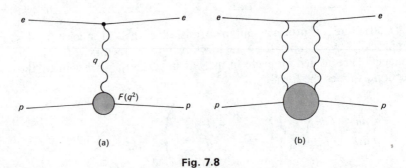

Fig. 7.7 The electron-proton scattering cross section plotted for fixed q^2 and different scattering angles θ (Rosenbluth plot). (After Weber 1967.)

Fig. 7.8

photon exchange may also occur. The equality of the cross-sections for e^+p and e^-p scattering (for which the two-photon exchange terms should have opposite signs) indicates that two-photon exchange is unimportant. This is reinforced by the observed linearity of the Rosenbluth plot (Fig. 7.7).

We may note at this point that the basic Rutherford scattering formula (7.16) may be written down almost by inspection from Fig. 7.8(a). The matrix element will be the product of vertex functions (equal to the charges coupling to the photon) and the photon propagator term $1/q^2$:

$$M = e \times \frac{1}{q^2} \times e = \frac{e^2}{q^2}.$$

The element of phase space is proportional to dq^2, since

$$q^2 = 2p_0^2(1 - \cos\theta), \qquad \text{or} \quad dq^2 = p_0^2 \, d\Omega/\pi.$$

So the differential elastic cross-section is given by

$$\frac{d\sigma}{dq^2} = |M|^2 \propto \frac{e^4}{q^4}. \tag{7.33}$$

The quantity $1/q^4$ in the cross-section simply gives the probability that the virtual photon carries the 4-momentum q; it corresponds to the classical-physics statement that the electric field between the particles varies as $1/r^2$. The blob at the bottom of Fig. 7.8(a) is supposed to represent the nucleon form factor, left out in (7.33).

The experimental determination of the proton form factor has been carried out by directing high-energy (400-MeV to 16-GeV) electron beams at a hydrogen target, and making precision measurements of the momentum and angle of scattered electrons by means of magnetic spectrometers. Elastic events can be selected using the kinematic relation (7.7) between the energy and angle of the scattered electron (with M as the proton mass). For the neutron, scattering is observed with deuterium targets, and a subtraction procedure employed:

$$\frac{d\sigma}{d\Omega}(en) = \frac{d\sigma}{d\Omega}(ed) - \frac{d\sigma}{d\Omega}(ep) + \text{correction factor},$$

where the correction factor involves the nuclear physics of the deuteron. Because of this, the neutron data are less precise. The first experiments, demonstrating the deviation of the scattering from that expected for a point particle – and thus measuring the form factors – were carried out by Hofstadter and his collaborators in 1961, at Stanford, and have since been extended in numerous laboratories. The end result of the present experiments is that the form factors obey the simple *scaling law*

$$G_E^p(q^2) = \frac{G_M^p(q^2)}{|\mu_p|} = \frac{G_M^n(q^2)}{|\mu_n|} = G(q^2), \qquad G_E^n(q^2) = 0, \tag{7.34}$$

and the empirical *dipole formula*

$$G(q^2) = \left(1 + \frac{q^2}{M_V^2}\right)^{-2}, \qquad \text{with} \quad M_V^2 = (0.84 \, \text{GeV})^2. \tag{7.35}$$

The relations (7.34) hold over the experimental range so far investigated, $q^2 = 0$–$2 \, \text{GeV}^2$ (see Fig. 7.9). At higher q^2, only electron-proton scattering data exist, and, as (7.31) indicates, these refer essentially to the magnetic form factor $G_M^p(q^2)$, the electric contribution being small and unmeasurable. For this scattering, the dipole formula (7.35) fits to within 10% accuracy up to $q^2 = 25$ GeV^2, as shown in Fig. 7.10. Over this range, the form factor squared falls by a

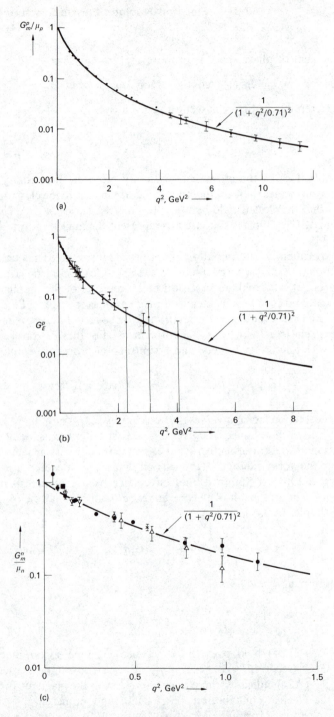

Fig. 7.9 Comparison of the magnetic and electric form factors of neutron and proton. They are consistent with the scaling law (7.34). (a) Proton magnetic form factor; (b) proton electric form factor; (c) neutron magnetic form factor. (After Weber 1967.)

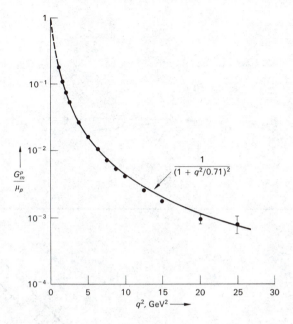

Fig. 7.10 The electromagnetic form factor of the proton, in the region of high q^2. The observations are fairly well fitted by the dipole formula. The data are principally from the SLAC laboratory. (After Panofsky 1968.)

factor of 10^6. Using Eq. (7.29), one finds that (7.35) can be interpreted in terms of an exponential charge–magnetic-moment distribution of the proton of density

$$\rho(R) = \rho_0 \exp(-M_V R),\tag{7.36}$$

with a root-mean-square radius

$$R_{\text{rms}} = \frac{\sqrt{12}}{M_V} = 0.80 \text{ fm.}\tag{7.37}$$

The elastic form factors reflect the spatial distribution of charge and magnetic moment in the nucleon. If we had a complete theory of hadron structure, these form factors could be calculated. At the present time no such theory exists and the subject must be treated on a largely empirical basis.

7.4 DEEP-INELASTIC ELECTRON-NUCLEON SCATTERING

As indicated above, at high momentum transfers, the elastic form factor is very small, and inelastic scattering of the incident electron is much more probable than elastic scattering. In a general inelastic scattering process, there is an extra variable because the space and time components $(\mathbf{q}, i\nu)$ of the momentum

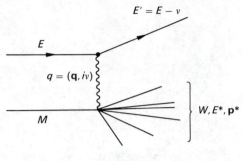

Fig. 7.11

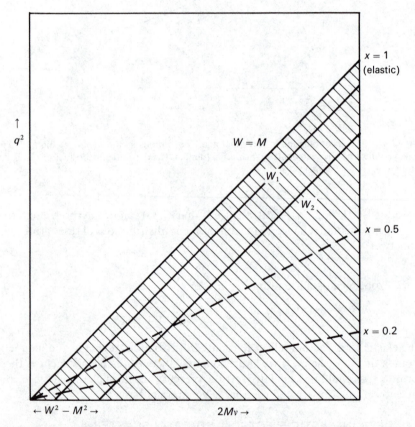

Fig. 7.12 Kinematic relations in inelastic lepton-nucleon scattering. The 4-momentum transfer squared, q^2, is plotted against $2Mv$, where M is the nucleon mass and v is the energy transfer measured in the rest frame of the target. The invariant mass of the final state of hadrons is given by $W^2 = M^2 + 2Mv - q^2$, and the variable $x = q^2/2Mv$. Elastic scattering corresponds to $x = 1$, $W = M$. Contours of fixed W are 45° lines; those of fixed x are radial lines from the origin with slope x.

transfer q are no longer related by $q^2 = 2Mv$, as in (7.18). If we denote the 3-momentum, energy, and invariant mass of the final hadron state by $\mathbf{p}^*$, E^*, and W, we obtain (see Fig. 7.11)

$$q^2 = (\mathbf{p}^* - 0)^2 - (E^* - M)^2,$$

$$v = E^* - M,$$

$$W^2 = E^{*2} - \mathbf{p}^{*2},$$

so

$$q^2 = 2Mv - W^2 + M^2. \tag{7.38}$$

$W = M$ corresponds to elastic scattering, with $q^2 = 2Mv$ as before. In terms of q^2, v the cross-section can be written down in analogy with (7.30) as

$$\frac{d^2\sigma}{dq^2\,dv} = \frac{4\pi\alpha^2}{q^4}\frac{E'}{EM}\left[W_2(q^2, v)\cos^2\frac{\theta}{2} + 2W_1(q^2, v)\sin^2\frac{\theta}{2}\right], \tag{7.39}$$

where E and E' are the incident and emergent electron energies (E, $E' \gg mc^2$). W_1, W_2 are arbitrary structure functions corresponding to the two possible polarization states, transverse and longitudinal, of the mediating photon. The validity of this formula has been established from the linearity of the Rosenbluth plot, just as in the case of elastic scattering. From such a plot, the ratio W_1/W_2 of magnetic to electric scattering can be determined.

The variables q^2, $2Mv$ are plotted in Fig. 7.12, where the shaded area represents the physical region accessible by inelastic scattering, and we have defined the ratio

$$x = \frac{q^2}{2Mv} \qquad (1 > x > 0). \tag{7.40}$$

Lines of constant W are indicated at $45°$ to the axes; those of constant x diverge from the origin with a slope equal to x. $q^2 < 2Mv$, $x < 1$ corresponds to the inelastic-scattering region, while the line $x = 1$ defines the kinematics for elastic scattering.

It is instructive to compare the dependence of $d^2\sigma/dq^2\,dv$ on the energy transfer $v = E - E'$ in different regions of q^2, for electron-nucleus and electron-nucleon scattering. Figure 7.13 shows typical results for scattering of energetic electrons by nuclei. At low q^2 [Fig. 7.13(a)] one observes a strong elastic-scattering peak, at a unique value

$$E' = E - v = E - \frac{q^2}{2M_{\text{nucleus}}}. \tag{7.41}$$

At larger v-values, other, broader peaks are observed, which are due to inelastic resonances in the cross-section when the nucleus is left in an excited state. At $q^2 > (50\ \text{MeV}/c)^2$, as shown in Fig. 7.13(c), elastic and resonance peaks are

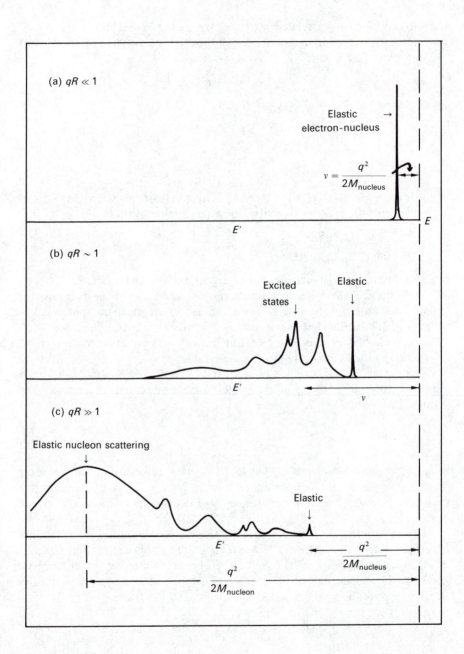

Fig. 7.13 Electron-nucleus scattering as a function of energy transfer $v = E - E'$, where E, E' are the energies of incident and scattered electrons. Three regions of qR are shown, where q is the momentum transfer and R the nuclear radius.

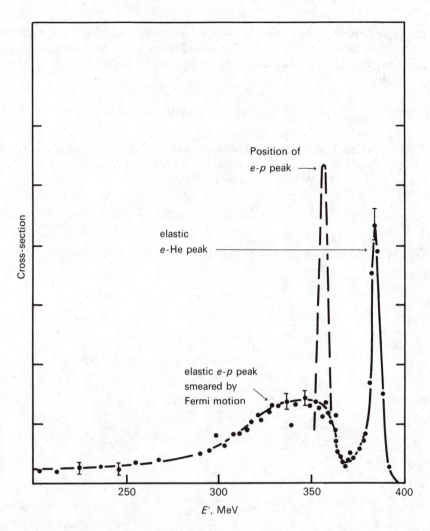

Fig. 7.14 Scattering of 400-MeV electrons in helium at 45°, showing a sharp elastic helium peak, and a smeared peak due to elastic scattering from individual nucleons (after Hofstadter 1956).

suppressed — in other words, the nuclear form factor, measuring the probability amplitude that the nucleus (in an excited state or not) will recoil coherently under the impact q, is small. At still larger values of v, however, the cross-section increases and has a smooth v-dependence. This is the "continuum" excitation. The maximum cross-section is observed to occur in the region

$$v \simeq \frac{q^2}{2M_{\text{nucleon}}}, \tag{7.42}$$

corresponding to *quasielastic scattering off individual nucleons*. If the nucleons were free, a sharp spike at $v = q^2/2M$ would be seen. The target nucleons are, however, bound in a potential well, of radius R, and thus have a Fermi momentum

$$p_F \sim \frac{h}{R} = 200 \, \text{MeV}/c.$$

The nuclear binding energy of 10 MeV or so is negligible in comparison with v and hardly affects the elastic kinematics; but the Fermi momentum smears the elastic peak, with a symmetrical spread of order

$$\frac{\Delta v}{v} = \pm \frac{p_F}{M} \simeq 10\%. \tag{7.43}$$

(The proof is left as an exercise.) Put another way, the elastic kinematic relation

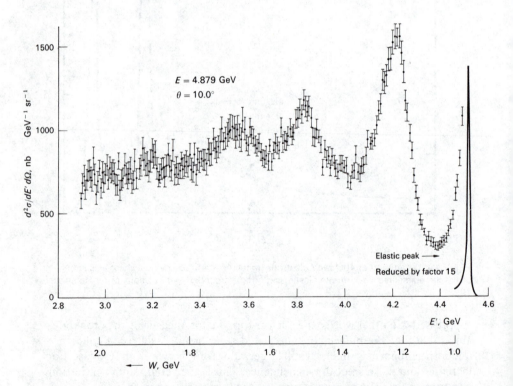

Fig. 7.15 Excitation curve of inelastic *ep* scattering, obtained at the DESY electron accelerator (Bartel *et al.* 1968). *E* and *E'* are the energies of the incident and the scattered electron, and *W* is the mass of the recoiling hadronic state. The peaks due to the pion-nucleon resonances of masses 1.24, 1.51, and 1.69 GeV are clearly visible.

$q^2 = 2M\nu$ defines the energy transfer ν in the rest frame of the struck nucleon. In the laboratory frame where the bound nucleon is moving, the measured energy transfer will be greater or less, depending on the direction of its velocity.

To summarize, in electron-nucleus scattering, as q^2 increases from low values, the coherent nuclear scattering dies away and the incoherent elastic scattering from individual nucleons becomes progressively more important. Increasing q^2 is equivalent to decreasing the wavelength or resolution of our "probe", and at sufficiently large q^2 we begin to "see" the constituents rather than the entire nucleus. An example of results from electron-helium scattering is given in Fig. 7.14.

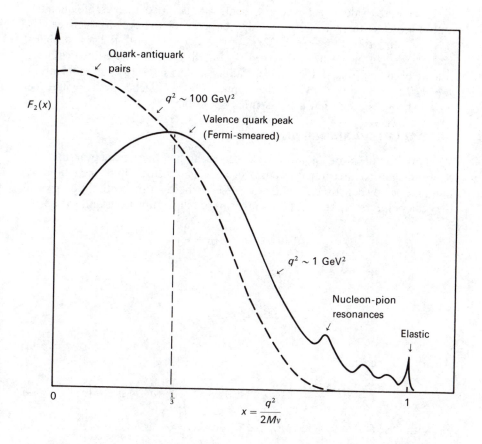

Fig. 7.16 Schematic diagram of the dependence of the cross-section, as measured by the structure function $F_2(x, q^2)$, on q^2 as a function of $x = q^2/2M\nu$. At low q^2, the elastic peak and those due to resonances, as in Fig. 7.15, appear at large x, followed by continuum excitation (lepton-quark elastic scattering) at lower x. At high q^2, the elastic and resonance peaks have been suppressed by form-factor effects and the whole distribution shifts to lower x.

Turning now to electron-proton scattering, we observe the *same phenomena* as in the nuclear case (only at much higher values of q^2). For $q^2 < 1$ GeV^2, with v increasing, we observe first a strong elastic peak at $v = q^2/2M$, or $x = q^2/2Mv = 1$, followed by successive peaks due to the broad nucleon resonances $\Delta(1238)$, $N(1450)$, $\Delta(1688)$, ... at $x = q^2/(q^2 + W^2 - M^2)$. In terms of the parameter x, as q^2 increases, these peaks move in towards $x = 1$ and are suppressed by the form factors, but the continuum excitation at smaller x-values remains large. In analogy with the nucleon case, we expect this to be due to *quasielastic scattering by nucleon constituents* which are more pointlike than the nucleon itself – and hence are not suppressed at large q^2 (see Figs. 7.15 and 7.16).

If one were to identify these constituents with pointlike quarks, of which the proton possesses three (u, u, d), one might expect each to have an effective mass $m = M/3$ and therefore $x = q^2/2Mv = (q^2/2mv)(m/M) = \frac{1}{3}$, that is, a sharp elastic peak. No such peak is observed. Again, the quarks are confined within a volume of dimensions R, the nucleon radius, and so have a Fermi momentum $p_F \sim \hbar/R \simeq 250$ MeV/c. Thus the spread of the elastic peak would be $\Delta x/x \sim p_F/mc \sim 1$, so that it is essentially invisible.

7.5 SCALE INVARIANCE AND PARTONS

We now pursue more quantitatively the idea that inelastic lepton-nucleon scattering can be interpreted in terms of the elastic scattering of the lepton by pointlike parton constituents, and how these can be identified with the quarks (if they can be). First we rewrite (7.39) in terms of related functions and variables:

$$F_2(q^2, v) = \frac{v W_2(q^2, v)}{M},$$

$$F_1(q^2, v) = W_1(q^2, v),$$

$$y = v/E,$$

$$E'/E = 1 - y,$$

$$q^2 = 2MExy.$$

Then (7.39) becomes

$$\frac{d^2\sigma}{dq^2\, dv} = \frac{4\pi\alpha^2}{q^4} \frac{E'}{Ev}\left[F_2(q^2, v)\cos^2\frac{\theta}{2} + \frac{2v}{M}F_1(q^2, v)\sin^2\frac{\theta}{2}\right], \tag{7.44}$$

or, with $\cos^2(\theta/2) = 1 - q^2/(4EE') \simeq 1$ and $dv/v = dx/x$,

$$\frac{d^2\sigma}{dq^2\, dx} = \frac{4\pi\alpha^2}{q^4}\left[(1 - y)\frac{F_2(x, q^2)}{x} + \frac{y^2}{2}\frac{2xF_1(x, q^2)}{x}\right]. \tag{7.45}$$

If the lepton-parton scattering is pointlike, then F_1 and F_2 cannot depend on q^2 and are purely functions of x. This assumption is known as the *Bjorken scaling*

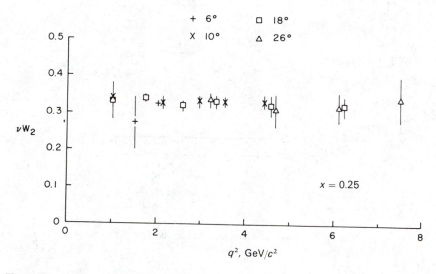

Fig. 7.17 νW_2 (or F_2) as a function of q^2 at $x = 0.25$. For this choice of x, there is practically no q^2-dependence, that is, exact "scaling". (After Friedman and Kendall 1972.)

hypothesis (Bjorken 1967). In simple terms, this states that if, in the limit $q^2 \to \infty$, $\nu \to \infty$, the function $F(q^2, \nu)$ remains finite, it can depend only on the dimensionless and finite ratio of these two quantities, that is on $x = q^2/2M\nu$. Since x is dimensionless, there is no scale of mass or length; hence the term scale invariance. Figure 7.17 shows the values of $F_2(x, q^2)$ at $x = 0.25$ from SLAC electron-scattering experiments, where there is essentially no q^2-dependence. Even at larger or smaller values of x, the q^2-dependence is weak (see Fig. 8.21). For example, for $x = 0.5$, F_2 falls by 50% as q^2 increases from 1 to 25 GeV^2; this is to be contrasted with the square of the elastic form factor (Fig. 7.10), which falls by a factor 10^6 over the same interval. The scale invariance proposed by Bjorken in the limit $q^2 \to \infty$ seems then to hold approximately in the region of q^2 of a few times M^2.

The interpretation of scale invariance is given in physical terms by the *parton model* of Feynman (1969). Imagine a reference frame in which the target proton has very large 3-momentum – the so-called infinite-momentum frame (Fig. 7.18). The proton mass can be neglected, so that it has 4-momentum $P = (p, 0, 0, ip)$ and is visualized as consisting of a parallel stream of partons, each with 4-momentum xP ($0 < x < 1$). Again, if P is large, masses and transverse momentum components of the partons can be neglected. Suppose now that one parton of mass m is scattered elastically by absorbing the current 4-momentum q from the scattered lepton. Then

$$(xP + q)^2 = -m^2 \simeq 0,$$

$$x^2 P^2 + q^2 + 2xP \cdot q \simeq 0. \tag{7.46}$$

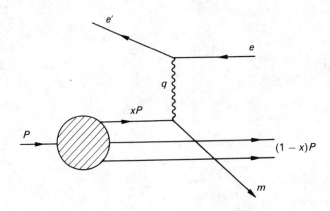

Fig. 7.18   The parton model of a deep-inelastic collision.

If $|x^2 P^2| = x^2 M^2 \ll q^2$, we obtain

$$x = \frac{-q^2}{2P \cdot q} = \frac{q^2}{2Mv}, \tag{7.47}$$

where the invariant scalar product $P \cdot q$ has been evaluated in the laboratory system in which the energy transfer is v and the nucleon is at rest. Thus, x in (7.40) represents the fractional 3-momentum of the parton in the infinite-momentum frame. It is as if we had a hypothetical parton of mass m, *stationary* in the lab system, with the elastic relation $q^2 = 2mv$, so that, provided always that $q^2 \gg M^2$,

$$x = \frac{q^2}{2Mv} = \frac{m}{M}. \tag{7.48}$$

Then x is also the fractional mass of the nucleon carried by such a hypothetical free parton initially at rest in the lab system. The cross-section, or equivalently $F(x)$ in (7.45), therefore gives, in a sense, a measure of the effective mass distribution of parton constituents.

Of course, we do not observe partons in the final state, but hadrons. Somehow the scattered and unscattered partons have to recombine to form hadrons, and at present no one knows the mechanism involved in this final-state parton interaction. The basic assumption is that the collision occurs in two independent stages. First, one parton is scattered, the collision time being that required to define the energy transfer, i.e. $t_1 \sim \hbar/v$. Over a much longer time, the partons recombine to form the final hadronic state, of mass W. Clearly the proper lifetime of this state must be $t_2 > \hbar/W$, or, transformed into the lab frame, $t_2 > \gamma\hbar/W = v\hbar/W^2$, so that since $W^2 \sim 2Mv$, we have finally $t_2 \sim \hbar/M$

$\gg t_1$ for $v \gg M$. The recombination therefore takes place over a long time scale and can be treated separately from the initial collision. It is also surmised that the cross-section will depend first and foremost on the dynamics of the initial stage, and only weakly or not at all on the complexities of the final-state interaction. This turns out to be a good guess except in the low-energy region ($v \sim M$), where there are significant resonance effects: even then, when integrated over all W-values at fixed E, as in the total neutrino cross-sections in Fig. 7.2, the combined effect of several resonances averages out to give a linear cross-section, that is, the pure parton-like behavior, even in the low-energy domain.

7.6 NEUTRINO-NUCLEON INELASTIC SCATTERING

The expression for the neutrino-nucleon inelastic scattering cross-section is obtained by replacing the electromagnetic coupling factor $4\pi\alpha^2/q^4$ in (7.45) with $G^2/2\pi$, where G is the Fermi constant, and including a third structure function $F_3(x)$. In this treatment we shall neglect effects due to strangeness- and charm-changing transitions, which are suppressed by the Cabibbo factor $\tan^2\theta_C$ relative to the $\Delta S = \Delta C = 0$ reactions. There are then three independent helicity states $(-1, +1, 0)$ for the mediating boson $W^\pm$, since in the weak interactions there is no conservation of parity compelling helicity -1 and $+1$ states to occur with equal probability as a coherent superposition, as in the electromagnetic case. Thus *three* structure functions are needed to describe vp scattering, and three each for $\bar{v}p$, vn, and $\bar{v}n$ scattering. Since many experiments are carried out on nuclear targets containing approximately equal numbers of protons and neutrons, we consider here neutrino-nucleon (i.e. proton-neutron average) and antineutrino-nucleon structure functions. These are equal via the principle of charge symmetry:

$$\left.\begin{matrix} F_i^{vn} = F_i^{\bar{v}p} \\ F_i^{vp} = F_i^{\bar{v}n} \end{matrix}\right\} \quad i = 1, 2, 3$$

so that

$$F_i^{vN} = \tfrac{1}{2}(F_i^{vp} + F_i^{vn}) = \tfrac{1}{2}(F_i^{\bar{v}p} + F_i^{\bar{v}n}) = F_i^{\bar{v}N}, \tag{7.49}$$

except that the *V-A* interference term changes sign for antineutrinos: $F_3^{\bar{v}N} = -F_3^{vN}$. One then obtains, for exact scaling,

$$\frac{d^2\sigma^{vN,\bar{v}N}}{dx\,dq^2} = \frac{G^2}{2\pi}\left[(1-y)\frac{F_2^{vN}(x)}{x} + \frac{y^2}{2}\frac{2xF_1^{vN}(x)}{x} \pm y\left(1 - \frac{y}{2}\right)\frac{xF_3^{vN}(x)}{x}\right],$$

$$\tag{7.50}$$

or, with $dq^2 = 2MEx\,dy$,

$$\frac{d^2\sigma^{vN,vN}}{dx\,dy} = \frac{G^2ME}{\pi}\left[(1-y)F_2^{vN}(x) + \frac{y^2}{2}2xF_1^{vN}(x) \pm y\left(1 - \frac{y}{2}\right)xF_3^{vN}(x)\right].$$

$$\tag{7.51}$$

Upon integration over x and y from 0 to 1, Eq. (7.51) gives total cross-sections σ^{vN}, $\sigma^{\bar{v}N}$ proportional to E, as noted previously.

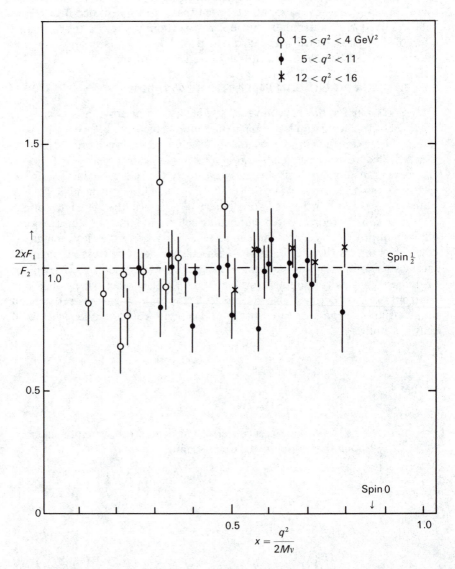

Fig. 7.19 The ratio $2xF_1/F_2$ measured in SLAC electron-nucleon scattering experiments. For spin-$\frac{1}{2}$ partons, with $g = 2$, a ratio of unity is expected in the limit of large q^2 – the Callan-Gross relation. (Data compiled from published SLAC data.)

7.7 LEPTON-QUARK SCATTERING

The nature of the partons can be established by relating the data on electromagnetic and weak structure functions of nucleons, (F_1^{eN}, F_2^{eN} and F_1^{vN}, F_2^{vN}, F_3^{vN} respectively) to the quantum numbers of the partons.

7.7.1 Parton Spin

First we compare the expression (7.39) with the Dirac cross-section for scattering from spin-$\frac{1}{2}$ pointlike particles of charge ze and mass m. We have from (7.15 and 7.30)

$$\left(\frac{d\sigma}{dq^2}\right)_{\text{Dirac}} = \frac{4\pi\alpha^2 z^2}{q^4}\left(\frac{E'}{E}\right)^2\left(\cos^2\frac{\theta}{2} + \frac{q^2}{2m^2}\sin^2\frac{\theta}{2}\right)$$

and

$$\left(\frac{d^2\sigma}{dq^2\,dx}\right)_{\text{inelastic}} = \frac{4\pi\alpha^2}{q^4}\frac{E'}{E}\left(F_2(x)\cos^2\frac{\theta}{2} + \frac{q^2}{2M^2x^2}2xF_1(x)\sin^2\frac{\theta}{2}\right)\frac{1}{x}.$$

Comparing coefficients of $\cos^2(\theta/2)$ and $\sin^2(\theta/2)$ and with the identification $m^2 = M^2x^2$ as in (7.48), we immediately obtain

$$\frac{2xF_1(x)}{F_2(x)} = 1. \tag{7.52}$$

This is the relation expected between F_1 and F_2 (i.e. magnetic and electric scattering) in the scaling region (q^2 large), if the scattering is to be interpreted as due to *pointlike constituents of spin $\frac{1}{2}$* and normal (Dirac) magnetic moments (i.e. $\mu = zeh/2mc$). Equation (7.52) is called the *Callan-Gross relation* (1968). Figure 7.19 shows the observed ratios $2xF_1/F_2$ measured in electron scattering experiments at SLAC for low values of q^2. They indicate spin $\frac{1}{2}$ for the partons. Zero-spin partons ($2xF_1/F_2 = 0$) are obviously excluded.

7.7.2 Parton Charges

Let us now compare the neutrino and antineutrino cross-sections in (7.51), for deep-inelastic scattering from nucleon targets, with those expected for elastic scattering from pointlike targets of spin $\frac{1}{2}$. In the *V-A* theory of weak interactions, neutrino-electron elastic scattering in reactions involving the so-called charged currents (i.e. via virtual $W^\pm$ boson exchange) is described by the following formulae:*

$$\frac{d\sigma^{ve}}{dy} = \frac{d\sigma^{\bar{v}e}}{dy} = \frac{2G^2mE}{\pi}, \tag{7.53a}$$

$$\frac{d\sigma^{\bar{v}e}}{dy} = \frac{d\sigma^{v\bar{e}}}{dy} = \frac{2G^2mE}{\pi}(1-y)^2, \tag{7.53b}$$

---

* See L. B. Okun, *Weak Interactions of Elementary Particles*, Pergamon, 1965, p. 61 or E. D Commins, *Weak Interactions*, McGraw-Hill, 1973, p. 59.

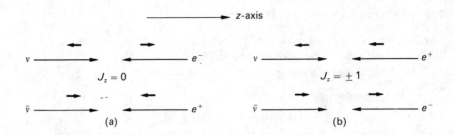

Fig. 7.20

where y is the fractional energy carried off by the recoiling electron target, and it is assumed that $E \gg m$, the electron mass. The cross-section (7.53a) corresponds to isotropic scattering in the CMS, for a particle-particle or antiparticle-antiparticle scattering, with $J_z = 0$ for the net spin component along the collision axis z [see Fig. 7.20(a)]. In (7.53b), however, as shown in Fig. 7.20(b), $J_z = \pm 1$ and $180°$ scattering is impossible for a pointlike (S-wave) interaction, by angular-momentum conservation. Thus, either neutrino-positron or anti-neutrino-electron scattering contains a term $(1 - y)^2$, leading to a factor of $\frac{1}{3}$ when integrated over y. In simple terms, the reactions (7.53b) proceed through a $J = 1$ state, but angular-momentum conservation only allows one of the 3 ($= 2J + 1$) substates, so the cross-section is reduced relative to that for (7.53a).

If we replace electrons by spin-$\frac{1}{2}$ partons and positrons by antipartons, the same formulae will apply, so that for neutrinos on a combination of partons Q and antipartons $\bar{Q}$ of mass $m = xM$ we get

$$\frac{d^2\sigma}{dy\,dx} = \frac{2G^2 ME}{\pi}[xQ(x) + x\bar{Q}(x)(1 - y)^2], \tag{7.54}$$

which is to be compared with (7.51) rewritten assuming $2xF_1 = F_2$:

$$\frac{d^2\sigma^{\nu N}}{dy\,dx} = \frac{G^2 ME}{2\pi}\{[F_2(x) + xF_3(x)] + [F_2(x) - xF_3(x)](1 - y)^2\}. \tag{7.55}$$

Thus

$$F_2^{\nu N}(x) = 2x[Q(x) + \bar{Q}(x)],$$
$$xF_3^{\nu N}(x) = 2x[Q(x) - \bar{Q}(x)]. \tag{7.56}$$

The structure functions $F_2^{\nu N}$ and $F_3^{\nu N}$ describing neutrino scattering on nucleons are thus proportional to the sums and differences of the parton and antiparton densities at x, weighted by the mass fraction x, i.e. the fractional nucleon mass (or, strictly, 3-momentum) carried by partons (antipartons) at x.

If we identify the partons with the quarks u and d, the elementary charged-current reactions for muon-type neutrinos will be

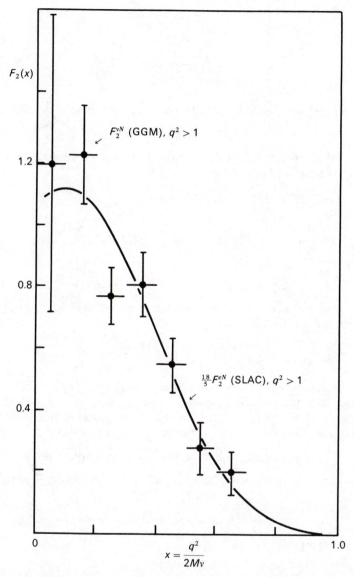

Fig. 7.21 Comparison of $F_2^{\nu N}$ measured in neutrino-nucleon scattering in the Gargamelle heavy-liquid bubble chamber in a PS neutrino beam at CERN, with SLAC data on F_2^{eN} from electron-nucleon scattering, in the same region of q^2. The data points are the neutrino results, and the curve is a fit through the electron data, multiplied by the factor $\frac{18}{5}$, which is the reciprocal of the mean squared charge of u- and d-quarks in the nucleon. This is a confirmation of the fractional charge assignments for the quarks. Note that the total area under the curve, measuring the total momentum fraction in the nucleon carried by quarks, is about 0.5. The remaining mass is ascribed to gluon constituents, which are the postulated carriers of the interquark color field (see Chapter 8).

$$v_\mu d \to \mu^- u, \tag{7.57a}$$

$$v_\mu \bar{u} \to \mu^- \bar{d}, \tag{7.57b}$$

$$\bar{v}_\mu u \to \mu^+ d, \tag{7.57c}$$

$$\bar{v}_\mu \bar{d} \to \mu^+ \bar{u}. \tag{7.57d}$$

If we let $u(x), d(x), \bar{u}(x), \bar{d}(x)$ stand for the quark densities in the proton, then

$$F_2^{vp}(x) = 2x[d(x) + \bar{u}(x)]. \tag{7.58}$$

From isospin invariance, we know that the quark populations in the neutron are given by $u(x)^n = d(x), d(x)^n = u(x)$ and thus

$$F_2^{vn}(x) = 2x[u(x) + \bar{d}(x)], \tag{7.59a}$$

or

$$F_2^{vN}(x) = x[u(x) + d(x) + \bar{u}(x) + \bar{d}(x)]. \tag{7.59b}$$

Similarly

$$xF_3^{vN}(x) = x[u(x) + d(x) - \bar{u}(x) - \bar{d}(x)]. \tag{7.59c}$$

The value of F_2 in electron scattering can be easily obtained. From (7.45) we note that, in the limit $y \to 0$ $(\theta \to 0)$,

$$\frac{d\sigma}{dq^2} = \frac{4\pi\alpha^2}{q^4} \int \frac{F_2^{eN}(x)\, dx}{x}, \tag{7.60}$$

so that comparing with the formulae in Section (7.7.1), $\int F_2(x)\, dx / x$ is to be interpreted as the *sum of the squares of the charges of the partons*. Thus $F_2^{ep}(x)/x$ is given by the quark densities in the proton, weighted by the squares of the charges; i.e., if we consider u, d, and s quarks,

$$F_2^{ep}(x) = x\{\tfrac{4}{9}[u(x) + \bar{u}(x)] + \tfrac{1}{9}[d(x) + \bar{d}(x) + s(x) + \bar{s}(x)]\}. \tag{7.61}$$

F_2^{en} is obtained by replacing the symbols u with d and $\bar{u}$ with $\bar{d}$, so that for a nucleon

$$F_2^{eN}(x) = x\{\tfrac{5}{18}[u(x) + \bar{u}(x) + d(x) + \bar{d}(x)] + \tfrac{1}{9}[s(x) + \bar{s}(x)]\}. \tag{7.62}$$

Thus, combining with (7.59b), we find

$$F_2^{vN}(x) \leqslant \tfrac{18}{5} F_2^{eN}(x), \tag{7.63}$$

where the number results directly from the fractional quark-charge assignment, and the equality holds if $s, c, \ldots$ quarks can be neglected (they in fact make only a few per cent contribution). Figure 7.21 shows early neutrino data on F_2^{vN} compared with those from electron scattering. The comparison shows that neutrinos and electrons "see" the same type of substructure in the nucleon, and that the parton constituents do indeed have the fractional charges previously ascribed to quarks.

7.7.3 Antiquark Content of the Nucleon

According to the V-A theory of the weak interactions, the ratio of cross-sections for antineutrinos and neutrinos on spin-$\frac{1}{2}$ pointlike particles should be

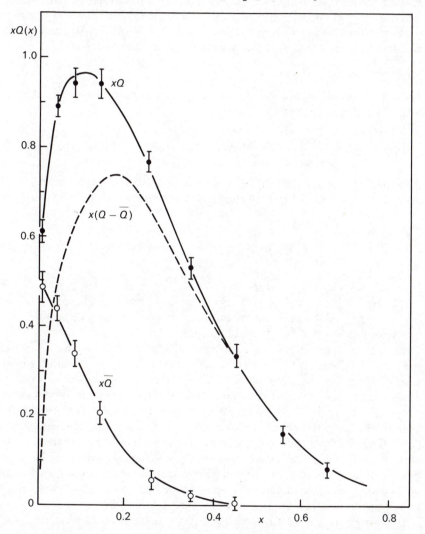

Fig. 7.22 Momentum distributions of quarks (Q) and antiquarks ($\overline{Q}$) in the nucleon, at a value of q^2 of order 10 GeV², obtained from results on neutrino and antineutrino scattering in experiments at CERN and Fermilab. The neutrino and antineutrino differential cross-sections measure the structure functions F_2 and F_3 in Eq. (7.50), and the difference and sum of these, through Eq. (7.56), give the quark and antiquark populations weighted by the momentum fraction x. The antiquarks ($\overline{Q}$) are concentrated at small x, the region of the so-called quark-antiquark "sea". The "valence" quarks of the static quark model ($Q - \overline{Q}$) are concentrated towards $x = 0.2$.

$R = \frac{1}{3}$ exactly. As pointed out above, this factor arises because in the antineutrino scattering on a fermion, the initial state has $J = 1$ (and $J_z = +1$), but in the final state, only one of the three possible $J = 1$ substates is allowed by angular momentum conservation. However, if antipartons are involved as well, then $R > \frac{1}{3}$, and the ratio of antiparton to parton densities will be simply $\bar{Q}/Q = (3R - 1)/(3 - R)$. Figure 7.2 shows that $R \simeq 0.45$ in the energy range 10–100 GeV, so $\bar{Q}/Q \sim 0.1$. The x-dependence is shown in Fig. 7.22.

7.7.4 Gluon Constituents

The identification of the partons with quarks leads us to expect that the structure-function integrals

$$\frac{18}{5} \int F_2^{eN}(x)\,dx = \int F_2^{vN}(x)\,dx = \int [u(x) + \bar{u}(x) + d(x) + \bar{d}(x)]x\,dx \simeq 1, \quad (7.64)$$

since the total fractional momentum summed over all constituents is necessarily unity. Ten years ago, the early electron and neutrino scattering experiments gave (see Fig. 7.21)

$$\frac{18}{5} \int F_2^{eN}(x)\,dx \simeq \int F_2^{vN}(x)\,dx = 0.50 \pm 0.05 \quad (7.65)$$

for values of q^2 in the region 1–10 GeV^2. The partons responsible for the scattering of leptons, via their electric and weak charges, only account for *half* the nucleon mass. We must either abandon the model entirely or postulate that there is another type of constituent besides the quarks, which is inert to leptons. These were subsequently identified with so-called *gluon* constituents. As we shall see later, the gluons are massless vector bosons and the specific quanta of the strong (color) force between quarks. They interact only with the strong charges of the quarks, and have no interaction with electromagnetic or weak charges. There is, incidentally, no *a priori* reason why the fractional energy-momentum content of the gluons should be just equal to that of the quarks, as indicated in (7.65). The fact that the mass fraction just happens to be $\sim 50\%$ in the region of q^2 a few GeV^2 is an indication of the mass scales associated with the strong quark-quark interaction, and the number of degrees of freedom carried by the quarks and gluons respectively. We merely note that we have postulated gluons as an important constituent of matter inside the nucleon.

At this point it is worth trying to summarize the results so far from deep-inelastic lepton-nucleon scattering:

(a) The nucleon contains pointlike constituents, as evidenced from the approximate scale invariance of the structure functions, $F_2(x, q^2) \simeq F_2(x)$.

(b) The constituents have spin $\frac{1}{2}\hbar$; $[2xF_1(x) \simeq F_2(x)]$.

(c) Electromagnetic and weak cross-sections are consistent with the identification of the "active" partons with fractionally-charged quarks.

(d) The quarks account for only a fraction ($\simeq 50\%$) of the nucleon mass; the remainder is ascribed to gluon constituents which are responsible for the interquark binding.

7.8 ELECTRON-POSITRON ANNIHILATION TO HADRONS

We have already discussed in Section 5.5 the decay of the vector mesons ϕ, ω, ρ, ψ and Υ to lepton pairs, and indeed seen that the ψ and Υ have been observed in the reactions $e^+e^- \rightarrow$ hadrons, in electron-positron colliding-beam accelerators. The pioneer experiments on production of the lighter vector mesons ρ, ω, and ϕ were carried out at Novosibirsk and Orsay in the late 1960s. Figures 7.23 and 7.24 show these early results on the production of ρ- and ω-mesons, with the characteristic Breit-Wigner behavior of the cross-section that we already discussed in the case of the ψ [see Eq. (5.45) and Fig. 5.10]. Near resonance the cross-section has the form (for ρ-production)

$$\sigma(e^+e^- \rightarrow \rho \rightarrow \pi^+\pi^-) = \frac{\pi\lambda^2}{4} \frac{(2J+1)\Gamma^2 B}{(E-M_\rho)^2 + \Gamma^2/4}, \qquad (7.66)$$

where Γ is the total width ($\Gamma_\rho \simeq 150$ MeV) and B is the branching ratio for $\rho \rightarrow e^+e^-$. The measurements yield $B = 6 \times 10^{-5}$. This is the order of magnitude expected from purely qualitative arguments.* In Fig. 7.25(a) we note that

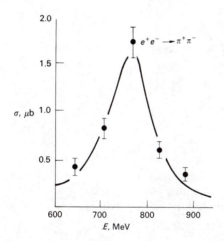

Fig. 7.23 Cross-section for the process $e^+e^- \rightarrow \pi^+\pi^-$ as measured in the electron-positron colliding-beam experiments at Orsay (Augustin *et al.* 1969). The dominant feature is the production of the ρ-meson resonance. The curve is from the Breit-Wigner formula, using best-fit parameters; central mass $M_\rho = 765$ MeV, width $\Gamma_\rho = 150$ MeV.

---

* The coupling of hadrons to photons via an intermediate vector meson with the same spin, parity, and C-parity as the photon ($J^{PC} = 1^{--}$) is called the hypothesis of *vector-meson dominance*.

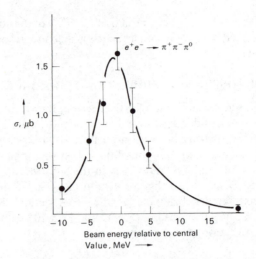

Fig. 7.24 Cross-section for $e^+e^- \to \pi^+\pi^-\pi^0$ as a function of beam energy. The curve corresponds to a Breit-Wigner resonance of width $\Gamma_\omega = 16$ MeV.

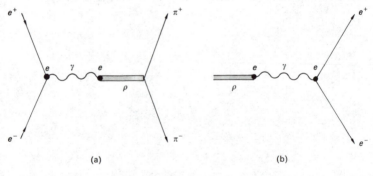

Fig. 7.25

the electromagnetic coupling constant $\sqrt{\alpha} = e$ (in units $\hbar = c = 1$) appears twice, coupling the photon to the e^+e^- pair and the ρ-meson respectively. Thus, we expect that, in order of magnitude, $B \sim (\sqrt{\alpha}\sqrt{\alpha})^2 = (137)^{-2} \sim 10^{-4}$. In the colliding-beam experiments, the normalization of the cross-section is made by reference to the Bhabha scattering $e^+e^- \to e^-e^+$, for which the cross-section is calculable from quantum electrodynamics.

It is also possible to produce ρ-mesons, either in strong interactions or in the inelastic collisions of energetic electron or photon beams with nucleons, and to observe the decay into (instead of formation from) an e^+e^- pair, as in Fig. 7.25(b). The value of B obtained is consistent with that above.

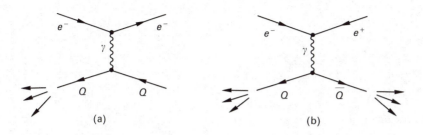

Fig. 7.26

Let us turn now to electron-positron annihilation at high energy, that is, above 10-GeV CMS energy, clear of the resonance region associated with ρ, ω, ϕ, and the ψ and Υ series of levels. The success of the parton model in deep-inelastic lepton-nucleon scattering naturally leads to the expectation that other processes involving hadrons and leptons can be described in similar fashion. The process of e^+e^- annihilation to hadrons can be considered as the two-stage reaction

$$e^+e^- \to Q\bar{Q}, \qquad Q, \bar{Q} \to \text{hadrons,} \tag{7.67}$$

in close analogy with lepton-quark scattering (Fig. 7.26).

The pointlike process (7.67) can be compared with that for lepton-lepton scattering

$$e^+e^- \to \mu^+\mu^-. \tag{7.68}$$

If the incident beams are unpolarized, this has a cross-section of magnitude

$$\sigma(e^+e^- \to \mu^+\mu^-) = \frac{4\pi\alpha^2}{3s}, \tag{7.69}$$

where

$$s = E_{\text{CMS}}^2 = 4E_1 E_2 \tag{7.70}$$

is the square of the total CMS energy, where E_1 and E_2 are the energies of the e^+ and e^- beams, measured in the laboratory system, and where we have assumed $E \gg m_\mu c^2$. The form (7.69) follows from the assumption that the interaction is mediated by single photon exchange, with 4-momentum q. This introduces the square of the photon propagator, $1/q^4$, and the coupling constant squared, α^2. The phase-space factor for the final-state leptons is q^2, if we neglect the lepton masses. The factor $4\pi/3$ results from the integration over solid angle in the final state and the averaging over initial-state spins. With $|q^2| = s$, the formula follows. Purely dimensional arguments also show that, when s is large compared with any of the masses concerned, $\sigma \propto s^{-1}$.

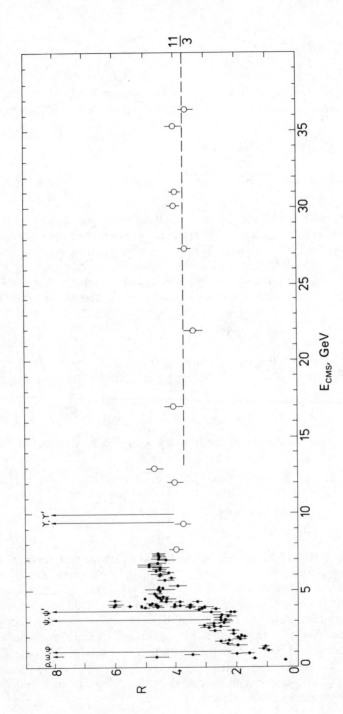

Fig. 7.27 The ratio R of the cross-section for $e^+e^- \to$ hadrons, divided by that that for $e^+e^- \to \mu^+\mu^-$. The fact that R is constant above 10-GeV CMS energy is a proof of the pointlike nature of hadron constituents. The predicted value of R, assuming that the primary process is formation of a quark-antiquark pair, is $\frac{11}{3}$ if pairs of u, d, s, c, b quarks are excited and they have three color degrees of freedom. The data come from many storage-ring experiments. At high energy (> 10 GeV CMS) it is from the PETRA ring at DESY, Hamburg.

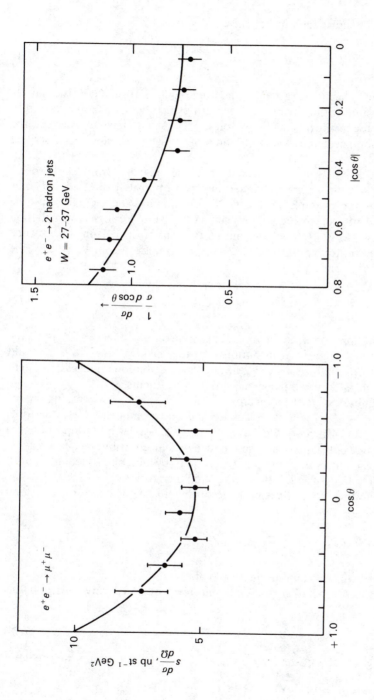

Fig. 7.28 Angular distribution, relative to the e^+e^- beam direction, of muon pairs (at left) and hadronic jets (as in Fig. 2.20). A $1 + \cos^2\theta$ distribution fits both data sets, as expected if the fundamental constituents (muons or quarks) have half-integral spin. Data compiled from several experiments.

Figure 7.27 shows the ratio of the cross-sections

$$R = \frac{\sigma(e^+e^- \to \text{hadrons})}{\sigma(e^+e^- \to \mu^+\mu^-)} \qquad (7.71)$$

as a function of beam energy. Over the range $E_{\text{CMS}} = 10\text{–}35$ GeV, the ratio is indeed a constant within errors, demonstrating once again the pointlike nature of the hadron constituents in the parton model. At intermediate energies, the situation is complicated by the excitation of heavy $Q\bar{Q}$ resonances as described above.

At high energies, it is observed that the hadrons from the process $e^+e^- \to Q\bar{Q} \to$ hadrons are collimated into two oppositely directed "jets", as illustrated in the example in Fig. 2.20. Presumably the jet axis must be approximately that of emission of the primary $Q\bar{Q}$ pair. The angular distribution in the process $e^+e^- \to Q\bar{Q}$ will depend on the spin of the parton constituents. Again, in the prototype process $e^+e^- \to \mu^+\mu^-$ involving spin-$\frac{1}{2}$ particles, the angular distribution has the form

$$\frac{dN}{d\Omega} \propto (1 + \cos^2\theta), \qquad (7.72)$$

where θ is the angle between the $\mu^\pm$ and beam directions. Figure 7.28 shows the results for the angular distributions in the process $e^+e^- \to \mu^+\mu^-$ and $e^+e^- \to$ hadrons. In the latter case, the angular distribution is indeed consistent with the form as expected for spin-$\frac{1}{2}$ quark constituents.

The magnitude of R, the total cross-section for $e^+e^- \to$ hadrons, in terms of that for $e^+e^- \to \mu^+\mu^-$, is easily calculated in the quark model. The (Rutherford) cross-section for electron-quark scattering [Fig. 7.26(a)] will be proportional to $\sum e_i^2$, the square of the quark charge summed over all contributing quark flavors i. If we replace incoming (outgoing) fermions by outgoing (incoming) anti-fermions on the legs of Fig. 7.26(a), we obtain the process $e^+e^- \to \gamma \to Q\bar{Q}$, Fig. 7.26(b). It follows that we expect, according to the Rutherford scattering formula,

$$R = \frac{\sum e_i^2}{1} \qquad (7.73)$$

summed over all contributing quark flavors.

At low s-values, below $c\bar{c}$ threshold, only u, d, s quarks are involved, and thus we expect

$$R_{\text{th}}(\sqrt{s} < 3\,\text{GeV}) = (\tfrac{1}{3})^2 + (\tfrac{1}{3})^2 + (\tfrac{2}{3})^2 = \tfrac{2}{3}, \qquad (7.74)$$

while at high s-values, where u, d, s, c, b quarks can contribute, we predict

$$R_{\text{th}}(\sqrt{s} > 10\,\text{GeV}) = (\tfrac{2}{3})^2 + (\tfrac{2}{3})^2 + (\tfrac{1}{3})^2 + (\tfrac{1}{3})^2 + (\tfrac{1}{3})^2 = \tfrac{11}{9}. \qquad (7.75)$$

It is obvious that these predictions are a long way below the data, and indeed this conflict was one of the primary reasons for introducing the *color* degree of freedom mentioned in Section 5.2. Quarks are endowed with three possible values of the strong color charge: red, green, and blue (R, G, B). Thus, any given $Q\bar{Q}$ flavor combination occurs in three substates $R\bar{R}$, $G\bar{G}$, and $B\bar{B}$, and the expected R-values must be multiplied by the factor 3, in quite good agreement with experiment (Fig. 7.27).

7.9 LEPTON PAIR PRODUCTION IN HADRON COLLISIONS – THE DRELL-YAN PROCESS

The process of lepton pair production in hadron-hadron collisions, as first described by Drell and Yan in 1970, can also be simply interpreted in terms of the quark-parton model. Consider the reaction

$$p + p \rightarrow \mu^+ \mu^- + X, \tag{7.76}$$

where X is any hadronic state. On the basis of the lepton-quark scattering diagram of Fig. 7.29(b), or that of $e^+ e^-$ annihilation to hadrons in Fig. 7.29(c), as discussed in the previous sections, it is natural to draw the related diagram of Fig. 7.29(a), in which a quark from one proton annihilates with an antiquark from the other, and the virtual photon produces a lepton pair. Note that, in both (a) and (c), the 4-momentum of the photon is timelike (i.e. $q^2 < 0$), while in (b) it is spacelike ($q^2 > 0$).

In analogy with Fig. 7.29(c) and Eq. (7.69) we can write down the cross-section in terms of pointlike scattering:

$$\sigma(Q\bar{Q} \rightarrow l^+ l^-) = \frac{4\pi\alpha^2}{3q^2} e_i^2, \tag{7.77}$$

where e_i, $-e_i$ are the electric charges of the $Q\bar{Q}$ pair. The differential cross-section is then

$$\frac{d\sigma}{dq^2}(Q\bar{Q} \rightarrow l^+ l^-) = \frac{4\pi\alpha^2}{3q^4} e_i^2. \tag{7.78}$$

The value of $-q^2$, the mass squared of the virtual photon, is related to the proton momenta and the momentum fractions x_i, $\bar{x}_i$ carried by the constituents. Let P_1, P_2 be the 4-momenta of the colliding protons: those of the constituents are $k_i = x_i P_1$ and $\bar{k}_i = \bar{x}_i P_2$ so that the invariant masses squared of the photon and of the dilepton are

$$m^2 = -q^2 = -(k_i + \bar{k}_i)^2 = -(x_i P_1 + \bar{x}_i P_2)^2$$
$$= -(2P_1 P_2 x_i \bar{x}_i + x_i^2 P_1^2 + \bar{x}_i^2 P_2^2). \tag{7.79}$$

The total CMS energy squared of the proton-proton system is

$$s = -(P_1 + P_2)^2 = -(2P_1 P_2 + P_1^2 + P_2^2).$$

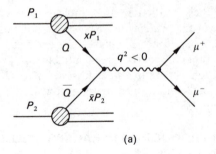

(a)

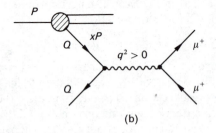

(b)

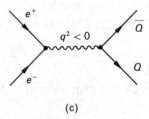

(c)

Fig. 7.29 Diagrams for different types of quark-lepton interaction. (a) Production of a muon pair in a hadron-hadron collision (the Drell-Yan process) with $q^2 < 0$. (b) Muon-nucleon deep-inelastic scattering, with $q^2 > 0$. (c) Electron-positron annihilation to a quark-antiquark pair, with $q^2 < 0$.

Now $P_1^2 = P_2^2 = - M^2$, and thus if $s \gg M^2$, $m^2 \gg x^2 M^2$, we obtain

$$m^2 = x_i \bar{x}_i s. \tag{7.80}$$

In this analysis, components of momentum of the constituents transverse to the proton beam direction have been neglected.

Introducing the dimensionless parameter

$$\tau = \frac{m^2}{s} = x_i \bar{x}_i, \tag{7.81}$$

we obtain for the lepton-pair cross-section in pp collisions

$$\frac{d\sigma}{dq^2} = \frac{4\pi\alpha^2}{3q^4} F(\tau),$$ (7.82)

where, writing a δ-function to ensure the equality (7.81),

$$F(\tau) = \sum_i e_i^2 \frac{1}{N_c} \int_0^1 x_i \, dx_i \int_0^1 \bar{x}_i \, d\bar{x}_i \, f_i^1(x_i) f_i^2(\bar{x}_i) \, \delta(x_i \bar{x}_i - \tau)$$ (7.83)

and where the integration is performed over the densities $f_i^1(x_i), f_i^2(\bar{x}_i)$ of quarks and antiquarks of flavor i in the protons, weighted by their fractional momenta x_i, $\bar{x}_i$ and by their squared charges e_i^2, in exact analogy with Eq. (7.61) for the structure function. The color factor $N_c = 3$ reduces the cross-section by the probability of pairing (for example) a red quark with an anti-red antiquark. According to (7.82), the cross-section

$$2q^4 \frac{d\sigma}{dq^2} = m^3 \frac{d\sigma}{dm} \propto F(\tau)$$ (7.84)

should scale, that is, depend only on the dimensionless ratio m^2/s. Direct tests of this prediction are difficult, since experiments tend to cover only a limited range of momentum of the dimuon pair. One can define the so-called rapidity of the dimuon pair,

$$y = \frac{1}{2} \ln \left(\frac{E + p_L}{E - p_L} \right),$$ (7.85)

where E and p_L are the total energy and net longitudinal momentum of the muon pair in the laboratory system (so $E^2 = p_L^2 + m^2$). Since y is dimensionless, we expect that the cross-section with respect to y will also scale; in particular, the dimensionless double differential cross-section

$$m^3 \frac{d^2\sigma}{dm \, dy}, \quad \text{or} \quad s \frac{d^2\sigma}{dy \, d\sqrt{\tau}},$$

should be a function of τ only. Figure 7.30 shows data on the invariant-mass distribution of muon pairs, obtained in an experiment at Fermilab by Yoh *et al.* (1978) using protons of 200–400 GeV ($\sqrt{s} = 20$–28 GeV) on a stationary target. Within the errors, the cross-section is a function of $\tau = m^2/s$ only. From such data it is possible to extract $F(\tau)$ defined in (7.82) and fit it to an analytical form, for example

$$F(\tau) = \text{const} \cdot \exp(-18.6\sqrt{\tau}) \quad (0.2 < \sqrt{\tau} < 0.5).$$ (7.86)

This may then be used to predict the absolute cross-sections at the much higher energies of the CERN ISR, where oppositely circulating proton beams of up to 30 GeV collide head on ($\sqrt{s}$ up to 62 GeV). Upon comparing the ISR data with

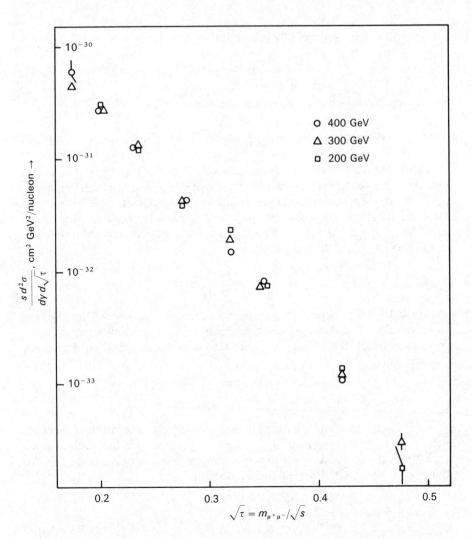

Fig. 7.30 Cross-section for muon pair production measured by the BFS collaboration (Yoh *et al.* 1978) in Fermilab experiments at three different proton energies. The cross-sections appear to "scale", being a function of the dimensionless variable $\tau = m_{\mu\mu}^2/s$, where s is the square of the CMS energy.

the predictions of (7.86), it is found that the cross-sections do scale even though their absolute magnitudes in the two experiments differ by up to 2 orders of magnitude; this verification of the parton model holds for values of $q = m_{\mu\mu} > 3$ GeV.

The assumption that lepton pair production proceeds via quark-antiquark annihilation to a virtual photon is verified in these experiments by observing the

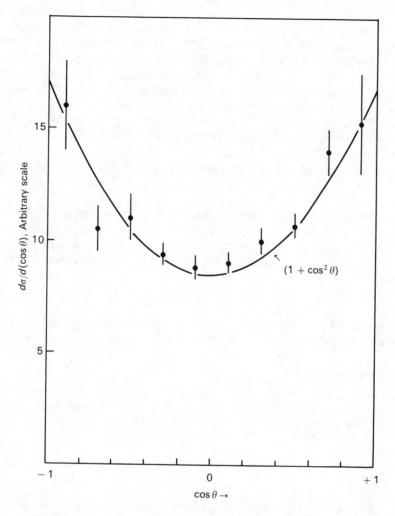

Fig. 7.31 Angular distribution of muons, measured in the dimuon rest frame, relative to the incident beam direction, in the process of dimuon production in hadron-hadron collisions (after Anderson *et al.* 1978). If the process follows the Drell-Yan mechanism of Fig. 7.29(a), involving spin-$\frac{1}{2}$ quarks ($Q\bar{Q} \to \mu^{+}\mu^{-}$), a distribution of the form $1 + \cos^2 \theta$ is expected.

polar angular distribution of either lepton in the dilepton rest frame, relative to the incident proton beam direction. As in the inverse process $e^{+}e^{-} \to Q\bar{Q}$ via single photon exchange, we expect an angular distribution of the form

$$\frac{d\sigma}{d\Omega} = \text{const} (1 + \cos^2 \theta),$$

as is indeed observed (Fig. 7.31).

Equation (7.83) shows that $F(\tau)$ is related to integrals over the quark and antiquark distribution functions, which have been in principle determined independently in the process of lepton-nucleon scattering. We recall that in neutrino scattering, the antiquark distributions $\bar{u}(x)$, $\bar{d}(x)$, $\bar{s}(x)$, ... are relatively difficult to measure, as their content in the nucleon is typically 5–10% only, and has to be determined using the neutrino-antineutrino cross-section differences (see for example Fig. 7.22). On the contrary, the dilepton pair cross-section in hadronic interactions is directly proportional to the product of quark and antiquark distributions. The antiquark distributions obtained from such experiments are in reasonable agreement with those found from lepton scattering; in comparable regions of $|q^2|$, both for example give $x\bar{d}(x) \simeq 0.5(1 - x)^7$.

A further important feature of massive dilepton production in hadronic reactions is that, using incident pion rather than proton beams, the cross-section is proportional to the antiquark distribution in the pion, where $\bar{u}_\pi(x) = u_\pi(x)$, from symmetry. From such experiments, the pion structure function has been determined as

$$xu_\pi(x) \simeq 0.25(1 - x).$$

It is thus much harder than the valence or sea-quark momentum distributions in the nucleon, varying as $(1 - x)^3$ and $(1 - x)^7$ respectively.

PROBLEMS

7.1 Show that if, in the parton model relation (7.46), we do not neglect the mass M of the nucleon or the mass m of the scattered parton, the fractional momentum of the nucleon carried by the initial parton is given by

$$\xi = x\left[1 - \frac{M^2x^2 - m^2}{q^2} + \cdots\right],$$

where $q^2 \gg M^2x^2$ or m^2, and as usual, $x = q^2/(2Mv)$.

7.2 Assuming that the quark and antiquark distributions in the nucleon and pion are of the form $xQ(x) = A(1 - x)^3$ and $x\bar{Q}(x) = B(1 - x)$ respectively, find an expression for the invariant cross-section for production of muon pairs of mass m in a pion-nucleon collision, in the limit where $\tau = m^2/s$ is small, where s is the square of the CMS energy.

7.3 The ratio of total cross-sections for antineutrinos and neutrinos on nucleon targets is found to be $R = 0.5$ in a particular experiment. Deduce the ratio of antiquark to quark momentum content in the nucleon and compute the average value of the quantity y [defined in (7.44)] in neutrino and antineutrino events.

7.4 From Fig. 7.25 it is seen that the process $e^+e^- \to \pi^+\pi^-$ measures the electromagnetic form factor of the pion in the timelike region of q^2. If this process is dominated by the ρ intermediate state, show that the rms radius of the pion is 0.64 fm (for $M_\rho = 765$ MeV).

7.5 In an e^+e^- colliding-beam experiment, the ring radius is 10 m and each beam is of 10 mA, with a cross-sectional area of 0.1 cm². Assuming the electrons and positrons are bunched and the two bunches meet head-on twice per revolution, calculate the luminosity in $cm^{-2} sec^{-1}$ (a luminosity L provides a reaction rate of σL per second for a process of cross-section σ). The cross-section for the reaction $e^{+}e^{-} \to \pi^{+}\pi^{-}\pi^{0}$ at the ω-peak is 1.5 μb. What would be the event rate per hour with the above luminosity?

7.6 A 10-GeV electron collides with a proton and emerges from the collision with a 10° deflection and an energy of 7 GeV. Calculate the rest mass W of the recoiling hadronic state.

7.7 Show that an exponential charge-density distribution of the form (7.36) leads to a form-factor corresponding to the dipole formula (7.35), and verify that for $M_V = 0.84$ GeV the rms radius is 0.8 fm.

7.8 A pencil electron beam of energy 15 GeV and intensity 10^{14} particles sec^{-1} impinges on a liquid hydrogen target of length 1 m parallel to the beam and of cross-section sufficient to cover the beam. Estimate the number of electrons per second scattered elastically through 0.1 rad and into a solid angle of 10^{-4} sr, for (a) pointlike spinless protons, (b) protons with the form factors of Eq. (7.35). (Hydrogen density = 0.06.)

BIBLIOGRAPHY

Close, F. E., "The quark parton model", *Rep. Prog. Phys.* **42**, 1285 (1979).

Close, F. E., *An Introduction to Quarks and Partons*, Academic Press, New York, 1979.

Feynman, R. P., *Photon-Hadron Interactions*, Benjamin, New York, 1972.

Francis, W. F., and T. B. W. Kirk, "Muon scattering at Fermilab", *Phys. Rep.* **54**, 307 (1979).

Friedman, J. I., and H. W. Kendall, "Deep inelastic electron scattering", *Ann. Rev. Nucl. Science* **22**, 203 (1972).

Harari, H., "Quarks and leptons," *Phys. Rep.* **42**, 235 (1978).

Hofstadter, R., "Nuclear and nucleon scattering of high-energy electrons", *Ann. Rev. Nucl. Science* **7**, 231 (1957).

Jacob, M., and P. Landshoff, "Inner structure of the proton", *Sci. Am.* **243**, 46 (Mar. 1980).

Lederman, L. M., "Lepton production in hadron collisions", *Phys. Rep.* **26**, 149 (1976).

Perkins, D. H., "Inelastic lepton nucleon scattering", *Rep. Prog. Phys.* **40**, 409 (1977).

Schwitters, R. F., and K. Strauch, "The physics of $e^{+}e^{-}$ collisions", *Ann. Rev. Nucl. Science* **26**, 89 (1976).

Tung-Mow Yan, "The parton model", *Ann. Rev. Nucl. Science* **26**, 199 (1976).

West, G. B., "Electron scattering from atoms, nuclei and nucleons", *Phys. Rep.* **18c**, 264 (1975).

Chapter 8

Fundamental Interactions and Their Unification

As indicated in Chapter 1, we are faced in nature with several types of fundamental interaction or field between particles. Each field has its distinct characteristics, such as space-time transformation properties (vector, tensor etc.), the set of conservation rules which are obeyed by the interaction, the characteristic coupling constant determining reaction cross-sections, and so forth (see Tables 1.4, 1.5 in Chapter 1).

The fact that the strength of the gravitational interaction between two protons, for example, is only 10^{-38} of their electrical interaction has always been a puzzle and a challenge, and many attempts have been made to try to understand the interrelation between the different fundamental fields.* In recent years it has become fashionable to believe that the strong, weak, electromagnetic, and gravitational interactions are different aspects of a single universal interaction, which would be manifest at some colossally high energy. At much lower energies it is necessary to assume that this symmetry is for some reason badly broken, at mass or energy scales which are puny relative to the unification energy, but large enough to result in dramatic differences in the characteristics of the interactions which can be measured in laboratory experiments.

The first successful attempt to unify two interactions was achieved by Clerk Maxwell in 1865. He showed that electricity and magnetism could be unified into a single theory involving a vector field (the electromagnetic field) interacting between charges and currents. The Maxwell equations involve the introduction of one arbitrary constant – called the velocity of light, c – which is not predicted by the theory and has to be determined by experiment. In the late 1960s Weinberg, Salam, and Glashow described how it might be possible to treat electromagnetic and weak interactions as different aspects of a single electroweak interaction, with the same coupling, e. This symmetry between electromagnetic and weak interactions would be manifest at very large momentum transfers ($q^2 > 10^4 \text{ GeV}^2$). At low energies it is a *broken symmetry*;

---

* The first attempt to relate gravity and electromagnetism seems to have been made by Weyl in 1921.

of the four vector bosons involved, one (the photon) is massless and the others $- W^+, W^-, Z^0 -$ are massive. Thus, compared with electromagnetism, weak interactions are short-range and apparently feeble.

More recently, ambitious attempts have been made to carry unification further, to include strong interactions as well. Although we discuss these so-called grand unified theories briefly in the closing pages of this chapter, the ideas are very speculative and have very little experimental support at the present time. Unification of gravity with the other interactions has also been considered.

8.1 RENORMALIZABILITY IN QUANTUM ELECTRODYNAMICS

In Chapters 1 and 3 we have noted the desirability of formulating quantum field theories which have the property of *renormalizability*, meaning that the amplitudes for different processes associated with an interaction should be well behaved, i.e. nondivergent at high energy and to high orders in the coupling constant. The prototype field theory, that of quantum electrodynamics (QED), does in fact contain divergent terms associated with integrals over intermediate states, but it is found that these divergences can always be absorbed into a redefinition of the "bare" lepton charges and masses, which are in any case arbitrary, as being equal to the physically measured values. All other predicted physical quantities, such as level shifts in atoms, magnetic moments of leptons, or collision cross-sections, are then finite. Thus, we can say that a theory is renormalizable if, at the cost of introducing a *finite* number of arbitrary parameters (to be determined from experiment) the predicted amplitudes for physical processes remain finite at all energies and to all orders in the coupling constant. Quantum electrodynamics is an example of such a theory (and for many years, was the only one); in it there are two arbitrary constants, the charge and mass of the electron.

In contrast, early theories of weak interactions, while well behaved at low energy and to first order, involved divergences in higher orders, which could be canceled only at the price of introducing an indefinitely large number of arbitrary constants, thus losing essentially any predictive power. Good high-energy behavior and cancellation of divergent terms in higher order are thus sensible demands for any physical theory. The subject of renormalizability is highly technical and outside the scope of this text, but we just remark that it appears to be connected with the property of invariance of the interaction under gauge transformations, as discussed below. Not surprisingly therefore, great emphasis has been placed in recent years on the development of generalized gauge theories, which also lead in a natural way to unification of the various fundamental interactions.

Quantum electrodynamics is the best-studied field theory, and it has been impressively successful. To illustrate this fact we now discuss the work on the magnetic moments of electrons and muons.

8.2 QUANTUM-ELECTRODYNAMIC PREDICTIONS OF ELECTRON AND MUON MAGNETIC MOMENTS

Some of the most stringent tests of QED have been obtained by comparing the observed and predicted values of the gyromagnetic ratio of the charged leptons.

The magnetic moment of the electron or muon is given to good approximation by the Dirac theory, describing pointlike spin-$\frac{1}{2}$ particles. According to this theory, the interaction of a stationary charged fermion of mass m with a magnetic field B is described by the energy eigenvalues $E = \mu_z B$, corresponding to an intrinsic magnetic moment associated with the spin which has two components

$$\mu_z = \pm \mu_B,$$

where

$$\mu_B = \frac{e\hbar}{2mc} \tag{8.1}$$

is called the Bohr magneton. Generally, the magnetic moment $\boldsymbol{\mu}$ is related to the spin vector $\mathbf{s}$ by

$$\boldsymbol{\mu} = g\mu_B \mathbf{s}, \tag{8.2}$$

where g is called the Landé g-factor, and $g\mu_B = \mu/s$ is the gyromagnetic ratio, i.e. the ratio of magnetic to mechanical moment. Thus the Dirac theory predicts for particles of spin $\frac{1}{2}$

$$g = 2. \tag{8.3}$$

The actual g-values of the electron and muon have been determined with great precision and differ by a small amount (0.2%) from the value 2. Thus the Dirac picture of a structureless, point particle is not exact. Similar departures are encountered in atomic levels: for example, the famous Lamb shift is a 1% correction to the $2P_{3/2}$-$2S_{1/2}$ level separation in the fine structure of hydrogen.

The magnetic moment (8.1) of a charged particle depends on the ratio e/m and thus, classically, for a rotating structure, on the spatial distributions of charge and mass. (If the two distributions are the same, a value $g = 1$ is obtained on classical arguments.) Deviation from the g-value of 2 for a spin-$\frac{1}{2}$ particle therefore argues that processes are taking place which distort the relative charge and mass distributions. For example, the proton has a g-value of 5.59, which must arise from its composite structure and different values of e/m for the components (quarks).

In QED, an electron interacts with other charges via the exchange of virtual photons. At any instant of time therefore, an electron has a certain probability of consisting of a "bare" or pointlike object, surrounded by a cloud of one or more virtual photons, which are continually being emitted and reabsorbed. Qualitatively, we can see that although the charge e resides on the electron, part

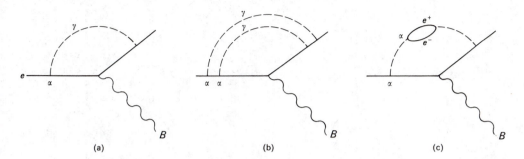

Fig. 8.1   Radiative corrections to the g-factor of electron and muon.

of the mass energy is carried by the photon cloud, and hence the value of e/m for the electron itself will be slightly increased. This increased value would obtain if one measures the magnetic moment of the electron with an external magnetic field B. The amount of the correction will be proportional to the photon emission probability, determined by the coupling constant $\alpha = e^2/4\pi\hbar c$. It is obvious that correction terms of order α, α^2,... correspond to emission of $1, 2, \ldots$ virtual photons at the moment the field B is applied.

Figure 8.1 gives examples of Feynman diagrams depicting these corrections. Figure 8.1(a) and (b) correspond to single and double photon emission respectively, while (c) corresponds to emission of a virtual photon which transforms to a virtual e^+e^- pair. Clearly, B can interact with either the "parent" electron or either member of the pair, and the correction, like that in (b), will be of order α^2. The process (c) is called "vacuum polarization", because the e^+ and e^- of the pair will be polarized and tend to shield the charge of the parent e^-.

The calculations of magnetic moments in QED have been improved in accuracy over many years. An indication of the labor involved can be glimpsed from Fig. 8.2, showing the 72 Feynman graphs which have to be evaluated to determine the coefficient of the α^3 term. The present QED prediction for the electron is

$$\left(\frac{g-2}{2}\right)_e^{\text{QED}} = 0.5\frac{\alpha}{\pi} - 0.32848\left(\frac{\alpha}{\pi}\right)^2 + 1.19\left(\frac{\alpha}{\pi}\right)^3 + \cdots$$

$$= (1\,159\,652.4 \pm 0.4) \times 10^{-9}, \tag{8.4}$$

while that for the muon is

$$\left(\frac{g-2}{2}\right)_\mu^{\text{QED}} = 0.5\frac{\alpha}{\pi} + 0.76578\left(\frac{\alpha}{\pi}\right)^2 + 24.45\left(\frac{\alpha}{\pi}\right)^3 + \cdots$$

$$= (1\,165\,851.7 \pm 2.3) \times 10^{-9}. \tag{8.5}$$

Fig. 8.2 The Feynman graphs which have to be evaluated in computing the α^3 corrections to the lepton magnetic moments (after Lautrup *et al.* 1972).

Note that the leading correction $0.5(\alpha/\pi)$ is obviously the same for electron and muon, while the α^2 term differs in both magnitude and sign in the two cases. This arises because the momenta of the particles in the intermediate virtual states scale in proportion to the parent particle mass. The vacuum-polarization term for the muon is much larger.

The QED result (8.5) differs significantly from the observed value by some 9 standard deviations (Table 8.1). There are however additional effects which arise because a virtual photon can transform not only into a lepton pair but also into hadrons. This contribution can be computed directly from the results of electron-positron colliding-beam experiments, $e^+e^- \to \gamma \to$ hadrons. It is negligible for the $g - 2$ of the electron, but for the muon the correction results in revision of (8.5) to

$$\left(\frac{g-2}{2}\right)_{\mu}^{\text{theory}} = (1\,165\,918 \pm 10) \times 10^{-9}, \tag{8.6}$$

in perfect agreement with experiment. Contributions from other effects (e.g. weak interactions, QED production of pairs of heavy $\tau^\pm$ leptons) can be neglected at the present level of accuracy.

8.3 EXPERIMENTAL DETERMINATION OF THE g-FACTORS OF ELECTRONS AND MUONS

Experiments to measure the magnetic moments of free electrons and muons started more than 20 years ago and form a saga to which we cannot hope to do justice here. We describe briefly the two most recent experiments on the $g - 2$ value of the electron and muon. Both employ a version of the so-called Penning trap, consisting of an electric quadrupole field with a vertical axis of symmetry, and a superposed uniform magnetic field $\mathbf{B}$ also in the vertical direction z. The effect of the electric field is to confine the electron or muon in a potential well in which they make small vertical oscillations. Due to the magnetic field, the particles perform circular cyclotron orbits in the horizontal plane, and the particle spin also precesses about the field direction. The orbital (cyclotron) frequency is given by

$$\omega_c = \frac{eB}{mc\gamma} = \frac{2\mu_B B}{\hbar\gamma}, \tag{8.7}$$

where m is the particle rest mass, $\gamma = (1 - v^2/c^2)^{-1/2}$. Equation (8.7) follows from the usual expression for the radius of curvature ρ in terms of the momentum p, namely $p = Be\rho$. The spin-precession frequency is given by*

$$\omega_s = \frac{eB}{mc\gamma}\left[1 + \left(\frac{g-2}{2}\right)\gamma\right]. \tag{8.8}$$

---

* For a derivation of this formula, see V. Bargmann, L. Michel, and V. L. Telegdi, *Phys. Rev. Lett.* **2**, 435 (1959).

Let us consider first the case of a nonrelativistic electron in a uniform magnetic field B. The cyclotron frequency is then $\omega_0 = 2\mu_B B/\hbar$. The energy eigenvalues have the form

$$E = (2n + 1 + gm_s)\mu_B B + \hbar(n_z + \tfrac{1}{2})\omega_z, \tag{8.9}$$

where the second term is the contribution from vertical oscillations in the electric field, of frequency ω_z determined by the magnitude and shape of the potential well. n, $n_z = 0, 1, 2, \ldots$, are the quantum numbers of the orbital and axial motion, and $m_s = \pm \tfrac{1}{2}$. Setting the anomaly

$$a = (g - 2)/2, \tag{8.10}$$

the two possible values of m_s give, from the first term in (8.9), two distinct sets of levels:

$$E_1 = [2(n + 1) + a]\mu_B B, \qquad E_2 = [2n - a]\mu_B B. \tag{8.11}$$

Thus if we can induce simultaneous transitions flipping the electron spin ($\Delta m_s = 1$) and changing n by one unit, these will occur at the difference

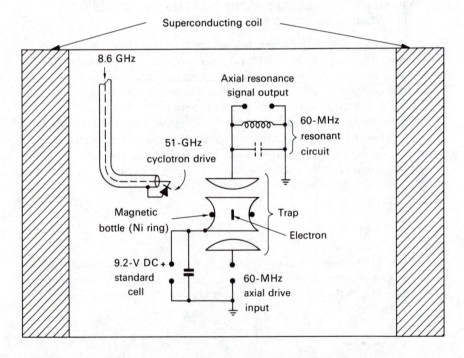

Fig. 8.3 The Penning trap employed by Van Dyck, *et al.* (1976) at the University of Washington in their very precise measurements of the electron magnetic moment.

frequency

$$\omega_a = 2a\mu_B B/\hbar = a\omega_0 \qquad (8.12)$$

measuring the anomaly $(g - 2)/2$ directly.

The experiment by Van Dyck, Dehmelt, *et al.* (1976, 1977) is illustrated in Fig. 8.3. A uniform vertical magnetic field of 1.83 T is supplied by a superconducting coil, and the quadrupole electric field by electrodes of parabolic shape, the whole system being in high vacuum. A pulsed beam of 1-keV electrons was used to liberate slow electrons from gas molecules in the trap. By applying an RF voltage to one electrode, an electron will be driven into vertical oscillation, passing through a sharp maximum at the resonant frequency ω_z for free oscillations. By increasing the RF amplitude, electrons can be driven out of the trap, so that only a single one remains — which can be observed for

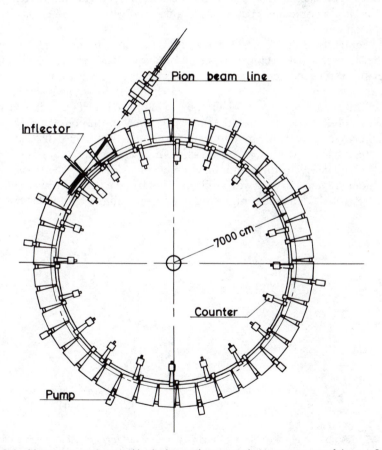

Fig. 8.4 Muon storage ring used in the last and most precise measurement of the $g - 2$ value of the muon at the CERN laboratory (after Bailey *et al.* 1977).

several days. By introducing a nickel ring around the trap, a very weak extra component of field could be introduced, so that $B = B_0 + B_1 z^2$. This couples vertical and horizontal oscillations. Using an auxiliary axial RF drive field of frequency near ω_a, the spin-flip transition rate was measured as a function of minute changes (~ 10 Hz in 60 MHz) in the axial frequency ω_z, and the resonance curve mapped out. Typically, electrons followed horizontal cyclotron orbits of 1-μm radius and vertical oscillations of amplitude 100 μm. In this experiment, precise determination of ω_a is achieved because a single electron can be observed for very long periods of time. The final result of the experiment gave

$$a_e = \frac{g-2}{2} = (1\,159\,652.4 \pm 0.2) \times 10^{-9}, \tag{8.13}$$

in astonishingly good agreement with theory [see Eq. (8.4) and Table 8.1]. Note that the experimental error is smaller than that in the prediction, which has to rely on the experimentally determined value of α (with its error) from the a.c. Josephson effect.

In contrast with the electron experiment, the most recent $g-2$ experiment on the muon involves cyclotron orbits of radius 7 m and vertical oscillations of order 10 cm. The method employs a muon storage ring, consisting of a vacuum tube threading a ring of 40 electromagnets (Fig. 8.4) providing a very uniform vertical magnetic field $B\,(= 1.47\text{ T})$. A beam of pions from a secondary target at the CERN PS was injected into the ring. After a few revolutions the pions decay in flight into muons. Those muons projected close to the forward direction in the pion rest frame have nearly the same momentum as the pions—3.098 GeV/c—and can be trapped into an equilibrium orbit. Trapping (focusing) of the muons is again achieved with an electric quadrupole field. "Forward" muons within 1% of the maximum momentum are nearly fully (97%) polarized, with

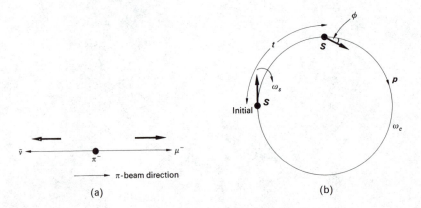

(a)   (b)

Fig. 8.5   (a) Spin polarization sense of muon emitted in the "forward" direction in π-decay in flight. (b) For a particle of $g \neq 2$ in a uniform magnetic field, the spin vector **s**, initially aligned with the momentum **p**, will "lead" by a phase angle ϕ at later times.

spin vectors pointing initially along the direction of motion [see Fig. 8.5(a) and the discussion in Section 6.8.1]. Counters around the ring were biased to detect the highest-energy electrons from muon decays, i.e. those projected forward in the muon rest frame. Thus, the electron angular distribution in the laboratory follows closely the precession of the muon spin [frequency ω_s in (8.8)], while the muons themselves execute cyclotron orbits of frequency ω_c in (8.7). Since $\omega_s > \omega_c$, the spin vector, initially aligned with the momentum, will lead at later times, and the counting rate of decay electrons is modulated by the frequency $(\omega_s - \omega_c) = \omega_a$ — see Fig. 8.5(b).

So far, we have neglected the effects of the quadrupole electric field. Since a charged particle of velocity βc traversing an electric field E "sees" an additional magnetic field $\boldsymbol{\beta} \times \mathbf{E}$, there is an extra precession given by

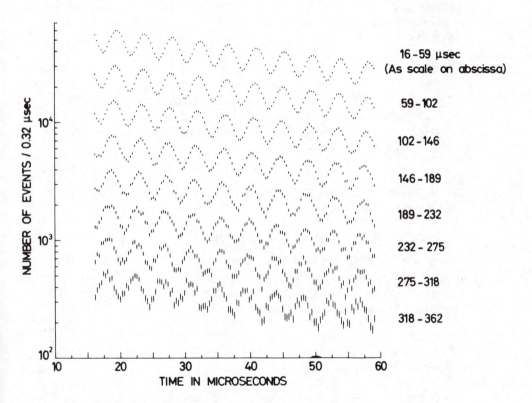

Fig. 8.6 Time dependence of the electron counting rate from the decay of muons in the CERN $g - 2$ experiment. The general exponential decrease corresponds to the loss of muons by radioactive decay, with a mean lifetime dilated by the relativistic γ-factor of 30. (It is of interest to remark that this experiment provides the most precise (0.1% accuracy) check of Einstein's time-dilation formula. The overall decrease in count rate with time is modulated by the $g - 2$ frequency $\omega_s - \omega_c$.

$$\Delta\omega_a = \frac{e}{mc}\left(\frac{1}{\gamma^2 - 1} - a\right)\boldsymbol{\beta} \times \mathbf{E}. \tag{8.14}$$

This effect can be exactly canceled by the choice of a "magic" muon momentum, given by $\gamma^2 = 1 + 1/a$, or $p = 3.098$ GeV/c. Thus the electric-field corrections were reduced to second-order effects (a few parts per million).

TABLE 8.1 $(g - 2)/2$ anomaly for leptons

	$10^9(g - 2)/2$	
	Electron	Muon
QED prediction	1 159 652.4 ± 0.4	1 165 851.7 ± 2.3
With strong-interaction correction	–	1 165 918 ± 10
Observed value	1 159 652.4 ± 0.2	1 165 924 ± 9

Figure 8.6 shows an example of the decay-electron count rate as a function of time from injection of the pions, consisting of a slow exponential decay modulated by the frequency ω_a. The final result of this experiment was, for the weighted average of μ^+ and μ^-,

$$a_\mu = (1\ 165\ 924 \pm 9) \times 10^{-9}, \tag{8.15}$$

again in agreement with theory when hadronic contributions are included (see Table 8.1).

8.4   DIVERGENCES IN THE WEAK INTERACTIONS

In contrast with QED, the *V-A* (Fermi) theory of weak interactions is badly divergent. We recall that in the Fermi theory of β-decay, the four fermions involved are assumed to have a contact interaction specified by the Fermi constant G. An example is the process

$$\nu_e + e^- \rightarrow e^- + \nu_e. \tag{8.16}$$

The cross-section has the pointlike form given in (7.53) using (7.48) for the substitution $q^2 = 2M\nu = 2MEy$:

$$\frac{d\sigma}{dq^2} = \frac{G^2}{\pi}, \tag{8.17}$$

where q^2 is the momentum transfer squared. The total cross-section is then

$$\sigma_{\text{tot}}(\nu e) = \frac{G^2}{\pi}q^2_{\text{max}} = \frac{2G^2mE}{\pi} = \frac{G^2s}{\pi}, \tag{8.18}$$

where m is the electron mass, E the incident neutrino energy, and s is the CMS energy squared ($s = 2mE$). The cross-section depends only on G and the phase space, which rises with CMS momentum p^* as $p^{*2} = s/4$ (we assume $E \gg m$).

The cross-section for such an elastic scattering process can also be found from wave theory (Eq. (4.41) and (4.51)) which yields, for $\eta = 0$, spin $s = \frac{1}{2}$ for the target electron and pointlike scattering ($l = 0$),

$$\sigma_{\text{max}} = \pi \lambda^2 (2l + 1)/(2s + 1) = \pi \lambda^2/2 \qquad (8.19)$$

where $\lambda = \hbar/p^*$ is the CMS wavelength. So Eq. (8.18) predicts a value for $\sigma_{\text{tot}} > \sigma_{\text{max}}$ when

$$\frac{4G^2 p^{*2}}{\pi} > \frac{\pi}{2p^{*2}}$$

or when

$$p^* > (\pi/G\sqrt{8})^{\frac{1}{2}} \simeq 300 \text{ GeV}/c. \qquad (8.20)$$

At sufficiently large energy, the Fermi theory therefore predicts a cross-section exceeding the wave-theory limit, which is determined by the condition that the scattered intensity cannot exceed the incident intensity in any partial wave: frequently this is called the unitarity limit. The basic reason for the bad high-energy behaviour of the V-A theory (which becomes steadily worse if processes of higher order in G — i.e. G^2, G^4, etc. — are considered) is that G has the dimensions of an inverse power of the energy. Somehow, we have to redefine the weak interaction in terms of a dimensionless coupling constant.

The proposed intermediate vector boson $W^{\pm}$ of the weak interactions was introduced in Section 6.10. Its effect is to introduce a propagator term $(1 + q^2/M_W^2)^{-1}$ into the scattering amplitude, which "spreads" the interaction over a finite range, of order M_W^{-1}, so that σ_{tot} in (8.18) will tend to a constant value $G^2 M_W^2/\pi$ at high energy. The unitarity limit in a given partial wave is still broken, although only logarithmically. Quadratic divergences however still appear in other, more esoteric, processes; for example, $\nu\bar{\nu} \rightarrow W^+ W^-$ is a conceivable reaction for which $\sigma_{\text{tot}} \propto s$. What is needed therefore is a mechanism to cancel the weak-interaction divergences systematically and to all orders in G.

8.4.1 Introduction of Neutral Currents

Figure 8.7 shows in (a) one of the recalcitrant diagrams giving a quadratic divergence (for longitudinally polarized Ws). One can in principal cancel this divergence by introducing ad hoc a neutral boson Z^0 or a new heavy lepton E^+ (with the same lepton number as e^- and ν_e) — or both — with suitably chosen couplings. Note that diagram (b) involves a neutral current, for which there is experimental evidence, as discussed in Section 6.10. There is no present evidence for a third, electron-type lepton E^+, and we do not consider this possibility (c) further.

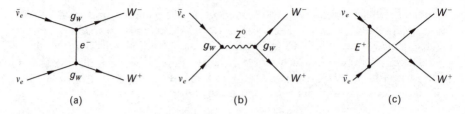

Fig. 8.7

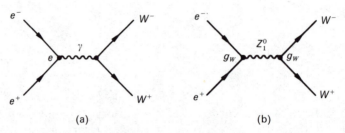

Fig. 8.8

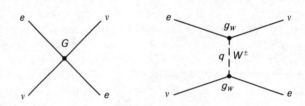

Fig. 8.9

Similarly the electromagnetic process $e^+e^- \rightarrow W^+W^-$ shown in Fig. 8.8(a) is also divergent, and cancellation in this order may again be effected by postulating a neutral vector boson Z_1^0 as in (b). Although in principle the cancellations in Figs. 8.7 and 8.8 could be effected with two different particles, Z^0 and Z_1^0, it is more economical to have just one. Nature seems to have chosen this course. In that case, it is clear that the Z^0 weak coupling, g_W, will have to be similar in magnitude to the electromagnetic coupling, e. In other words, the *two fields are unified*, with (apart from numerical factors of order unity) the same intrinsic coupling strength $g_W \sim e$ to leptons of the mediating bosons W, Z^0, and γ. Recalling from (6.63) and Fig. 8.9 that in the low-q^2 limit

$$\lim_{q^2 \rightarrow 0} \frac{g_W^2}{q^2 + M_W^2} \equiv \frac{G}{\sqrt{2}}, \tag{8.21}$$

one finds, setting $g_W = e$,

$$M_W^2 = \frac{g_W^2 \sqrt{2}}{G} = \frac{e^2 \sqrt{2}}{G}. \tag{8.22}$$

Inserting $e^2 = 4\pi/137$, $G = 10^{-5}/M_p^2$ (in units $\hbar = c = 1$) gives

$$M_{W,Z^0} \simeq 100\,\text{GeV}. \tag{8.23}$$

This is the approximate mass required for the W and Z^0 particles if they are to have the same coupling as in electromagnetism and to give an effective four-fermion coupling of magnitude G in weak interactions at low energy.

Actually we implied that by introducing a Z^0 with the right coupling, all the divergences in $e^+e^- \rightarrow W^+W^-$ would disappear. This is only true if the electron mass can be neglected. For a finite electron mass, a residual divergence exists and has to be canceled by introduction of further scalar particles ϕ, with special couplings proportional to the lepton mass. Such particles are called *Higgs scalars*.

8.5 GAUGE INVARIANCE IN QED

In order to understand the construction of the electroweak unified theory, we have to discuss the subject of gauge invariance, since this is at the center of renormalizable models.

In quantum mechanics, the physical observables of a system are described by the expectation values obtained when some operator Q works on the wavefunction $\psi(x)$ of the system (where x represents the space-time coordinate)

$$\langle Q \rangle = \int \psi^*(x) Q \psi(x)\, d\tau. \tag{8.24}$$

Clearly $\langle Q \rangle$ is invariant under a *global* phase transformation (i.e. the same over all space) of the form

$$\psi(x) \rightarrow e^{i\alpha}\psi(x), \tag{8.25}$$

where α is any scalar. In other words, the phase of the wavefunction is arbitrary and unobservable. The equations of motion of the system in fact involve derivatives of the wavefunction, via terms like $\psi^*(x)\partial_\mu\psi(x)$, and this is also invariant. Here ∂_μ represents a space and time gradient ($\mu = 1, \ldots, 4$).

Suppose however that we wish to choose α *differently at each space-time point*. Since α is arbitrary, it is clearly desirable and esthetic to let the phase be chosen at will in different places – a so-called local phase transformation. This turns out to be possible, provided we introduce a massless (i.e. long-range) vector field with which the particle (described by ψ) interacts. In the case of a charged particle, this will be the 4-vector electromagnetic field A_μ. The local phase transformations on the particle wavefunction have then to be accompanied by local gauge transformations on the field A_μ, and actually specify the form of the particle-field interaction.

Consider a phase rotation on the charged-particle wavefunction $\psi(x)$ of the form

$$\psi(x) \to e^{ie\theta(x)}\psi(x), \tag{8.26}$$

where e is the electric charge of the particle. Then the gradient

$$\partial_\mu \psi(x) \to e^{ie\theta(x)}[\partial_\mu \psi(x) + ie\psi(x)\partial_\mu\theta(x)]$$
$$\neq e^{ie\theta(x)}\partial_\mu\psi(x),$$

and $\psi^*(x)\partial_\mu\psi(x)$ is not invariant. However, let us write a local gauge transformation on the field A_μ:

$$A_\mu(x) \to A_\mu(x) + \partial_\mu\theta(x), \tag{8.27}$$

where the field equations of Maxwell are invariant if we add the derivative of any scalar $\theta(x)$ to the 4-vector potential (see Section 3.8). Let us also replace ∂_μ by the *covariant derivative*, defined by

$$D_\mu = \partial_\mu - ieA_\mu. \tag{8.28}$$

Under the simultaneous phase transformation on $\partial_\mu\psi(x)$ and gauge transformation on $A_\mu(x)$ we get

$$D_\mu\psi(x) \to e^{ie\theta(x)}[\partial_\mu\psi(x) + ie\psi(x)\partial_\mu\theta(x)$$
$$- ie\psi(x)A_\mu(x) - ie\psi(x)\partial_\mu\theta(x)]$$
$$= e^{ie\theta(x)}D_\mu\psi(x), \tag{8.29}$$

and $\psi^*(x)D_\mu\psi(x)$ is again invariant. The fact that one must use D_μ instead of ∂_μ is already specifying the form of the interaction (eA_μ) of the charged particle with the field. As indicated in Chapter 3, the gauge invariance of the electromagnetic interaction leads to a conserved current (charge conservation) and the masslessness of the bosons* (photons) associated with A_μ. Most importantly, the theory contains a high degree of symmetry, and as a result, it is renormalizable, with systematic cancellations of divergent terms, order by order in the coupling constant e.

8.6 GENERALIZED GAUGE INVARIANCE

The infinite set of phase transformations (8.26) form the unitary group called U(1). Since $\theta(x)$ is a scalar quantity, the U(1) group is said to be Abelian. More complex phase transformations are also possible, specified by noncommuting operators, and these belong to the so-called non-Abelian groups. Such gauge transformations were proposed by Yang and Mills in 1954 and involved fields containing both charged and neutral massless bosons. Specifically, they chose

---

* Gauge invariance fails if the bosons are massive, and can only be restored by including extra interactions (Higgs scalars).

the group SU(2) of isospin, which involves the *noncommuting* Pauli matrices $\tau = \tau_1, \tau_2, \tau_3$ (see Appendix B). Recall that the conservation of isospin in strong interactions implies invariance under an isospin rotation (see (3.16)):

$$\psi \to e^{ig\tau \cdot \Lambda}\psi, \tag{8.30}$$

where Λ is an arbitrary vector about which the rotation in "isospin space" takes place and g is a constant. Again, we can require that $\Lambda(x)$ be chosen arbitrarily at different space-time points x. For example, we could choose the two states ψ of the nucleon as proton or neutron independently at different places. In the same way as for electromagnetism, a gauge-invariant description can be obtained by introducing a massless isovector field $\mathbf{W}_\mu$ with charged and neutral components, and a strong coupling constant g, analogous to e. Invariance of $\psi^* D_\mu \psi$ under the transformation (8.30) is obtained by introducing a covariant derivative of the form

$$D_\mu = \partial_\mu - ig\tau \cdot \mathbf{W}_\mu, \tag{8.31}$$

where an infinitesimal gauge transformation of the field $\mathbf{W}_\mu$ is given by

$$\mathbf{W}_\mu \to \mathbf{W}_\mu + \partial_\mu \Lambda - g\Lambda \times \mathbf{W}_\mu. \tag{8.32}$$

The extra term, as compared with (8.27), is associated with the fact that the isospin matrices $\tau = \tau^{(1)}, \tau^{(2)}, \tau^{(3)}$ do not commute*. The transformations (8.30) and (8.32) can be verified by writing down the expression for $\psi^* D_\mu \psi$ in the case where Λ is infinitesimal. Then, neglecting second-order terms in Λ,

$$\psi^* D_\mu \psi = \psi^*(\partial_\mu - ig\tau \cdot \mathbf{W})\psi$$

$$\to \psi^*(1 - ig\tau \cdot \Lambda)[\partial_\mu - ig\tau \cdot (\mathbf{W} + \partial_\mu \Lambda - g\Lambda \times \mathbf{W})(1 + ig\tau \cdot \Lambda)]\psi$$

$$= \psi^*(\partial_\mu - ig\tau \cdot \mathbf{W})\psi$$

$$+ g^2\psi^*[-(\tau \cdot \Lambda)(\tau \cdot \mathbf{W}) + (\tau \cdot \mathbf{W})(\tau \cdot \Lambda) + i\tau \cdot \Lambda \times \mathbf{W}]\psi. \tag{8.33}$$

The second term, written out in isovector components i, j, k, is

$$\sum \tau^i W^i \sum \tau^j \Lambda^j - \sum \tau^j \Lambda^j \sum \tau^i W^i + i \sum \tau^k \sum \Lambda^j W^i$$

$$= \sum W^i \Lambda^j \sum (\tau^i \tau^j - \tau^j \tau^i) - i \sum W^i \Lambda^j \sum \tau^k = 0. \tag{8.34}$$

Hence, inclusion of the vector-product term in (8.32) leads to local gauge invariance. It implies an interaction of $\mathbf{W}_\mu$ with all particles carrying isospin and hence with itself. Thus the bosons $\mathbf{W}_\mu$ are at one and the same time the carriers and a part of the source of the isospin field. A parallel example is that of gravity, where the field quanta are the massless gravitons, mediating the interaction between masses; but since the gravitons carry energy and momentum, they are

---

* Boldface type indicates isovector quantities, and superscripts the Cartesian components in "isospin space". The subscript μ stands for the space-time components of a 4-vector ($\mu = 1, \ldots, 4$).

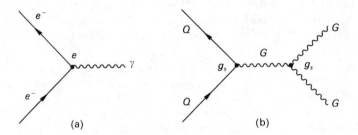

Fig. 8.10

themselves part of the source of gravitational field. Similarly, we shall see that in the gauge theory of strong interactions, involving the non-Abelian gauge group of color, SU(3), the quanta (gluons) of the color field themselves carry color and therefore are self-coupled (Fig. 8.10).

We note, however, that the hypothetical charged bosons of the isospin field are massless, just like the photon, and we infer that isospin symmetry in hadronic interactions cannot be (and is not) exact, since no massless charged particles exist.

8.7   THE WEINBERG-SALAM SU(2) × U(1) MODEL

In 1967–1968 Weinberg and Salam proposed a gauge theory unifying weak and electromagnetic interactions – the so-called electroweak interactions, based on an SU(2) group of "weak isospin" I and a U(1) group of "weak hypercharge" Y (phase transformations). They invoked a process called "spontaneous symmetry breaking" which endows the gauge bosons with mass, without spoiling the renormalizability of the theory. This may be achieved with the help of an isospin doublet of scalar mesons called Higgs scalars, which generate mass as a result of self-interaction, and which we mentioned above in connection with removal of divergences in $e^+e^- \rightarrow W^+W^-$. This topic is outside the scope of this text, and the reader is referred to the chapter bibliography for details. At the price of extra scalar particles, the quanta of the "weak isospin" field are given a mass and the theory remains renormalizable. The crucial question of renormalizability was not settled until 1971 by 't Hooft, and until that time the theory was not taken too seriously. The fundamental vector bosons are a massless isovector triplet $\mathbf{W}_\mu = W_\mu^{(1)}, W_\mu^{(2)}, W_\mu^{(3)}$ [for SU(2)] and a massless isosinglet B_μ [for U(1)]. As a result of spontaneous symmetry breaking, three bosons (denoted W_μ^+, W_μ^-, and Z_μ^0) acquire mass, and one (A_μ, the photon) remains massless. These four bosons are combinations of $\mathbf{W}_\mu$ and B_μ, as discussed below. There are no new leptons in this model, and the cancellation of divergences in Fig. 8.7 is accomplished by the neutral boson Z^0.

As stated previously, the interaction energy, usually represented by the so-called Lagrangian density $\mathscr{L}$, of fermions ψ with the fields $\mathbf{W}_\mu$, B_μ will be of the form $\bar{\psi}\gamma^\mu D_\mu\psi$, where γ^μ is the Dirac operator appropriate to a 4-vector field and where D_μ will now contain, apart from the gradient term, additional terms of the type shown in (8.28) and (8.31). Thus the interaction to first order in the coupling can be written (see also Eq. (6.29)):

$$\mathscr{L} = g\mathbf{J}_\mu \cdot \mathbf{W}_\mu + g'J^Y_\mu B_\mu, \tag{8.35}$$

where $\mathbf{J}_\mu$ and J^Y_μ represent respectively the isospin and hypercharge currents of the fermions (leptons or quarks), and g and g' are their couplings to $\mathbf{W}_\mu$ and B_μ, analogous to the charge e in (8.28).

Defining $Y = Q - I_3$ from the Gell-Mann–Nishijima formula, we can write

$$J^Y_\mu = J^{e.m.}_\mu - J^{(3)}_\mu, \tag{8.36}$$

where $J^{e.m.}_\mu$ is the electromagnetic current, coupling to the charge Q, and $J^{(3)}_\mu$ is the third component of the isospin current $\mathbf{J}_\mu$. The physical bosons consist of the charged particles $W^\pm_\mu$ and the neutrals Z_μ and A_μ (the photon). The latter are taken as linear combinations of $W^{(3)}_\mu$ and B_μ. Thus we can set [see Eqs. C4 and C5(c), Appendix C]

$$W^\pm_\mu = \frac{1}{\sqrt{2}}(W^{(1)}_\mu \pm iW^{(2)}_\mu) \tag{8.37}$$

and

$$W^{(3)}_\mu = \frac{gZ_\mu + g'A_\mu}{\sqrt{g^2 + g'^2}}, \tag{8.38}$$

$$B_\mu = \frac{-g'Z_\mu + gA_\mu}{\sqrt{g^2 + g'^2}}, \tag{8.39}$$

where $W^{(3)}_\mu$ and B_μ are orthogonal as required ($\langle W^{(3)}_\mu | B_\mu \rangle = 0$). Hence

$$\mathscr{L} = g(J^{(1)}_\mu W^{(1)}_\mu + J^{(2)}_\mu W^{(2)}_\mu) + g(J^{(3)}_\mu W^{(3)}_\mu) + g'(J^{e.m.}_\mu - J^{(3)}_\mu)B_\mu$$

$$= (g/\sqrt{2})(J^-_\mu W^+_\mu + J^+_\mu W^-_\mu) + J^{(3)}_\mu(gW^{(3)}_\mu - g'B_\mu) + J^{e.m.}_\mu g'B_\mu,$$

where $J^\pm_\mu = J^{(1)}_\mu \pm iJ^{(2)}_\mu$.

Inserting the expressions for $W^{(3)}_\mu$ and B_μ from (8.38) and (8.39) and setting

$$g'/g = \tan\theta_w \tag{8.40}$$

gives the result

$$\mathscr{L} = \frac{g}{\sqrt{2}}(J^-_\mu W^+_\mu + J^+_\mu W^-_\mu) + \frac{g}{\cos\theta_w}(J^{(3)}_\mu - \sin^2\theta_w J^{e.m.}_\mu)Z_\mu + g\sin\theta_w J^{e.m.}_\mu A_\mu.$$

$$\qquad\qquad \uparrow \qquad\qquad\qquad\qquad\qquad \uparrow \qquad\qquad\qquad\qquad \uparrow$$

$$\text{weak CC} \qquad\qquad\qquad\qquad \text{weak NC} \qquad\qquad\qquad \text{e.m. NC}$$

$$\tag{8.41}$$

This equation shows that the interaction contains the weak charge-changing current, a weak neutral current, and the electromagnetic neutral current, for which we know [see (8.28)] the coupling to be e. Hence

$$e = g \sin \theta_w. \tag{8.42}$$

The angle θ_w is called the weak mixing angle (or Weinberg angle). We now have to relate the coupling g to the Fermi constant G. Let us consider the leptonic current. In the theory, the leptons as well as the gauge bosons are endowed with isospin and hypercharge. The charged current (V-A) weak interactions (mediated by $W^\pm$) connect v_e and e^-, which are assigned to a left-handed doublet ψ_L, while the right-handed electron state belongs to a singlet, ψ_R:

$$\psi_L = \frac{1 + \gamma_5}{2} \begin{pmatrix} v_e \\ e^- \end{pmatrix} \quad \begin{matrix} I = \tfrac{1}{2}, I_3 = +\tfrac{1}{2}, Q = 0 \\ I = \tfrac{1}{2}, I_3 = -\tfrac{1}{2}, Q = -1 \end{matrix} \left.\vphantom{\begin{matrix}a\\a\end{matrix}}\right\} \quad Y = -\tfrac{1}{2},$$

$$\psi_R = \frac{1 - \gamma_5}{2} (e^-) \quad I = 0, Q = -1, \qquad\qquad Y = -1. \tag{8.43}$$

The lepton wavefunctions in (8.43) are normalized to give unity for the electron state (both LH and RH components). The form of the charged current interaction in Fig. 8.9 is defined as $\mathcal{L}_{cc} = g_w \cdot J(\text{lepton}) \cdot W^+$, where $J(\text{lepton}) = \bar{v}_e \gamma_\mu (1 + \gamma_5)e$ as in (6.33). So from (8.22) and (8.41) we can make the identification

$$\frac{g}{2\sqrt{2}} = g_w = \left(\frac{GM_W^2}{\sqrt{2}} \right)^{1/2}$$

and from (8.42) it follows that

$$M_{W^\pm} = \left(\frac{g_w^2 \sqrt{2}}{G} \right)^{1/2} = \left(\frac{e^2 \sqrt{2}}{8G \sin^2 \theta_w} \right)^{1/2} = \frac{37.4}{\sin \theta_w} \text{ GeV} \tag{8.44}$$

to be compared with our first rough guess, (8.23). It can also be shown that, in the simplest gauge model (of Weinberg and Salam) with a single doublet of Higgs scalars:

$$M_{Z^0} = \frac{M_{W^\pm}}{\cos \theta_w} = \frac{75}{\sin 2\theta_w} \text{ GeV}. \tag{8.45}$$

In this model, there is therefore one free parameter, θ_w, determining the ratio of the couplings g' and g or, equivalently, the relative magnitude of charged- and neutral-current reaction cross-sections.

To summarize therefore: a renormalizable gauge-invariant description of the electroweak interactions of leptons may be obtained by postulating an "isospin" triplet vector boson $\mathbf{W}_\mu$ and an isosinglet vector boson B_μ. This model predicted neutral weak currents, which were subsequently observed. The masslessness of the bosons is destroyed by spontaneous breaking of the gauge

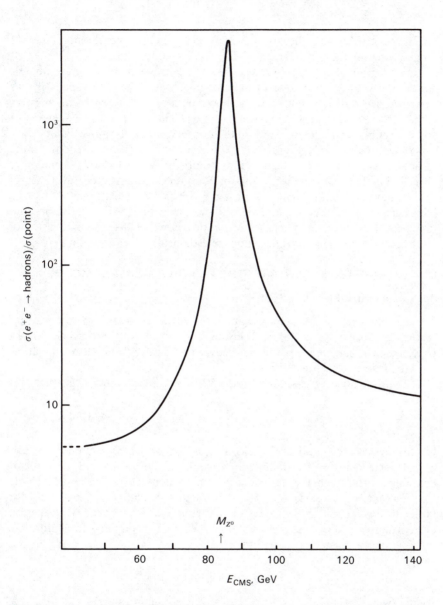

Fig. 8.11 Predicted dependence of the cross-section for the process $e^+e^- \to$ hadrons, compared with the pointlike cross-section ($M_{Z^0} = \infty$), near the Z^0 resonance, for $\sin^2 \theta_W = 0.25$. For the current world average of 0.23, $M_W \sim 80$ GeV and $M_{Z^0} \sim 90$ GeV.

symmetry, endowing three of the four bosons with large masses. The intrinsic coupling of the bosons to leptons has basically the same value (e), and the weakness of the weak interactions — compared with electromagnetic interactions — is attributed to the large boson masses and consequently short effective range of the interaction.

The theory makes several predictions. All electroweak interactions must be described by a unique value of the parameter θ_w, and experimentally, as described below, this so far indeed seems to be the case. Secondly, the massive bosons $W^\pm, Z^0$, with known masses, must exist. These have well-defined widths and decay characteristics, and indeed their observation is a major goal of future experiments (as of 1981). Figure 8.11 shows the anticipated cross-section for the resonant process $e^+e^- \to Z^0$, the main motivation for the LEP collider at CERN. Finally, one (or more) "isospin" doublets of particles, called Higgs scalars, have been postulated to endow the $W^\pm, Z^0$ with large masses, but the Higgs masses are unknown, and we still await evidence for their existence and very peculiar properties. There is no question however that the unification of electromagnetic and weak interactions is esthetically very appealing.

8.8 EXPERIMENTAL TESTS OF THE WEINBERG-SALAM MODEL

8.8.1 Neutrino-Electron Scattering

The pioneer experiment demonstrating the existence of the neutral current scattering process $\bar{\nu}_\mu e \to \bar{\nu}_\mu e$ has been described in Section 6.10. The various processes involving neutral (and charged) currents in neutrino-electron scattering are illustrated in Fig. 8.12.

The differential cross-section, in the Weinberg-Salam model, has the form

$$\frac{d\sigma}{dE} = \frac{G^2 m}{2\pi} \left[(g_V + g_A)^2 + (g_V - g_A)^2 \left(1 - \frac{E}{E_\nu} \right)^2 - \frac{mE}{E_\nu^2}(g_A^2 - g_V^2) \right], \quad (8.46)$$

where E is the energy of the recoil electron, E_ν that of the incident neutrino, and m the electron mass. The values of g_V and g_A, the vector and axial-vector coupling constants, are given in Table 8.2 for reactions (i)–(iv) of Fig. 8.12. (Note that for charged currents, $g_V = g_A = 1$ for neutrinos and $g_V = 1, g_A = -1$ for antineutrinos, as in (7.53).) Consider first reaction (i). Only the weak neutral current coupling of (8.41) can contribute. The V and A couplings for leptons are

$$g_V = g_L + g_R,$$

$$g_A = g_L - g_R,$$

where, from (8.41) and (8.43), the LH and RH couplings are

$$g_L = I_3 - Q \sin^2 \theta_w,$$

$$g_R = - Q \sin^2 \theta_w.$$

So, for the coupling of the neutrino to Z_0, $g_V = g_A = \frac{1}{2}$, and it has the usual V-A form, $\gamma_\mu(g_V + \gamma_5 g_A) = \gamma_\mu(1 + \gamma_5)/2$, as in the case of charged currents. For the electron coupling,

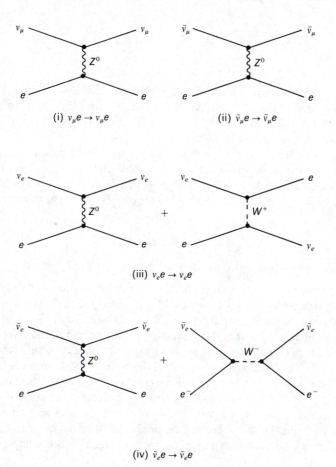

Fig. 8.12 Diagrams for the scattering of electron-neutrinos and muon-neutrinos and the corresponding antineutrinos from electron targets.

TABLE 8.2 Neutrino-electron couplings in the Weinberg-Salam model

	g_V	g_A
(i) $\nu_\mu e \to \nu_\mu e$	$-\frac{1}{2} + 2\sin^2\theta_w$	$-\frac{1}{2}$
(ii) $\bar{\nu}_\mu e \to \bar{\nu}_\mu e$	$-\frac{1}{2} + 2\sin^2\theta_w$	$+\frac{1}{2}$
(iii) $\nu_e e \to \nu_e e$	$+\frac{1}{2} + 2\sin^2\theta_w$	$+\frac{1}{2}$
(iv) $\bar{\nu}_e e \to \bar{\nu}_e e$	$+\frac{1}{2} + 2\sin^2\theta_w$	$-\frac{1}{2}$

$$g_V = -\tfrac{1}{2} + 2\sin^2\theta_w,$$

$$g_A = -\tfrac{1}{2}.$$

For the antineutrino reaction (ii), the signs of g_A are reversed. Reactions (iii) and (iv) can proceed via both charged and neutral couplings, the former being $g_V = g_A = 1$ for (iii) and $g_V = 1$, $g_A = -1$ for (iv). Adding these two contributions, we obtain the results in the table.

The cross-sections for reactions (i) and (ii) have been measured in several accelerator neutrino experiments, using bubble-chamber and counter techniques. Based on only about 100 events altogether, which simply reflects the very low value of the cross-sections ($\sigma/E_\nu \sim 10^{-42}\,\mathrm{cm^2\,GeV^{-1}}$), the data provide an estimate

$$\sin^2\theta_w = 0.24 \pm 0.04. \tag{8.47}$$

8.8.2 Deep-Inelastic Neutrino-Nucleon Scattering

The electroweak model is primarily one of the interactions of fundamental "isospin" triplet and singlet bosons, $W^\pm$, Z^0 and γ with leptons, which are assigned to an "isospin" doublet (ψ_L) and a singlet (ψ_R). To describe the electroweak interactions of hadrons, recourse must be had to the quark-parton model of hadron constituents. Again the quarks are organized into left-handed doublets and right-handed singlets, but as described in Section 6.11, the quark states taking part in the electroweak interaction are the "Cabibbo-rotated" combinations of (6.68), thus ensuring that the neutral weak current does not contain a $|\Delta S| = 1$ or $|\Delta C| = 1$ part. The neutral current cross-sections involving neutrinos and quarks can therefore be predicted in terms of $\sin^2\theta_w$ and the measured quark distributions in the nucleon, as described in Chapter 7.

TABLE 8.3 Ratios of neutral- to charged-current total cross-sections for neutrinos (R) and antineutrinos ($\bar{R}$) on nucleons (after Sacton 1978 and Winter 1979)

Accelerator and experiment	Mean neutrino energy (GeV)	Target	R	$\bar{R}$
CERN PS	2	CF_3Br (Gargamelle)	0.26 ± 0.04	0.39 ± 0.06
FNAL (HPWF)	60	Liquid scintillator	0.30 ± 0.04	0.33 ± 0.09
FNAL (CITF)	50	Fe calorimeter	0.27 ± 0.02	0.40 ± 0.08
CERN SPS (ABCLOS)	100	Ne-H_2 (BEBC)	0.33 ± 0.05	0.36 ± 0.07
CERN SPS (CDHS)	100	Fe calorimeter	0.31 ± 0.01	0.37 ± 0.03
CERN SPS (CHARM)	100	Marble calorimeter	0.30 ± 0.02	0.39 ± 0.02

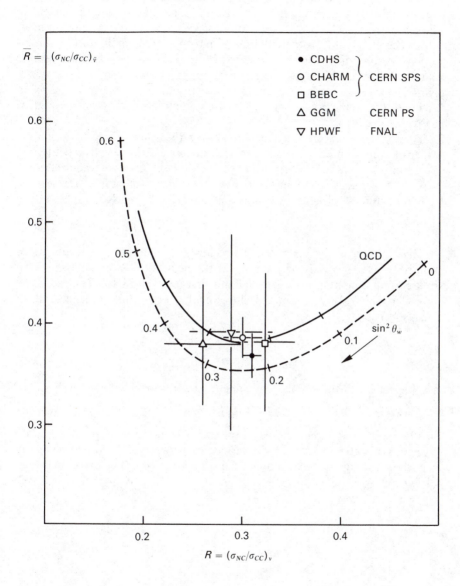

Fig. 8.13 Ratios of neutral-current to charged-current total cross-sections on nucleons, for antineutrinos ($\bar{R}$), and neutrinos, (R), as found in various experiments (see Table 8.3). The curve shows the prediction of the standard (Weinberg-Salam) model, based on quark distributions in the nucleon, with (full line) and without (dashed line) QCD corrections.

We do not discuss here the theory or data in detail (but see Problem 8.3). From the theory one can predict, for example, the ratio of neutral-current to charged-current inelastic cross-sections for incident neutrinos and antineutrinos. The *ratio* is more useful (though less informative) than the absolute cross-sections, because it is easier to measure a cross-section ratio than the absolute values, and because small uncertainties in the exact form of the quark distributions will tend to cancel out. The early results have already been given in Section 6.10, and the present world data are summarized in Table 8.3, as well as in Fig. 8.13.

There is clearly excellent agreement between the results in Fig. 8.13, relating to neutrino-nucleon scattering, and the value from neutrino-electron scattering, (8.47). In addition, elastic neutral-current processes involving hadrons (e.g. $v_\mu + p \rightarrow v_\mu + p$), as well as inelastic processes on neutron and proton targets, have been studied in detail. All lead to a value (Liede and Roos 1980)

$$\sin^2 \theta_w = 0.233 \pm 0.009 \qquad (\rho = 1). \tag{8.48a}$$

The analysis can also be carried out in more general terms, by introducing an extra parameter ρ specifying the ratio of the neutral (Z^0) and charged ($W^\pm$) weak current couplings in the second and first terms of (8.41). This would be allowed in the SU(2) × U(1) model with more than one doublet of Higgs scalars. The result is then

$$\rho = 1.002 \pm 0.015,$$

$$\sin^2 \theta_w = 0.234 \pm 0.013, \tag{8.48b}$$

in agreement with the simplest (Weinberg-Salam) model with $\rho = 1.0$.

8.8.3 Asymmetries in the Scattering of Polarized Electrons by Deuterons

Finally we discuss a very delicate experiment to detect tiny parity-violation effects (asymmetries) due to the interference between Z^0 and γ-exchange in inelastic scattering of polarized electrons by deuterons. The experiment was carried out with beams of electrons of 16–22-GeV/c momentum at SLAC,

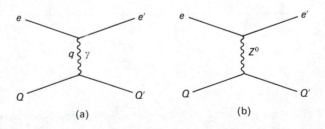

(a)   (b)

Fig. 8.14 Weak and electromagnetic neutral-current couplings of electrons to quarks, relevant in the SLAC polarized electron-deuteron scattering experiment.

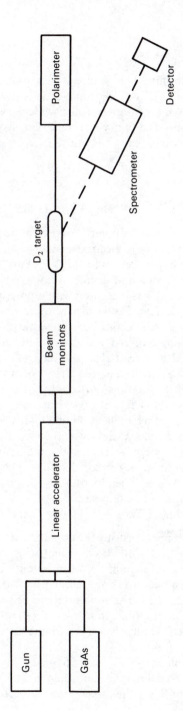

Fig. 8.15 Schematic layout of the SLAC experiment on scattering of polarized electrons on deuterons (after Prescott *et al.* 1978).

the reaction being

$$e^-_{L,R} + d_{\text{unpolarized}} \rightarrow e^- + X,$$

where X is any final hadron state. The weak and electromagnetic exchanges are depicted in Fig. 8.14(a) and (b). The electromagnetic scattering amplitude is of order e^2/q^2, where q is the 4-momentum transfer, while the weak amplitude is of order G, the Fermi constant. The parity-nonconserving asymmetry, measured by the difference of cross-sections for LH and RH electrons, will then be

$$A = \frac{\sigma_R - \sigma_L}{\sigma_R + \sigma_L} \sim \frac{Gq^2}{e^2} = \frac{137 \times 10^{-5}}{4\pi} \frac{q^2}{M_p^2}$$

$$\sim 10^{-4} q^2 \qquad (q^2 \text{ in GeV}^2). \tag{8.49}$$

The method employed to measure polarization asymmetries as small as 10^{-5} is illustrated in Fig. 8.15. A source of electrons, either polarized or unpolarized, is accelerated in the SLAC linear accelerator to 16–22 GeV/c and impinges on a liquid deuterium target. The sign and degree of polarization $|P_e|$ of the beam (37% for the polarized source) was measured with a polarimeter, in which one observed the left-right asymmetry in the Møller (electron-electron) scattering from a magnetized iron foil (see Section 6.5). Inelastically scattered electrons were focused and momentum-analyzed in a spectrometer, and recorded in a gas Čerenkov counter followed by a lead-glass shower counter.

The polarized-electron source consisted firstly of a dye laser producing linearly polarized light, which was transformed into circularly polarized light by means of a Pockels cell – a birefringent crystal in which the sign of circular polarization could be switched by application of a high-voltage electric field of either sign. This circularly polarized light was then used to optically pump a gallium arsenide crystal, between valence and conduction bands, and thus provide a source of longitudinally polarized electrons. The magnitude and sign of the polarization could also be varied by rotating the plane of polarization of the laser light with a rotatable calcite prism.

The steps in the experiment were:

(a) Measurement of asymmetry using the unpolarized source (electron gun), yielding $A/|P_e| = (-2.5 \pm 2.2) \times 10^{-5}$, consistent with zero and proving the sensitivity of the method and reliability of beam monitoring.

(b) Setting the calcite prism at azimuthal angles of $0°$, $45°$, and $90°$. For $\phi = 45°$, the electrons from the GaAs source should be unpolarized, while for $\phi = 0°$ or $90°$, the polarization should be R (L) or L (R) respectively, according as the sign of the Pockels-cell voltage is $+$ ($-$). Figure 8.16 shows the measured asymmetry as a function of ϕ.

(c) Varying the electron energy E_0 from 16 to 22 GeV/c. The beam, before hitting the target, has suffered a magnetic bending of $24.5°$, and in this process, the electron spin will "lead" over the momentum vector by an angle [see Eqs.

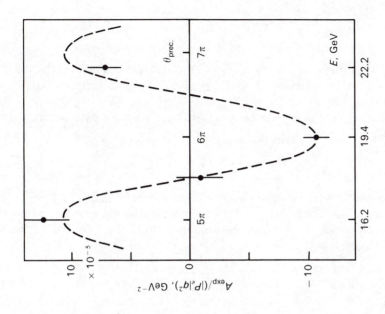

Fig. 8.17 Variation of asymmetry with electron-beam energy in the SLAC experiment of Prescott *et al.*, showing the $g - 2$ rotation of the electron spin.

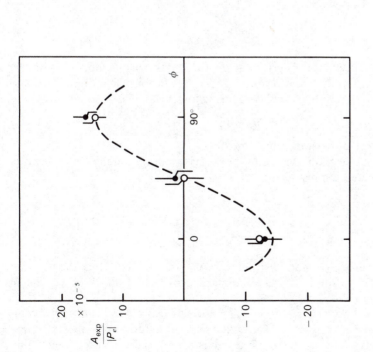

Fig. 8.16 Asymmetry, as defined in Eq. (8.49), as a function of the azimuth of the calcite prism in the experiment of Fig. 8.15.

(8.7) and (8.8)]

$$\theta_{\text{precession}} = \frac{E_0}{mc^2} \frac{g-2}{2} \theta_{\text{bend}}. \qquad (8.50)$$

Thus, as the beam energy E_0 is changed, the asymmetry should vary as the degree of longitudinal polarization is varied by the $g-2$ effect. This is demonstrated by the results in Fig. 8.17.

The final results of this experiment was that a clear asymmetry was observed, of magnitude

$$\frac{A}{q^2} = -(9.5 \pm 1.6) \times 10^{-5} \, (\text{GeV}/c)^{-2}. \qquad (8.51)$$

In succeeding experiments, the variation of A was measured as a function of $y = (E_0 - E)/E_0$, the fractional energy loss of the electron in the collision [see Eq. (7.45)]. The y-dependence as well as the magnitude of A depends on the weak angle θ_w, according to

$$\frac{A}{q^2} = -\frac{9G}{20\sqrt{2}\,\pi\alpha} \left\{ a_1 + a_2 \frac{1 - (1-y)^2}{1 + (1-y)^2} \right\}, \qquad (8.52)$$

where

$$a_1 = 1 - \tfrac{20}{9} \sin^2 \theta_w,$$

$$a_2 = 1 - 4 \sin^2 \theta_w.$$

These coefficients are derived from the quark-parton model, using the appropriate values of I_3 and Q in Eq. (8.41) – see Problem 8.6. The observed y-dependence was consistent with the prediction (8.52). The final result was

$$\sin^2 \theta_w = 0.22 \pm 0.02, \qquad (8.53a)$$

agreeing with the previous estimates.

However, if the ratio ρ of neutral- to charged-current couplings is retained as a free parameter, the results are

$$\rho = 1.74 \pm 0.36,$$

$$\sin^2 \theta_w = 0.293 \begin{array}{l} +\,0.033 \\ -\,0.100 \end{array}, \qquad (8.53b)$$

because the values of ρ and $\sin^2 \theta_w$ are strongly correlated. Thus, the neutrino results (8.48) provide by far the most stringent test of the Weinberg-Salam model.

Parity violation (arising from γ-Z^0 interference) in electromagnetic transitions has also been searched for, and found, in the form of optical rotation effects in heavy atoms. At the present time (1981), different experiments find somewhat different results. For a review, see Fortson and Wilets, 1981. They are in approximate agreement with the Weinberg-Salam model, but the comparison with theory is not straightforward because of atomic shielding effects (see Section 3.7). Neutral weak current effects at high energy have also been investigated by comparing the absolute cross-sections for $e^+e^- \to e^+e^-, \mu^+\mu^-,$ $\tau^+\tau^-$ with the pure QED predictions at the PETRA collider (DESY, Hamburg), up to CMS energies of 35 GeV. The γ-Z^0 interference term is proportional to the vector coupling of charged leptons to the Z^0, and the absence of any measurable deviations confirms that $\sin^2\theta_w \sim 0.25$, so that this coupling is very small (see Table 8.2).

8.9 QUANTUM CHROMODYNAMICS AND QUARK-QUARK INTERACTIONS

8.9.1 The Color Interaction between Quarks

The *color* quantum number has already been introduced as an extra degree of freedom in the quark model of hadrons (Section 5.2.2) and in lepton-quark interactions, particularly the cross-section for the process $e^+e^- \to$ hadrons (Section 7.8). Recall that the introduction of three colors for quarks increases the expected e^+e^- cross-section by a factor 3 and brings it into line with experiment. Quantum chromodynamics (QCD) is the formal gauge theory of the strong color interactions between quarks. The color charge of a quark has three possible values — say red, blue, or green. Antiquarks carry anticolor. The interquark interactions are assumed to be invariant under color interchange; in other words, they are described by the symmetry group SU(3). Since a quark can carry one of three possible colors, we can say that the quarks belong to the triplet (3) representation of SU(3). The bosons mediating the quark-quark interactions are called *gluons* and are postulated to belong to an octet (8) representation of SU(3). In analogy with the (flavor) octet of mesons in Section 5.3 we can write the color-anticolor states of the 8 gluons as follows:

$$r\bar{b}, r\bar{g}, b\bar{g}, b\bar{r}, g\bar{r}, g\bar{b}, \frac{r\bar{r} - b\bar{b}}{\sqrt{2}}, \frac{r\bar{r} + b\bar{b} - 2g\bar{g}}{\sqrt{6}}. \tag{8.54}$$

With 3 colors and 3 anticolors, we expect $3^2 = 9$ combinations, but one of these is a color singlet and has to be excluded. As an example, the color interaction between a red quark and a blue quark can proceed via exchange of a single $r\bar{b}$ gluon (Fig. 8.18).

The color charge of the strong quark interactions is analogous to the electric charge in electromagnetic interactions. Both forces are mediated by a massless, vector particle (a gluon or a photon). However, electromagnetism is an Abelian U(1) gauge theory, with two types of charge and an uncharged mediating boson. QCD is a non-Abelian gauge theory, with six types of charge (color or anticolor) and a charged (i.e. colored) mediating boson. This difference turns out to be crucial in understanding the features of the quark interactions at short distances and the success of the parton model.

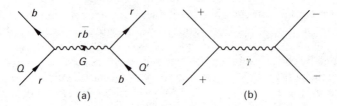

Fig. 8.18

The color quantum number does not enter our description of hadrons, so that both baryons and mesons must be colorless, i.e. be singlets of the SU(3) of color. If we write down the various contributions due to exchange of gluons [Eq. (8.54)] between quarks, the quark configurations of lowest energy are found to consist of the color singlet QQQ state (baryon) and the color singlet $Q\bar{Q}$ state (meson). Other quark combinations (e.g. the $Q\bar{Q}$ color octet) have weaker binding, or even a repulsive interaction (see Appendix G). Thus, QCD correctly predicts that only two of all the possible multiquark combinations should exist in nature.

The actual form of the strong-interaction potential between quarks is unknown. At small distances, the interaction is assumed to be of the Coulomb type, in analogy with electromagnetism, while at larger distances, the potential must increase indefinitely, so as to confine the quarks inside a hadron. Thus

$$V = -\frac{K_1}{r} + K_2 r \qquad (8.55)$$

is a possible form, which has been used previously [Eq. (5.47)] in discussing the energy levels of charmonium.

8.9.2 The Effects of Quark Interactions in Deep-Inelastic Lepton-Nucleon Scattering

In our discussion of deep-inelastic lepton-nucleon scattering in Chapter 7 we saw that the data could be accounted for, at least approximately, by postulating that the interactions consisted of the elastic scattering of the lepton by an effectively pointlike, free constituent or parton in the nucleon; these partons were identified with the quarks.

Suppose that in a deep inelastic scattering process, we vary q^2 (the 4-momentum transfer squared) between the incoming and the outgoing electron. q^2 is defined as a positive quantity and is equal to $-m^2$, where m is the imaginary value of the mass of the virtual photon exchanged between lepton and nucleon. The spatial resolution of this virtual photon "probe" is $\Delta r = h/q$. At low q^2, such that $\Delta r > R$, the target-proton radius ($R \sim 0.8$ fm), the entire proton recoils elastically from the collision; in other words it behaves like a pointlike particle [Fig. 8.19(a)]. As q^2 is increased the virtual photon begins to "see" the parton

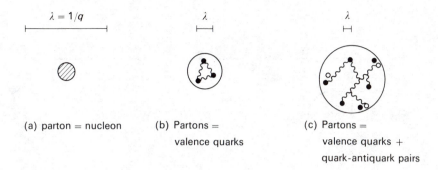

(a) parton = nucleon

(b) Partons =
valence quarks

(c) Partons =
valence quarks +
quark-antiquark pairs

Fig. 8.19 Symbolic picture of the nucleon, as probed in lepton scattering, as the wavelength $\lambda = 1/q$ of the probe is decreased and more structure is revealed.

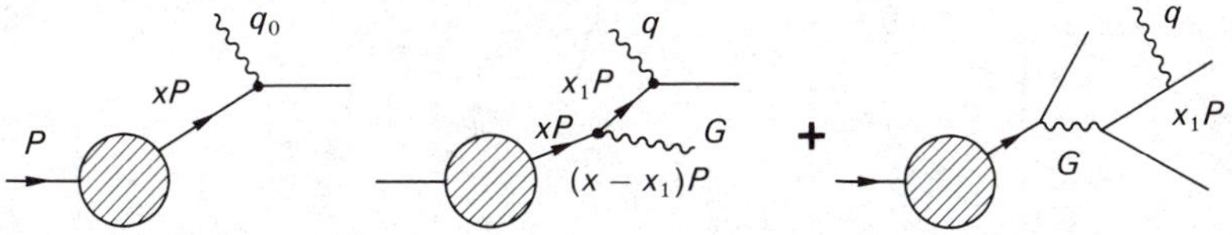

Fig. 8.20

structure of the nucleon, as three pointlike "valence" quarks. In the parton model, where the quarks are assumed to be both pointlike and free, further increase in q^2 does not lead to revelation of any finer structure [see Fig. 8.19(b)].

On the other hand, in the interacting-quark model, significant changes are expected as q^2 is increased. The exchange of gluons between quarks implies that the quarks have a structure. As our virtual photon probes smaller and smaller distances, what appeared as a single quark at low resolution is revealed as, for example, a quark which has emitted a gluon which in turn has formed a quark-antiquark pair. This process is analogous to that of vacuum polarization (virtual e^+e^- pair production) in QED (see Fig. 8.1). The pictures for the interacting quark model are shown in Fig. 8.19(b) and (c).

The effects of quark interactions on the behavior of nucleon-structure functions can easily be understood qualitatively from Fig. 8.20. In (a), a quark carrying a momentum fraction x of the nucleon momentum P absorbs a photon of momentum q_0. For $q^2 \gg q_0^2$ however, the quark has dissociated into a quark of momentum fraction $x_1 < x$, and a gluon with fraction $x - x_1$. Thus, if x is

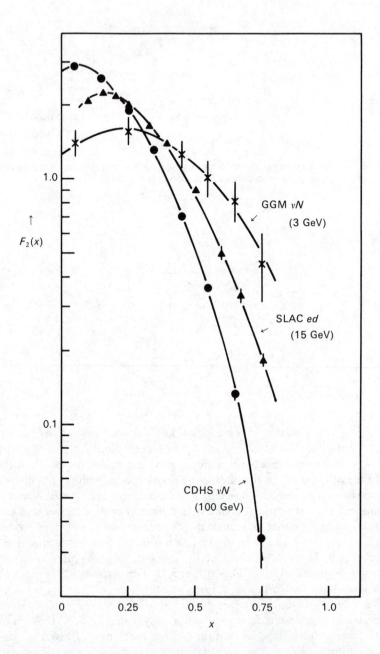

Fig. 8.21(a)

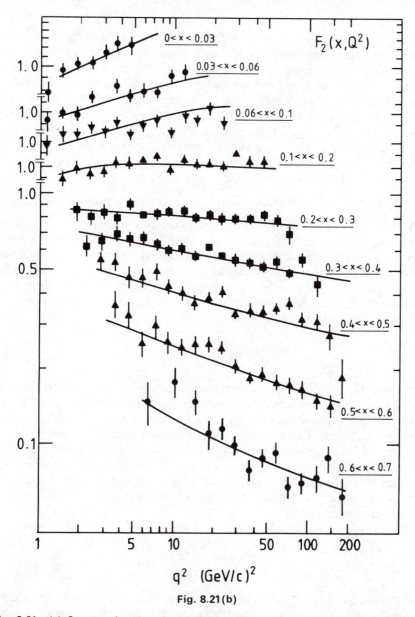

Fig. 8.21(b)

Fig. 8.21 (a) Structure function $F_2(x)$ measured in neutrino and electron scattering on nucleons at different beam energies. The shrinkage towards smaller x with increasing average energy and increasing average q^2 is illustrated. (b) $F_2^{\nu N}(x, q^2)$ plotted as a function of q^2 in different regions of x, from a counter experiment at CERN (see Fig. 2.21). Compare with Fig. 7.17, from electron scattering, at $x = 0.25$. The increase with q^2 at small x is shown, as well as the stronger decrease at large x. The curves are from empirical fits to quark momentum distributions, incorporating a q^2-dependence à la QCD, with $\Lambda = 0.3$ GeV. (After de Groot *et al.* 1979.)

large (> 0.3 say), the effect of increasing q^2 is to lead to a shrinkage of the quark distribution (i.e. structure function) towards small x. At very small x, on the other hand, the increasing number of gluons can form quark-antiquark pairs, in the domain of small x, and one of these can absorb the photon and thus lead to an increase of the structure function in this region [Fig. 8.20(b)]. This sort of behavior would be expected in any field theory of interacting quarks, and is not peculiar to QCD. Figure 8.21 shows practical examples of the evolution of structure functions with increasing q^2, bearing out the shrinkage in x as q^2 increases. Such deviations from the scale-invariant parton model were first observed in 1973 (Fox *et al.* 1974).

8.9.3 Running Coupling Constant: Quantitative Predictions of QCD

In Section 8.2 we described the radiative corrections to the electron and muon magnetic moments in QED. The expression for $(g - 2)/2$ appeared as a power series in the fine-structure constant α. We have already mentioned the labor involved in evaluating the higher-order corrections, but we note that their effect is equivalent to a first-order correction (i.e. of order α), with α depending however on the masses or momentum transfers in the virtual processes involved. Thus, we can always write for the anomaly

$$\frac{g - 2}{2} = \frac{0.5}{\pi} \alpha_{\text{eff}}, \tag{8.56}$$

where α_{eff} is larger for the muon than for the electron. In fact we are treating α as a "running coupling constant" dependent on the masses or momentum transfers involved in any particular case. This is not so arbitrary as it sounds. We noted previously that attempts to define "bare" couplings, charges, or masses $-\alpha_0, e_0, m_0 -$ in QED lead to infinities associated with self-energy terms, and we have to replace such quantities by the renormalized, physically measured values. The running coupling constant expresses the value of α at one value of q^2 in terms of that at another q^2 (incidentally avoiding bare couplings at $q^2 = \infty$). The relation can be approximated by

$$\alpha(q^2) = \frac{\alpha(q_0^2)}{1 - \frac{\alpha(q_0^2)}{3\pi} \ln\left(\frac{q^2}{q_0^2}\right)}, \tag{8.57}$$

so that the change in the value of α, as we go from a process involving a typical momentum transfer q_0^2 to one involving $q^2 > q_0^2$, depends logarithmically on the ratio q^2/q_0^2. Equation (8.57) is obtained by summing the higher-order corrections, involving terms of the general form $\alpha^n[\ln(q^2/q_0^2)]^m$, but retaining only the leading logarithms, i.e. $m = n$. We note that as q decreases, or the typical distance $r \sim 1/q$ increases, the effective coupling α gets smaller. This is a well-known effect in a polarizable (dielectric) medium. A test charge immersed in the dielectric exerts a potential, at distances larger than or comparable with

Fig. 8.22

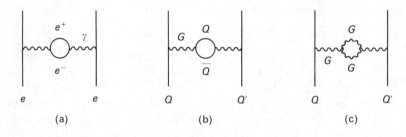

Fig. 8.23

molecular dimensions, which is smaller than the Coulomb potential in free space; in other words, the dielectric introduces a *shielding* effect (Fig. 8.22). Actually, the medium is unnecessary. Even in a vacuum, the test charge continually emits and reabsorbs virtual photons which can temporarily produce e^+e^- pairs, again producing a shielding effect – the so-called vacuum polarization. At extremely small distances, the shielding effect becomes small and one obtains the potential due to the bare charge.*

In QCD, the quark interactions can also be represented by a running coupling constant, $\alpha_s(q^2)$. Again, quark-antiquark pairs produce a shielding effect on the value of a test quark (color charge), Fig. 8.23(b). Because the field is non-Abelian, however, a gluon can also give rise to a gluon pair, and for certain states of polarization of this pair, an *antishielding* effect is produced [Fig. 8.23(c)]. $\alpha_s(q^2)$ has the form, in leading-logarithm approximation

$$\alpha_s(q^2) = \frac{\alpha_s(q_0^2)}{1 + B\alpha_s(q_0^2)\ln(q^2/q_0^2)} = \frac{1}{B\ln(q^2/\Lambda^2)}, \qquad (8.58)$$

---

* Virtual electron pairs, according to the Uncertainty Principle, will be mainly confined to distances less than $\lambda = h/mc$, so it is only for $r < \lambda$ that shielding effects are important. For $r \gg \lambda$, shielding effects decrease dramatically and the potential rapidly approaches the Coulomb value (see Gell-Mann and Low 1954).

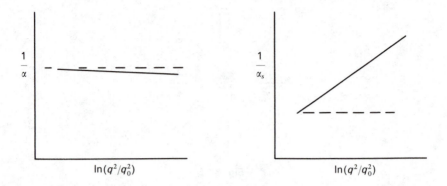

Fig. 8.24 Reciprocal of the running coupling constant as a function of q^2, for (left) QED and (right) QCD.

where $B = (33 - 2f)/12\pi$, and in the second expression we have made the substitution $\Lambda^2 = q_0^2 \exp[-1/B\alpha_s(q_0^2)]$. Thus, provided the number of quark flavors $f \leqslant 16$, it follows that $\alpha_s(q^2)$ decreases as q^2 increases. The effect of the self-coupling of the gluons is to "spread out" the color charge, and this effect becomes larger as q^2 increases. This decrease of α_s at large q^2 (or small distances) was first discussed in detail by Gross and Wilczek (1973) and by Politzer (1974). Clearly, (8.58) shows that at asymptotically large q^2 we have $\alpha_s(q^2) \to 0$, that is, the quarks behave as if free — a phenomenon called *asymptotic freedom*, and precisely what is expected in the parton model of quasifree quarks. At $q^2 \ll q_0^2$, such that $q \sim \Lambda$, $\alpha_s(q^2)$ becomes very large. Although at this point the theory becomes meaningless, large quark coupling may well be connected with the confinement of quarks at large distances, as for the potential (8.55) where $K_1 = \frac{4}{3}\alpha_s$. The contrasting q^2 behavior of α and α_s is shown in Fig. 8.24.

8.9.4 q^2 Evolution of Structure Functions

The theory makes quantitative predictions about the q^2 evolution of the quark distributions (structure functions), in the limit where $\alpha_s(q^2) \ll 1$ so that the approximate expression in (8.58) may hold, requiring $q^2 \gg \Lambda^2$. Suppose $u(x, q_0^2)$ represents the quark distribution at $q^2 = q_0^2$. Then, as explained before, at $q^2 > q_0^2$ the quark distribution can be modified by the radiation of a gluon. Since there is no scale or mass in the problem, the fractional change in $u(x)$ will be proportional to the fractional change in q^2 and we can write (Altarelli and Parisi 1977)

$$\frac{q^2 \, du(x, q^2)}{dq^2} = \frac{du(x, q^2)}{d \ln q^2}$$

$$= \alpha_s(q^2) \int_{y=x}^{y=1} u(y, q^2) P_{QQ}\left(\frac{x}{y}\right) dy. \qquad (8.59)$$

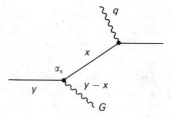

Fig. 8.25

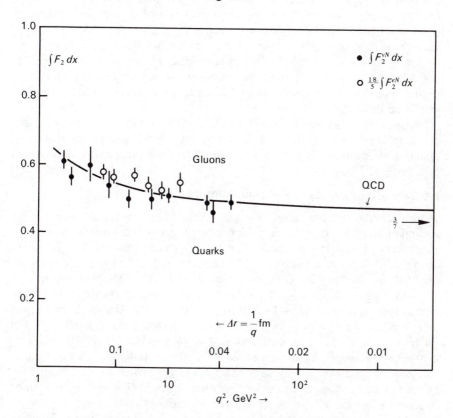

Fig. 8.26 The integral of $F_2(x, q^2)$ from $x = 0$ to $x = 1$ is a measure of the total momentum fraction in the nucleon accounted for by quarks and antiquarks [see Eqs. (7.59) and (7.61)]. This plot shows results from neutrino experiments in the CERN chambers BEBC and Gargamelle (Fig. 2.12 and Fig. 7.1) and from electron scattering at SLAC, where the integral could be evaluated over nearly the whole range of x. According to QCD, the quark mass fraction should decrease and tend asymptotically to a value of $\frac{3}{7}$, when the quark-gluon "gas" is in equilibrium. Nonasymptotically free field theories in general predict an increase in the quark momentum fraction with q^2 (Glück and Reya (1979)). Because of other noncalculable effects (varying as M^2/q^2), one can only say that the results are consistent with QCD and the non-Abelian nature of this theory (gluon-gluon coupling), but do not prove it.

The meaning of this equation is evident from Fig. 8.25. A quark with momentum fraction x which absorbs the current q is shown originating from a quark with fraction $y > x$ which has radiated a gluon with momentum fraction $y - x$. The probability for radiation to occur is measured by α_s, and the probability that a quark which radiates retains a fraction $z = x/y$ of its original momentum is described by the "splitting function" $P_{QQ}(z)$. The form of P_{QQ}, describing the radiation of a vector particle by a fermion, is known from QED to have the form $P_{QQ}(z) = (1 + z^2)/(1 - z)$. Thus, given the measured quark distribution $u(x, q^2)$ over the range x to 1, the logarithmic derivative with respect to q^2 of $u(x, q^2)$ measures α_s. A typical example of quark distributions (structure functions) from a recent experiment is given in Fig. 8.21(b). From such data, the value of α_s can be estimated, and hence the parameter Λ from Eq. (8.58). Typically, values of $\Lambda \simeq 0.1\text{--}0.5$ GeV are obtained. The smallness of Λ, together with the range of q^2 under experimental investigation, shows that α_s is not large ($\simeq 0.2$) and that the corresponding q^2-dependence of the structure functions is small and difficult to measure.

A possible test of the non-Abelian nature of QCD (i.e. the gluon-gluon self-coupling) is provided by the q^2-dependence of the total momentum fraction of quarks plus antiquarks, $\int F_2(x, q^2)\, dx$. This appears to show a slow decrease, as expected (see Fig. 8.26).

8.9.5   Multijet Events in e^+e^- Annihilation

More dramatic demonstrations of quark substructure are obtained in e^+e^- annihilation to hadrons at very high energy. As noted in Section 7.8, the elementary process is the annihilation to a $Q\bar{Q}$ pair, followed by "fragmentation" of the quarks to hadrons. At CMS energies of 30 GeV or more, typically about 10 hadrons (mostly pions) are produced. The average hadron momentum along the original quark direction is therefore large compared with its transverse momentum p_T — see p. 165 — limited to $p_T < 0.5$ GeV/c, that is, a magnitude $\sim 1/R_0$ where R_0 is a typical hadron size (~ 1 fm). Hence, the hadrons appear in the form of two "jets" collimated around the $Q\bar{Q}$ axis [see Fig. 8.27(a) and Fig. 2.20]. Occasionally, one might expect a quark to radiate a "hard" gluon, carrying perhaps half of the quark energy, at a large angle [Fig. 8.27(b)], the gluon and quark giving rise to separate hadronic jets. Such processes seem to be observed (Fig. 8.28). The measured rate of three-jet compared with two-jet events is clearly determined by α_s, and gives $\alpha_s \simeq 0.2$ at CMS energy 30 GeV. This estimate corresponds to a Λ-value in agreement with that quoted above. In particular, the distribution in angle between the two jets on one side of the event depends on the spin of the gluon and is consistent with their assumed vector nature (spin 1). In any case, the observed systematic and approximately linear increase of the transverse momentum p_T of hadrons relative to the event axis, with increasing CMS energy, is a sure indication of the existence of pointlike, hard scattering processes, due for example to radiation by the quark of one (or more) gluons.

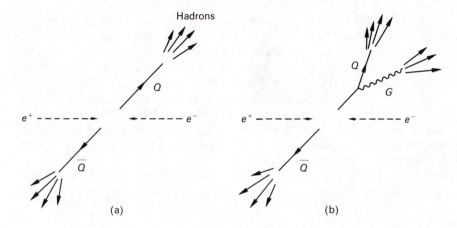

Fig. 8.27

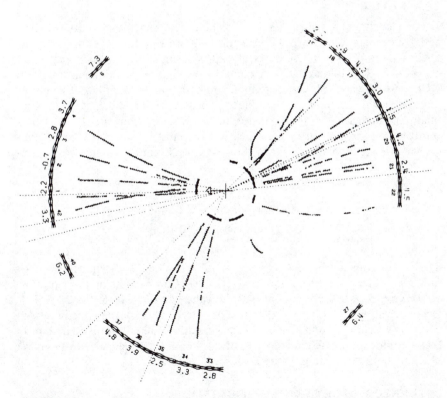

Fig. 8.28 Example of a three-jet event observed in the JADE detector at the PETRA e^+e^- collider (DESY, Hamburg). The total CMS energy in the event is 31 GeV. Such events (in comparison with the more common "two-jet" events of Fig. 2.20, p. 69) are suggestive of the process in Fig. 8.27(b). (Courtesy DESY)

A similar effect is observed in the Drell-Yan process of dilepton production in hadron collisions (Section 7.9): the p_T value of the dilepton pair, relative to the beam axis, increases with the mass of the pair.

The results which we have described from lepton-nucleon scattering and e^+e^- experiments seem to be consistent with the predictions of QCD. We have already seen that this theory gives, both qualitatively and quantitatively, a good account of the static properties of hadrons, including, for example, quite successful predictions for baryon magnetic moments and the mass splittings in hadron multiplets. QCD as a theory, however, has not yet been proved quantitatively in the detail which has been possible for QED or the electroweak model, and still has to be treated as tentative. In particular, the mechanism of the confinement of quarks in hadrons is not yet accounted for and is a major problem for the theory.

8.10 QCD AND THE WIDTHS OF CHARMONIUM AND UPSILON STATES

We previously remarked (in Section 5.11) on the narrow total widths of the ψ and Υ states, associated with "Zweig rule" suppression for unconnected quark lines as in Fig. 5.14(b). These states, as well as the mesons into which they decay, are color singlets, so that the connection between initial and final quark states in QCD must be via a color-singlet gluon combination. Thus, at least two gluons must be exchanged. Since ψ, Υ couple to photons, they have charge conjugation quantum number $C = -1$, and therefore, by the same argument as was used for positronium, the number of gluons exchanged must be odd. Thus three-gluon exchange is the simplest possibility. For the ψ, for example, we have from (5.30)

$$\Gamma(\psi \to \gamma \to e^+e^-) = \frac{16\pi\alpha^2}{M^2}|\chi(0)|^2|Q_i|^2 \simeq 5\,\text{keV},$$

while for hadronic decay via three gluons it is found that

$$\Gamma(\psi \to 3G \to \text{hadrons}) = \frac{160(\pi^2 - 9)}{81M^2}\alpha_s^3|\chi(0)|^2 \simeq 70\,\text{keV}.$$

The last equation is the same as that for $e^+e^- \to 3\gamma$ [Eq. (3.43)] apart from extra numerical factors due to color for $Q\bar{Q} \to 3G$. Comparison of the equations above gives $\alpha_s \simeq 0.2$, only about half the value, $\alpha_s \simeq 0.4$, from the $2\,^3S_1$–$1\,^3S_1$ level separation [see Eq. (5.47)]. The fairly large values of the effective strong coupling constant found from these arguments probably implies that more complex gluon exchanges are also involved; but at least the mechanism gives a qualitative understanding of the Zweig-rule suppression.

8.11 THEORIES OF GRAND UNIFICATION

In this final section we touch on the most speculative and, at the same time, probably the most exciting and far-reaching development in our present ideas and concepts in both high-energy physics and cosmology.

A major problem to our understanding of the fundamental interactions is in their number (four, if we include gravity together with strong, electromagnetic, and weak interactions) and disparate strengths and properties. The electroweak theory postulated a single interaction to describe electromagnetic and weak processes, and spontaneous symmetry breaking to account for their different apparent strengths in the energy domain well below the masses of the mediating bosons concerned. The so-called grand unified theories (called GUTs for brevity) appeal to further symmetry-breaking processes in order to reconcile the relatively great strength of strong interactions at low energies with a unique intrinsic coupling (approximately α, the fine-structure constant) for all three interactions at the unification energy.

The basic idea of the approach to a universal coupling is illustrated in Fig. 8.29. The graph indicates the dependence on momentum transfer, or mass scale, of the running coupling constants for the Abelian U(1) field described by g' in (8.35), for the non-Abelian SU(2) field denoted by g in (8.35), and for the non-Abelian SU(3) color field which we denote g_s (where $\alpha_s = g_s^2/4\pi$). As q increases, g' increases slightly while g decreases slightly and g_s decreases more quickly [compare (8.57) and (8.58)]. It is a remarkable fact that when all three interactions are extrapolated, they appear to meet at a single point at the colossal value of $q \sim 10^{14}$ GeV. (There are numerical constants of order unity multiplying g, g', and g_s, which we have omitted from Fig. 8.29.) Implicit in this enormous extrapolation is the bold assumption that Nature has run out of ideas

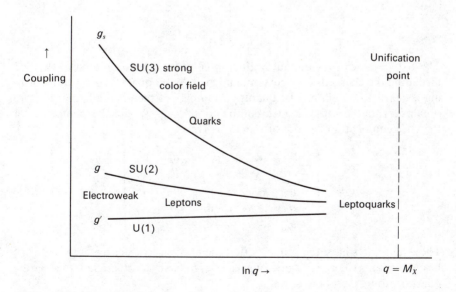

Fig. 8.29 Running coupling constants g, g', and g_s of the electroweak and strong interactions appear to extrapolate to a single value at $q \sim 10^{14}$ GeV.

and has no new surprises in the range from $q \sim 30$ GeV/c of present accelerator experiments, all the way to $q \sim 10^{14}$ GeV/c. There may instead be new internal degrees of freedom, quark or lepton substructure, only just around the corner and to be revealed at the next generation of accelerators. On the basis of past experience, one would think that this is rather likely.

There are many ways in which the SU(2), U(1), and SU(3) symmetries could be incorporated into a more global gauge symmetry. The simplest grand unifying symmetry is that of the group SU(5) (Georgi and Glashow 1974). This incorporates the known fermions (leptons and quarks) in multiplets, inside which quarks can transform to leptons, and quarks to antiquarks, via the mediation of very massive (10^{14}-GeV) bosons Y and X, with electric charges $-\frac{1}{3}$ and $-\frac{4}{3}$ respectively. There are a total of 24 gauge bosons in the model. These consist of the 8 gluons of SU(3) and the $W^{\pm}$, Z^0, and photon of SU(2) × U(1), plus the 12 varieties of X and Y boson (each carrying 3 colors and existing in particle and antiparticle states). The fermions (quarks and leptons) are assigned to different "generations". The first generation consists of 15 states: the u and d quarks, each in 3 color and 2 helicity states; the e^- with 2 helicity states; and the v_e, with 1 helicity only. By convention, we write them down as LH states, since for example the e_{RH}^- state and the e_{LH}^+ state are equivalent (by CP). The 15 states comprise "$\bar{5}$" and "10" representations as follows:

$$
\bar{5} = \begin{pmatrix} v_e \\ e^- \\ \bar{d}_R \\ \bar{d}_B \\ \bar{d}_G \end{pmatrix}_{\text{LH}} \quad \begin{matrix} \rfloor \; W^- \\ \leftarrow \\ \rfloor \; X \\ \\ \rfloor \; G_{B\bar{G}} \end{matrix} \; . \tag{8.60}
$$

The arrows indicate a gluon mediating the color force between quarks, the $W^{\pm}$ mediating the charged weak current, and an X "leptoquark" boson transforming a quark to a lepton. The quantum numbers of members of the "10" are obtained from the antisymmetric combinations $q_i q_j - q_j q_i$ of the members q_i of the "5" (conjugates of the $\bar{q}_i$ of the "$\bar{5}$"). Thus

$$
10 = \begin{pmatrix}
0 & e^+ & d_R & d_B & d_G \\
-e^+ & 0 & u_R & u_B & u_G \\
-d_R & -u_R & 0 & \bar{u}_G & \bar{u}_B \\
-d_B & -u_B & -\bar{u}_G & 0 & \bar{u}_R \\
-d_G & -u_G & -\bar{u}_B & -\bar{u}_R & 0
\end{pmatrix}_{\text{LH}} \tag{8.61}
$$

These multiplets have the property that the total electric charge $\sum Q_i = 0$. The heavier leptons (μ, v_μ, τ, v_τ) and quarks (s, c, b) have to be assigned to separate quark-lepton generations.

Among the attractive features of this (and other) grand unifying symmetries are:

(a) The fractional charges ($\frac{2}{3}$ and $\frac{1}{3}$) of the quarks occur because the quarks come in three colors and the electron is colorless (and $\sum Q_i = 0$).

(b) The equality of the baryon (proton) and electron charge – a historic puzzle – is accounted for.

(c) The strong similarity between the weak lepton and quark doublet patterns, for example $(v_e, e)_L$ and $(u, d_c)_L$, and the fact that $Q(v) - Q(e) = Q(u) - Q(d)$, occur as natural consequences of leptoquark unification.

The model makes a number of predictions, some at least of which are accessible in present or future experiments. We discuss here two of the predictions.

8.11.1 Prediction of the Weak Mixing Angle

The bringing together of quarks and leptons into multiplets allows one to estimate the weak mixing angle, as follows. Consider the diagram of Fig. 8.30 depicting the neutral boson Z^0 mixing with a photon, γ, via an intermediate fermion loop (analogous to the $K^0 \bar{K}^0$ mixing of Fig. 6.23). We know that Z^0 and γ are orthogonal states, $\langle Z^0 | \gamma \rangle = 0$; hence there should be no net coupling when summed over all fermions (leptons and quarks) which can contribute.

Fermions

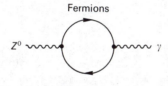

$$Z^0 \sim\!\!\sim\!\!\sim \bigcirc \sim\!\!\sim\!\!\sim \gamma$$

Fig. 8.30

The coupling of fermions to Z^0 is proportional to $I_3 - Q \sin^2 \theta_w$, and to the photon is simply Q – see (8.41). For the "5" representation, using (8.43), we obtain from (8.60)

$$
\begin{array}{ccc}
 & I_3 & Q \\
\begin{pmatrix} v_e \\ e^- \\ \bar{d}_R \\ \bar{d}_B \\ \bar{d}_G \end{pmatrix}_{LH} &
\begin{array}{c} +\frac{1}{2} \\ -\frac{1}{2} \\ 0 \\ 0 \\ 0 \end{array} &
\begin{array}{c} 0 \\ -1 \\ +\frac{1}{3} \\ +\frac{1}{3} \\ +\frac{1}{3} \end{array}
\end{array}
\qquad (8.62)
$$

where the LH states of antifermions (or RH states of fermions) have $I_3 = 0$, as explained previously. The net coupling, summed over all fermions, to both Z^0

and photon has to vanish, giving us the relation

$$\sum Q(I_3 - Q \sin^2 \theta_w) = 0$$

or

$$\sin^2 \theta_w = \frac{\sum Q I_3}{\sum Q^2} = \frac{3}{8}. \qquad (8.63)$$

A similar result is obtained for the "10" representation (8.61).

The above prediction for $\sin^2 \theta_w$ is clearly much larger than the observed value (8.48). However, it applies to the ratio $g'/g = \tan \theta_w \ (= \sqrt{3/5})$ at the *unification point*, $M_X \sim 10^{14}$ GeV. From Fig. 8.29 we see that corrections to the running coupling constants g and g' are required to bring them to the values expected at accelerator energies, and the effect of these corrections is clearly to decrease g'/g and $\sin^2 \theta_w$. When these corrections are made, a value $\sin^2 \theta_w = 0.21 \pm 0.01$ is obtained. Next, the experimental results of Section 8.8 must be amended to take account of radiative corrections to the quark cross-sections in both charged and neutral current processes, including effects of cuts in the data. When this is done, the best world average value is $\sin^2 \theta_w = 0.215 \pm 0.015$ (Marciano and Sirlin 1981, Llewellyn-Smith and Wheater 1981). Considering that, in principle, the prediction from grand unified theory could lie anywhere between 0 and 1, the near-agreement with the experimental results is quite spectacular.

8.11.2 Prediction of Proton Decay

There is no doubt that protons are very stable objects, and the mere existence of life on Earth sets a lower limit on the proton lifetime more than one million times longer than the age of the solar system (see Problem 8.4). Not surprisingly, baryon number was for many years considered to be an absolutely conserved quantity in all interactions, following the suggestion of Wigner and others. What we learn from gauge theories however is that absolute conservation laws are connected with gauge invariance and the existence of an appropriate long-range (massless) field. No fields are known which could be connected with baryon or lepton conservation (see Section 1.5).

The grand unified models, on the contrary, predict that protons must decay, violating baryon-number conservation. Some possible diagrams for proton decay via single X- or Y-boson exchange are given in Fig. 8.31. This exchange transforms a u-quark into a positron, and a d-quark into $\bar{u}$, for example [see (8.60)], corresponding to the decay $p \rightarrow e^+ \pi^0$.

The decay probability can be crudely estimated from such diagrams. The dominant factor will be the X-boson propagator, $(q^2 + M_X^2)^{-1}$. Since the momentum transfer is $q^2 \sim 1$ GeV^2 only, the decay rate will then be proportional to M_X^{-4}. Thus we can write for the lifetime

$$\tau = \frac{A}{\alpha^2} \frac{M_X^4}{M_p^5}, \qquad (8.64)$$

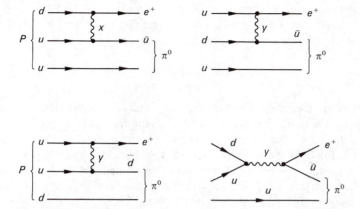

Fig. 8.31 Some diagrams relevant to nucleon decay, in which quarks transform to leptons and to antiquarks via massive vector bosons X and Y of charge $\frac{4}{3}$ and $\frac{1}{3}$.

where A is dimensionless and, in units $h = c = 1$, the right-hand side has dimensions of mass^{-1}. Our choice of the proton mass as the scale in the denominator is arbitrary but quite appropriate. The constant A involves a great deal of theory, which we do not discuss, but it turns out to be of order unity. If we take $A = 1$, then $M_X = 3 \cdot 10^{14}$ GeV gives $\tau_p = 10^{31}$ years. There are many uncertainties in this estimate, not the least of which is the exact value of M_X, or the energy scale of the grand unification point. This is determined mostly by g_s, the strong SU(3) color coupling, which has a much stronger q-dependence than either g or g'. g_s depends, through (8.58), on the strong scale parameter Λ, on which present experimental results vary by a factor $\simeq 4$, which propagates as a factor $4^4 = 256$ on the proton lifetime.

At the time of writing (1981), several experiments are under way to try to detect proton decay. The present lower limit, from underground experiments, is $\tau > 10^{30}$ years. If $\tau > 10^{33}$ years, background processes (cosmic-ray neutrino interactions) would essentially preclude any possibility of detection. Despite the formidable experimental difficulties, the prosecution of such experiments is important because they constitute a unique test of grand unification.

8.11.3 Cosmological Aspects of Grand Unification

The grand unified theory described above may be relevant to our understanding of the temporal development of, and present distribution of matter and radiation in, the universe. This is a vast subject, and the interested reader is referred to the chapter bibliography for recent reviews and papers. We only mention here a few of the ideas and facts, simply to indicate the relevance of results from particle physics to cosmology.

The isotropic cosmic background radiation in the microwave region, discovered in 1965, has a spectrum equivalent to that from a black body at 3°K and is interpreted as the cooled radiative remnant of the "Big Bang" preceding the expansion of the universe. From the present distribution of matter, the ratio of baryons to photons averaged over the universe is estimated to be

$$\frac{n_b}{n_\gamma} = 10^{-9 \pm 1}$$

and should be time-independent in an adiabatically and isotropically expanding universe. In the initial hot stage of the big bang ($t < 10^{-4}$ sec, $kT > 1$ GeV), baryon pairs should have been created with a number density comparable with that of photons. As the universe expanded and cooled, nucleon pair creation would no longer compensate annihilation to photons, and the baryon number would therefore fall catastrophically until a baryon residue was "frozen out" as the expansion rate exceeded the annihilation rate. This final ratio of baryons and antibaryons to photons can be calculated without too much uncertainty and is found to be

$$\frac{n_b}{n_\gamma} \sim 10^{-18},$$

in violent contrast with the observed ratio. So there are two problems: the observed baryon density is 9 orders of magnitude bigger than expected, and baryons preponderate over antibaryons. We know this is so in our own galaxy because heavy primary cosmic-ray nuclei – which are, so to speak, messengers from the most distant parts of the Milky Way – are invariably nuclei rather than antinuclei. Nor is there any evidence for the intense γ-ray emission which would follow annihilation between, for example, the matter of a galaxy and intergalactic antimatter. So, with all the usual reservations about the average isotropy of the universe, we can be sure that the antibaryon-to-baryon ratio, now, is $< 10^{-4}$.

It appears therefore that, in the early universe, some processes occurred which resulted in a net baryon-antibaryon asymmetry of order 10^{-9}. All but this tiny fraction of matter annihilated to photons, leaving a residue of baryons. Clearly, if one starts off with a symmetrical situation, an asymmetry can only arise through a baryon-number-nonconserving process, but there are two other requirements: the baryon-creating processes must be CP-nonconserving, and the baryons must be out of thermal equilibrium. (In thermal equilibrium, the baryon density can depend only on the temperature and on the particle mass, which is the same for particle and antiparticle, by the CPT theorem.) One possibility is the initial generation of the $X, \bar{X}$ bosons mentioned above, which by the CPT theorem will have identical total decay rates, but for which some partial decay rates into quarks, antiquarks, and leptons may violate CP, just as in the leptonic decay of K_L [Eq. (6.98)]. Such decays could generate a baryon asymmetry as a second-order process, and it thus appears possible to account for the preponderance of baryons in the universe as well as the observed baryon-to-photon ratio. The effects predicted in the SU(5) model of grand unification are too small, and more general and larger unification groups – involving more free parameters and hence less predictive power – seem to be required. Although, in principle, *CP*-violation is possible with only four quark flavors, the asymmetry becomes appreciable only with at least six flavors (see Section 6.12).

Astrophysical arguments can be used the other way round, to set interesting limits on the properties of elementary constituents. The density of light neutrinos (mass $\ll 1$ MeV) created in the big bang should be comparable with the photon density, now of order 10^9 times the nucleon density. Thus, a neutrino mass of only a few eV (i.e. 10^{-9} of the nucleon mass), would increase the gravitational potential energy by a substantial factor; indeed, this is one of the reasons for postulating nonzero neutrino masses, in order to correct the mismatch between motional and gravitational energy in the universe (Section 6.14). On the other hand, a mass as high as 100 eV might imply a degree of deceleration in the expansion of the universe in conflict with observation.

Limits on the number of flavors of the light neutrino have been set from considerations of the observed helium/hydrogen mass ratio (about 0.29), which must be a little greater (via stellar synthesis) than the primordial ratio attained in thermonuclear fusion following the big bang. The primordial helium content from nucleosynthesis (at $t \sim 100$ sec, $kT \sim 0.1$ MeV) depends on the value of the n/p ratio, which is determined by the reversible reactions $e^- + p \rightleftharpoons \nu_e + n$. The back reaction is exothermic by 0.8 MeV, so $n/p < 1$ at nucleosynthesis and is smaller (larger) the slower (faster) the cooling rate. The latter is in turn proportional to the number of fundamental fermions produced in the earlier stages, and each extra neutrino flavor is expected to increase the helium mass fraction by about 1%. These ideas suggest not more than four light-neutrino flavors; so perhaps the total eventual number of lepton (and presumably quark) flavors is finite and small, and not, as some people fear, embarrassingly large. The question of the number of possible types of neutrino will be independently and finally settled in a few years, from observation of the branching ratio for $Z^0 \to \nu\bar{\nu}$ at giant $e^+ e^-$ colliders.

Finally, we must stress that our search for the ultimate constituents of matter will never be completed. This book has been written as if quarks and leptons were the ultimate constituents, but there is absolutely no evidence for this. Perhaps quarks and leptons themselves possess a still more fundamental substructure, but at present this is pure speculation, and we may never know. In conclusion, we refer the reader to the quotations from Newton and Voltaire on the front page. Our view of the universe may not have changed so much in the past two or three hundred years.

PROBLEMS

8.1 Using Eq. (8.46) and the results from Table 8.2, plot the total cross-sections σ/E_ν for ν_e, $\bar{\nu}_e$, ν_μ, and $\bar{\nu}_\mu$ scattering from stationary electron targets as a function of $\sin^2 \theta_w$. Where do σ_{ν_μ}, $\sigma_{\bar\nu_\mu}$ have minimum values? For what value of $\sin^2\theta_w$ is the $\bar\nu_\mu$ coupling to electrons purely axial-vector?

8.2 By neglecting the third term in (8.46) and integrating over the electron recoil energy E, plot the relation between g_V and g_A assuming that $\sigma(\nu_\mu e) = 3 \times 10^{-42} E_\nu$ cm^2 GeV. Take $G = 1.02 \times 10^{-5}/M_p^2$. Show that, if $\sigma(\bar\nu_\mu e)$ is also known (without error), there are four possible solutions for g_A and g_V.

8.3 Using the results from Section 8.6, show that the left-handed and right-handed neutral current couplings of the u- and d-quarks are

$$g_L^u = \tfrac{1}{2} - \tfrac{2}{3}\sin^2\theta_w, \qquad g_L^d = -\tfrac{1}{2} + \tfrac{1}{3}\sin^2\theta_w,$$

$$g_R^u = -\tfrac{2}{3}\sin^2\theta_w, \qquad g_R^d = \tfrac{1}{3}\sin^2\theta_w.$$

From the relations $g_V = g_L + g_R$, $g_A = g_L - g_R$, and with the help of Eq. (8.46), derive expressions for the deep-inelastic neutral-current differential cross-sections for neutrinos and antineutrinos on nucleons (i.e. equal numbers of neutrons and protons in the target). Express the results as $d\sigma/dy$, where $y = E/E_\nu$ and E is the energy transferred to hadrons. The formulae you obtain will be in terms of the quark distributions $u(x)$ and $d(x)$ in the proton, as discussed in Chapter 7. By integrating the results over y and comparing the total cross-sections obtained with those for the charged current reactions, find the values of R and $\bar R$ as a function of $\sin^2\theta_w$ and compare with the results shown in Fig. 8.13. (Neglect the antiquark population in the nucleon.)

8.4 The unit of radiation dosage is the rad, which corresponds to an energy liberation (in ionization) of 100 erg g^{-1}. The annual permissible body dose for a human is cited as 5 rad. Assuming that 100 times this dose would lead to extinction of advanced life forms, what limit does this set on the proton lifetime? (Assume that in proton decay, a substantial fraction of the total energy released is deposited in body tissue.)

8.5 Verify Eq. (8.52) for the asymmetry in scattering of polarized electrons by deuterons. Use the expressions in Section 8.8.1 and Problem 3.3 for the couplings of electrons and quarks to the Z^0 and photon. The cross-section results from the sum of the photon and Z^0 exchange amplitudes. In summing these, only the pure photon exchange term and the Z^0-photon interference term will contribute, assuming $q^2 \ll M_Z^2$. Assume also equal numbers of u- and d-quarks in the deuteron, and neglect antiquarks. [In case of difficulty, consult for example R. N. Cahn and F. J. Gilman, *Phys. Rev. D* **17**, 1313 (1977), where the problem is fully worked out.]

8.6 Show that, at high q^2, the elastic form-factor of the nucleon in Eq. (7.35) has the form naively expected in QCD, if the "struck" quark and the two "spectator" quarks are to recoil coherently, and their interactions are mediated by single gluon exchanges.

8.7 (a) One expects the $W \pm$ boson to be produced in $p\bar p$ collisions at sufficiently high' energy. One decay mode would be $W^+ \to e^+ + \nu_e$. Taking the matrix element for the process $u + \bar d \to W^+ \to e^+ + \nu_e$ from nuclear β-decay, show that in order of magnitude, the cross-section $\sigma \sim G^2 M_w^2 \sim 10^{-34}$ cm^2 (assume the W produced nearly at rest in the CMS, and that $M_w = 80$ GeV).

(b) Including a factor $1/\sqrt{E}$ for boson wavefunction normalization (Eq. (D.3)) and using Eq. (1.27) and (8.22)—or from dimensional arguments—show that $\Gamma(W^+ \to e^+ + \nu_e) = AGM_w^3$ where A is a numerical constant. An exact calculation yields $A = 1/6\pi\sqrt{2}$. Show that the total width $\Gamma(W^+ \to$ anything) will be of order 2.5 GeV in the standard (6 quark) model.

BIBLIOGRAPHY

Aitchison, I. J., and A. J. Hey, "Gauge theories in particle physics" (Adam Hilger Ltd., Bristol 1981).

Bahcall, J. N., "Solar neutrino experiments", *Rev. Mod. Phys.* **50**, 881 (1978).

Barrow, J. D., "The baryon asymmetry of the universe", *Surveys in High En. Phys.* **1**, 182 (1980).

Beg, M. A., and A. Sirlin, "Gauge theories of weak interactions", *Ann. Rev. Nucl. Science* **24**, 379 (1974).

Combley, F., F. J. M. Farley, and E. Picasso, "The CERN muon $(g - 2)$ experiment", *Phys. Rep.* **68**, 93 (1981).

Combley, F., "$(g - 2)$ factors for the muon and electron and the consequences for QED", *Rep. Prog. Phys.* **42**, 1889 (1979).

Farley, F. J. M., and E. Picasso, "The muon $(g - 2)$ experiment", *Ann. Rev. Nucl. Science* **29**, 243 (1979).

Glashow, S. L., "Towards a unified theory: Threads in tapestry", *Rev. Mod. Phys.* **52**, 539 (1980).

Goldhaber, M., P. Langacker, and R. Slansky, "Is the proton stable?", *Science* **210**, 851 (1980).

't Hooft, G., "Gauge theories of the forces between elementary particles", *Sci. Am.* **243**, 90 (June 1980).

Hung, P. Q., and C. Quigg, "Intermediate bosons: weak interaction couriers", *Science* **210**, 1205 (1980).

Primakoff, H., and S. P. Rosen, "Baryon number and lepton number conservation laws", *Ann. Rev. Nucl. Science* **31**, 145 (1981).

Salam, A., "Gauge unification of fundamental forces", *Rev. Mod. Phys.* **52**, 525 (1980); *Science* **210**, 723 (1980).

Steigman, G., "Cosmology confronts particle physics", *Ann. Rev. Nucl. Science* **29**, 313 (1979).

Weinberg, S., "Conceptual foundations of the unified theory of weak and electromagnetic interactions", *Rev. Mod. Phys.* **52**, 515 (1980); *Science* **210**, 1212 (1980).

Weinberg, S., "Unified theories of elementary particle interactions", *Sci. Am.* **231**, 50 (July 1974).

Appendix A

Relativistic Kinematics

The relativistic relation between the total energy E, momentum p, and rest mass m of a particle is

$$E^2 = m^2 c^4 + p^2 c^2, \tag{A.1}$$

or, taking the velocity of light $c = 1$,

$$E^2 = p^2 + m^2.$$

The particle velocity in these units is

$$v = \beta c = \beta,$$

and the Lorentz factor

$$\gamma = (1 - \beta^2)^{-1/2},$$

so that

$$E = \gamma m,$$

$$p = \gamma \beta m = \sqrt{\gamma^2 - 1}\, m, \tag{A.2}$$

$$\beta = \sqrt{\gamma^2 - 1}\,/\gamma.$$

$E, \mathbf{p}$ can be written as components of a four-vector p_μ, where $\mu = 1, \ldots, 4$ and $p_1 = p_x$, $p_2 = p_y$, $p_3 = p_z$, the Cartesian space components. Then $p_4 = iE$, the time component. Thus the square of the "length" of this 4-vector is

$$p^2 = \sum_\mu p_\mu^2 = p_1^2 + p_2^2 + p_3^2 + p_4^2 = -m^2, \quad \text{invariant.} \tag{A.3}$$

Suppose the values $E, \mathbf{p}$ refer to properties of a particle measured in the laboratory frame Σ. In a different reference frame Σ', moving with constant velocity β^* along the x-axis, the new values $E', \mathbf{p}'$ can be found from the rules for transforming components of four-vectors. In matrix form the result is

$$
\begin{bmatrix} p_1' \\ p_2' \\ p_3' \\ p_4' \end{bmatrix}
=
\begin{bmatrix}
\gamma^* & 0 & 0 & i\beta^*\gamma^* \\
0 & 1 & 0 & 0 \\
0 & 0 & 1 & 0 \\
-i\beta^*\gamma^* & 0 & 0 & \gamma^*
\end{bmatrix}
\begin{bmatrix} p_1 \\ p_2 \\ p_3 \\ p_4 \end{bmatrix},
\tag{A.4}
$$

or in space time coordinates,

$$p'_x = \gamma^*(p_x - \beta^*E), \tag{A.4a}$$

$$p'_y = p_y, \tag{A.4b}$$

$$p'_z = p_z, \tag{A.4c}$$

$$E' = \gamma^*(E - \beta^*p_x), \qquad \text{where} \quad \gamma^* = (1 - \beta^{*2})^{-1/2}. \tag{A.4d}$$

Transformations of angle are found quite simply from these relations. Suppose a particle travels at angle θ relative to the x-axis in frame Σ, and at θ' in Σ'. Writing the transverse momentum component

$$p_T = \sqrt{p_y^2 + p_z^2} = p'_T,$$

then we find

$$\tan \theta' = \frac{p_T}{p'_x} = \frac{1}{\gamma^*} \frac{p_T/p}{(p_x/p) - \beta^* E/p} = \frac{1}{\gamma^*} \frac{\sin \theta}{\cos \theta - \beta^*/\beta}$$

$$\tan \theta = \frac{1}{\gamma^*} \frac{\sin \theta'}{\cos \theta' + \beta^*/\beta'}. \tag{A.5}$$

When $\beta' < \beta^*$, i.e. the particle velocity in Σ' is less than the velocity of Σ' in Σ, there is a *maximum angle* $\theta_{\max}$ at which the particle can be emitted in Σ. Differentiating (A.5), we have

$$\frac{\partial}{\partial \theta'}(\tan \theta) = 0 \qquad \text{when} \quad \theta = \theta_{\max},$$

yielding

$$(\cos \theta')_{\theta_{\max}} = -\beta'/\beta^*,$$

$$\gamma^* \tan \theta_{\max} = \beta'/(\beta^{*2} - \beta'^2)^{1/2}. \tag{A.6}$$

The equations (A.5), with $\beta' = \beta = 1$, give the familiar relativistic aberration formulae in optics, while the Doppler shift is given by equation (A.4d).

The 4-vector notation is extremely useful in threshold calculations in relativistic kinematics. Suppose we wish to compute the threshold energy for the reaction produced when an incident particle of mass m, total energy E, momentum $\mathbf{p}$, hits a stationary target of mass M, resulting in a final state M^*:

$$m + M \rightarrow M^* \rightarrow m_1 + m_2 + \cdots,$$

where $M^* > m + M$. The threshold energy E will correspond to the total energy of m and M in the center-of-momentum system just being equal to M^*.

The total 4-momentum squared of m and M, calculated in the laboratory frame, will be

$$P^2 = (\mathbf{p} + 0)^2 - (E + M)^2$$
$$= \mathbf{p}^2 - E^2 - M^2 - 2ME$$
$$= -(M^2 + m^2 + 2ME).$$

In the CMS, the total 3-momentum is $\mathbf{p}^* - \mathbf{p}^* = 0$, and the total energy is M^*. Thus, we also have that

$$(P^2)_{\text{threshold}} = -(M^*)^2.$$

Hence

$$E_{\text{threshold}} = \frac{M^{*2} - M^2 - m^2}{2M}. \tag{A.7}$$

Note that when $E \gg M, m$, the total CMS energy is

$$E^* \simeq \sqrt{2ME} \simeq 2p^*, \tag{A.8}$$

and thus rises as the square root of the product of the target mass and the energy of the incident particle.

Pauli Matrix Representation of the Isospin
of the Nucleon

Just as for the description of particles of spin $\frac{1}{2}$, we can formally express the isospin wavefunction of a nucleon as a two-component matrix

$$\chi = \begin{bmatrix} 1 \\ 0 \end{bmatrix} = p,$$

$$\chi = \begin{bmatrix} 0 \\ 1 \end{bmatrix} = n,$$

(B.1)

where, for brevity, p and n signify the proton (or isospin-up) state and neutron (or isospin-down) state, respectively. As in the formalism for spin $\frac{1}{2}$, we introduce an isospin operator τ which has Cartesian components given by the 2×2 matrix operators

$$\tau_1 = \frac{1}{2}\begin{bmatrix} 0 & 1 \\ 1 & 0 \end{bmatrix}, \quad \tau_2 = \frac{1}{2}\begin{bmatrix} 0 & -i \\ i & 0 \end{bmatrix}, \quad \tau_3 = \frac{1}{2}\begin{bmatrix} 1 & 0 \\ 0 & -1 \end{bmatrix}, \quad \text{(B.2)}$$

which, apart from the factor $\frac{1}{2}$, are identical with the Pauli spin matrices (see (F.7)) and obey the usual commutation relations of the angular momentum operators, as in (C.3). Thus

$$\tau_1\tau_2 - \tau_2\tau_1 = i\tau_3, \text{ etc.}$$

If we apply τ_3 to the wave functions (B.1), we obtain

$$\tau_3 p = \frac{1}{2}\begin{bmatrix} 1 & 0 \\ 0 & -1 \end{bmatrix}\begin{bmatrix} 1 \\ 0 \end{bmatrix} = \frac{1}{2}\begin{bmatrix} 1 \\ 0 \end{bmatrix} = \frac{1}{2}p,$$

(B.3)

and

$$\tau_3 n = -\frac{1}{2}n,$$

expressing the fact that the eigenvalues of τ_3 are $+\frac{1}{2}$ and $-\frac{1}{2}$ for the two charge states of the nucleon. When we say that a nucleon has an isospin $\tau = \frac{1}{2}$, we really mean that there are $2\tau + 1$ eigenvalues of the *operator* τ_3. The eigenvalue of τ^2

should then be $\tau(\tau + 1) = \frac{3}{4}$. We verify this from (B.2):

$$\tau^2 = \tau_1^2 + \tau_2^2 + \tau_3^2, \qquad \text{where} \quad \tau_1^2 = \tau_2^2 = \tau_3^2 = \frac{1}{4}\begin{bmatrix} 1 & 0 \\ 0 & 1 \end{bmatrix}, \qquad \text{(B.4)}$$

so

$$\tau^2\chi = \frac{3}{4}\begin{bmatrix} 1 & 0 \\ 0 & 1 \end{bmatrix}\chi = \tfrac{3}{4}\chi.$$

Other combinations of the Pauli operators are the "raising" and "lowering" operators (see (C.4)):

$$\tau_+ = \tau_1 + i\tau_2 = \begin{bmatrix} 0 & 1 \\ 0 & 0 \end{bmatrix},$$

$$\tau_- = \tau_1 - i\tau_2 = \begin{bmatrix} 0 & 0 \\ 1 & 0 \end{bmatrix}. \qquad \text{(B.5)}$$

Applying these to (B.1), one obtains

$$\begin{array}{ll} \tau_+ p = 0, & \tau_+ n = p, \\ \tau_- p = n, & \tau_- n = 0. \end{array} \qquad \text{(B.6)}$$

Thus the operator τ_+ transforms a neutron into a proton state, and τ_- transforms a proton into a neutron – in other words, the operators $\tau_\pm$ flip the sign of the third component of isospin of the nucleon.

Clebsch-Gordan Coefficients.
The Addition of Angular Momenta or Isospins

Suppose we have two particles of angular momenta $\mathbf{j}_1$ and $\mathbf{j}_2$ with z-components m_1 and m_2. The total z-component is

$$m = m_1 + m_2.$$

The total angular momentum is

$$\mathbf{j} = \mathbf{j}_1 + \mathbf{j}_2,$$

and may therefore lie anywhere inside the limits

$$|j_1 - j_2| \leqslant j \leqslant |j_1 + j_2|.$$

We wish to find the weights of the various allowed j-values contributing to the two-particle state, that is,

$$\phi_1(j_1m_1)\phi_2(j_2m_2) = \sum_j C_j\psi(j,m), \qquad \text{with} \quad m = m_1 + m_2. \qquad \text{(C.1)}$$

The C_j are called Clebsch-Gordan coefficients (or Wigner, or vector addition, coefficients). Alternatively, we may want to express $\psi(j,m)$ as a sum of terms of different j_1 and j_2 combinations. We can do this by the use of angular-momentum (or isospin) *shift operators* (also known as "raising" and "lowering" operators).

First let us recall the definition of the x-, y-, and z-component angular momentum operators, in terms of the differential Cartesian operators:

$$J_x = -\frac{ih}{2\pi}\left(y\frac{\partial}{\partial z} - z\frac{\partial}{\partial y}\right),$$

$$J_y = -\frac{ih}{2\pi}\left(z\frac{\partial}{\partial x} - x\frac{\partial}{\partial z}\right), \qquad \text{(C.2)}$$

$$J_z = -\frac{ih}{2\pi}\left(x\frac{\partial}{\partial y} - y\frac{\partial}{\partial x}\right).$$

It is readily verified that the operators

$$J_x, \quad J_y, \quad J_z \quad \text{and} \quad J^2 = J_x^2 + J_y^2 + J_z^2$$

obey the commutation rules

$$J^2 J_x - J_x J^2 = 0, \quad \text{etc.}$$

and

$$J_x J_y - J_y J_x = iJ_z,$$
$$J_y J_z - J_z J_y = iJ_x, \qquad \text{(C.3)}$$
$$J_z J_x - J_x J_z = iJ_y,$$

where we have used units $\hbar = c = 1$ for brevity. The eigenvalues of the operators J^2 and J_z are given in Eq. (C.5) below.

The *shift operators* are defined as

$$\left. \begin{matrix} J_+ = J_x + iJ_y \\ J_- = J_x - iJ_y \end{matrix} \right\} \quad \text{whence} \quad \begin{cases} J_z J_+ - J_+ J_z = J_+, \\ J_z J_- - J_- J_z = -J_-. \end{cases} \qquad \text{(C.4)}$$

Thus

$$J_z(J_- \phi) = J_z J_- \phi = J_-(J_z - 1)\phi = (m - 1)J_- \phi.$$

Similarly,

$$J_z(J_+ \phi) = (m + 1)(J_+ \phi).$$

This last equation shows that the wavefunction $J_+ \phi$ is an eigenstate of J_z with eigenvalue $m + 1$. We can therefore write it as

$$J_+ \phi(j, m) = C_+ \phi(j, m + 1),$$

where C_+ is an unknown (and generally complex) constant. If we multiply both sides of this equation by $\phi^*(j, m + 1)$, and integrate over volume, we get

$$\int \phi^*(j, m + 1)J_+ \phi(j, m)\, dV = C_+ \int \phi^*(j, m + 1)\phi(j, m + 1)\, dV$$

where * indicates complex conjugation.

We choose the normalization of ϕ so that the last integral is unity, and all allowed m-values have unit weight. So

$$C_+ = \int \phi^*(j, m + 1)J_+ \phi(j, m)\, dV.$$

Similarly,

$$C_- = \int \phi^*(j, m)J_- \phi(j, m + 1)\, dV.$$

$$= \int \phi^*(j,m) J_+^* \phi(j,m+1) \, dV$$

$$= C_+^*,$$

from Eq. (C.4).

If we neglect arbitrary and unobservable phases, we must have

$$C_+ = C_- = C \qquad \text{(a real number)}.$$

Also, from Eq. (C.4),

$$J_+ J_- = J_x^2 + J_y^2 - i(J_x J_y - J_y J_x) = J_x^2 + J_y^2 + J_z = J^2 - J_z^2 + J_z.$$

Then

$$J_+ J_- \phi(j,m+1) = [j(j+1) - m^2 - m]\phi(j,m+1) = C^2 \phi(j,m+1).$$

So

$$C = \sqrt{j(j+1) - m(m+1)}$$

is the coefficient connecting states $(j,m) \leftrightarrow (j,m+1)$.

To summarize, the angular-momentum operators have the following properties:

$$J_z \phi(j,m) = m\phi(j,m), \tag{C.5a}$$

$$J^2 \phi(j,m) = j(j+1)\phi(j,m), \tag{C.5b}$$

$$J_+ \phi(j,m) = \sqrt{j(j+1) - m(m+1)} \, \phi(j,m+1), \tag{C.5c}$$

$$J_- \phi(j,m) = \sqrt{j(j+1) - m(m-1)} \, \phi(j,m-1). \tag{C.5d}$$

Example (1)

As an example, we consider two particles of j_1, m_1, and j_2, m_2, forming the combined state $\psi(j,m)$, and we take the case where $j_1 = 1, j_2 = \frac{1}{2}$, and $j = \frac{3}{2}$ or $\frac{1}{2}$.

Obviously the states of $m = \pm \frac{3}{2}$ can only be formed in one way:

$$\psi(\tfrac{3}{2}, \tfrac{3}{2}) = \phi(1,1)\phi(\tfrac{1}{2}, \tfrac{1}{2}), \tag{C.6}$$

$$\psi(\tfrac{3}{2}, -\tfrac{3}{2}) = \phi(1,-1)\phi(\tfrac{1}{2}, -\tfrac{1}{2}). \tag{C.7}$$

Now we use the operators $J_\pm$ to form the relations:

$$J_- \phi(\tfrac{1}{2}, \tfrac{1}{2}) = \phi(\tfrac{1}{2}, -\tfrac{1}{2}), \qquad J_- \phi(\tfrac{1}{2}, -\tfrac{1}{2}) = 0$$

$$J_- \phi(1,1) = \sqrt{2}\phi(1,0), \qquad J_- \phi(1,0) = \sqrt{2}\phi(1,-1), \qquad J_- \phi(1,-1) = 0,$$

using (C.5c, d).

Now operate on (C.6) with J_- on both sides:

$$J_- \psi(\tfrac{3}{2}, \tfrac{3}{2}) = \sqrt{3}\psi(\tfrac{3}{2}, \tfrac{1}{2}) = J_- \phi(1,1)\phi(\tfrac{1}{2}, \tfrac{1}{2})$$

$$= \sqrt{2}\phi(1,0)\phi(\tfrac{1}{2}, \tfrac{1}{2}) + \phi(1,1)\phi(\tfrac{1}{2}, -\tfrac{1}{2}).$$

So

$$\psi(\tfrac{3}{2},\tfrac{1}{2}) = \sqrt{\tfrac{2}{3}}\,\phi(1,0)\phi(\tfrac{1}{2},\tfrac{1}{2}) + \sqrt{\tfrac{1}{3}}\,\phi(1,1)\phi(\tfrac{1}{2}, -\tfrac{1}{2}). \qquad (C.8)$$

Similarly, for (C.7),

$$\psi(\tfrac{3}{2}, -\tfrac{1}{2}) = \sqrt{\tfrac{2}{3}}\,\phi(1,0)\phi(\tfrac{1}{2}, -\tfrac{1}{2}) + \sqrt{\tfrac{1}{3}}\,\phi(1, -1)\phi(\tfrac{1}{2},\tfrac{1}{2}). \qquad (C.9)$$

The $j = \tfrac{1}{2}$ state can be expressed as a linear sum:

$$\psi(\tfrac{1}{2},\tfrac{1}{2}) = a\phi(1,1)\phi(\tfrac{1}{2}, -\tfrac{1}{2}) + b\phi(1,0)\phi(\tfrac{1}{2},\tfrac{1}{2})$$

with $a^2 + b^2 = 1$. Then

$$J_+\psi(\tfrac{1}{2},\tfrac{1}{2}) = 0 = a\phi(1,1)\phi(\tfrac{1}{2},\tfrac{1}{2}) + b\sqrt{2}\,\phi(1,1)\phi(\tfrac{1}{2},\tfrac{1}{2}).$$

Thus, $a = \sqrt{\tfrac{2}{3}}$, $b = -\sqrt{\tfrac{1}{3}}$, and so

$$\psi(\tfrac{1}{2},\tfrac{1}{2}) = \sqrt{\tfrac{2}{3}}\,\phi(1,1)\phi(\tfrac{1}{2}, -\tfrac{1}{2}) - \sqrt{\tfrac{1}{3}}\,\phi(1,0)\phi(\tfrac{1}{2},\tfrac{1}{2}). \qquad (C.10)$$

Similarly,

$$\psi(\tfrac{1}{2}, -\tfrac{1}{2}) = \sqrt{\tfrac{1}{3}}\,\phi(1,0)\phi(\tfrac{1}{2}, -\tfrac{1}{2}) - \sqrt{\tfrac{2}{3}}\,\phi(1, -1)\phi(\tfrac{1}{2},\tfrac{1}{2}). \qquad (C.11)$$

Expressions (C.6) to (C.11) give the coefficients appearing in Table III (p. 399), for the addition of $J = 1$ and $J = \tfrac{1}{2}$ states.

Example (2)

As a second example, suppose we wish to know the isospin states corresponding to combination of a Σ^0-hyperon and a π^0-meson. In other words we have to find which states of total isospin can be made by adding two states of $I = 1$ and $I_3 = 0$. We write for the amplitude

$$\phi(\Sigma^0)\chi(\pi^0) = A\psi(2,0) + B\psi(1,0) + C\psi(0,0). \qquad (C.12)$$

A, B, and C are to be determined, with $A^2 + B^2 + C^2 = 1$. $\psi(I, I_3)$ is the total isospin state, ϕ and χ the Σ^0 and π^0 isospin wavefunctions. We use the raising operator

$$I^+\psi(I, I_3) = \sqrt{I(I + 1) - I_3(I_3 + 1)}\,\psi(I, I_3 + 1).$$

Applying this to the left-hand side of (C.12) gives

$$I^+[\phi(1,0)\chi(1,0)] = [I^+\phi(1,0)]\chi(1,0) + [I^+\chi(1,0)]\phi(1,0)$$

$$= \sqrt{2}[\phi(1,1)\chi(1,0) + \chi(1,1)\phi(1,0)]. \qquad (C.13)$$

Applying it to the right-hand side gives

$$I^+[A\psi(2,0) + B\psi(1,0) + C\psi(0,0)] = A\sqrt{6}\,\psi(2,1) + B\sqrt{2}\,\psi(1,1) + 0. \qquad (C.14)$$

On multiplying each side by its complex conjugate state, and remembering that $\phi(1,1)\phi^*(1,1) = 1$ and that $\phi(1,0)\phi^*(1,1) = 0$ (orthonormality), we get

$$6A^2 + 2B^2 = 2(\sqrt{2})^2 = 4. \qquad (C.15)$$

Applying the operator I^+ once more to (C.13) and (C.14), we obtain

$$A\sqrt{6}\sqrt{4}\psi(2,2) + 0 = 4\phi(1,1)\chi(1,1),$$

or

$$A^2 = \tfrac{2}{3}.$$

From (C.15) and (C.12), one then finds $B = 0$ and $C^2 = \tfrac{1}{3}$. Thus

$$\phi(\Sigma^0)\chi(\pi^0) = \sqrt{\tfrac{2}{3}}\psi(2,0) \pm \sqrt{\tfrac{1}{3}}\psi(0,0),$$

which is the decomposition required. The fact that the $I = 1$ state does not contribute can also be understood from a simple vector model.

Lorentz-Invariant Phase Space

The transition-rate formula (1.27) is

$$W = \frac{2\pi}{\hbar} |M_{if}|^2 \rho_f, \tag{D.1}$$

where W is the probability, per second, of a transition between an initial state i and a particular final state f. M_{if} is the matrix element for the transition, and the factor ρ_f denotes the number of states available in momentum space, per unit interval of the total energy of the state f, for the particle or particles produced in the final state. The proof of (D.1) is given in most standard texts on quantum mechanics, in the framework of perturbation theory. In this case, the matrix element has the form $M = \int \psi_f^* H_I \psi_i \, dV$, where H_I is the operator corresponding to the interaction energy, such that $H_I \ll H_0$ and H_0 is the "free" Hamiltonian. H_0 has as eigenvalues the energy of the system i before the interaction is turned on. It is common to represent both i and f by free-particle plane wavefunctions (Born approximation), particularly in collision problems solved by perturbation methods (for example, electron scattering by nuclei, discussed in Chapter 7).

In general, and certainly in strong interactions, the form of M is not known explicitly – in fact (D.1) is effectively a definition of M. In any case, both M and ρ_f refer to some arbitrary normalization volume V (somewhere inside which the interaction occurs), and the wavefunctions of the individual particles must be normalized to this volume. Thus, if a particle is represented by a plane wave, with $\hbar = c = 1$,

$$\psi = V^{-1/2} \exp[i(\mathbf{p} \cdot \mathbf{r} - Et)], \tag{D.2}$$

so that $\int \psi^* \psi \, dV = 1$. In an actual problem, one usually sets $V = 1$ and forgets about it. V must drop out in the final result, since the physical quantities like cross sections or decay rates are expressed per particle and must be independent of V. We include it here because the normalization (D.2) is not correct if we want the matrix element and ρ_f to be in a relativistically invariant form, which is useful in many cases, particularly when one can only make general statements about the properties of M, as in S-matrix theory.

First let one of the particles in the interaction be at rest relative to a box of volume V. Then the particle density is $1/V$ in this frame. Now suppose the particle plus box moves at velocity $\mathbf{v}$ relative to the chosen reference frame, Σ, in which W is to be calculated. If the particle mass is m, its energy in Σ is E, where $E/m = (1 - v^2/c^2)^{-1/2}$. As a consequence of the Lorentz transformation, a length measured in the direction of $\mathbf{v}$ in the particle rest frame is reduced by a factor m/E when measured in the frame Σ, so that the volume of the box is contracted to a value $V' = Vm/E$. If the wavefunction is integrated over a volume V in Σ, then $\int \psi^* \psi \, dV = E/m$; i.e., the particle density is increased by a factor E/m. Therefore, to ensure that the particle density is correctly normalized in the same way in all frames, we should incorporate a factor $\sqrt{m/E}$ with the wavefunction ψ, for each particle, where E is the energy in the chosen frame Σ. Several conventions are in use: $\sqrt{m/E}$, $1/\sqrt{2E}$, etc. We shall take $1/\sqrt{2E}$. Setting the phase-space factor

$$\frac{p^2 \, dp \, d\Omega \, V}{h^3} = \frac{d^3\mathbf{p}}{(2\pi)^3},$$

i.e. with $V = 1$, and units $\hbar = c = 1$, the transition rate becomes

$$W = \frac{2\pi |M|^2}{\prod_{\text{initial}} 2E_j} \rho_f, \tag{D.3}$$

with

$$\rho_f = \frac{d}{dE_{\text{total}}} \frac{\int d^3\mathbf{p}_1 \, d^3\mathbf{p}_2 \cdots d^3\mathbf{p}_{n-1}}{(2\pi)^{3(n-1)} \prod_{\text{final}} 2E_k}. \tag{D.4}$$

The products of the E's have been factored into initial and final states. There are n particles in the final state, hence $n - 1$ independent three-momenta in phase-space. ρ_f in (D.4) is Lorentz-invariant. This is clearly so for the decay of a single particle ($j = 1$), since the decay rate per unit volume is proportional to γ^{-1} or E^{-1}, so that ρ_f must be invariant. To prove the general case, we first write

$$dN = \int d^3\mathbf{p}_1 \, d^3\mathbf{p}_2 \cdots d^3\mathbf{p}_{n-1} = \int d^3\mathbf{p}_1 \, d^3\mathbf{p}_2 \cdots d^3\mathbf{p}_n \, \delta(\mathbf{p}_i - \mathbf{p}_f),$$

where δ stands for the Dirac δ-function,* conserving the total momentum ($\mathbf{p}_i = \mathbf{p}_f$), so that $\int d^3\mathbf{p}_n \, \delta(\mathbf{p}_i - \mathbf{p}_f) = 1$. Also, energy conservation implies

$$\int \frac{dN}{dE} \delta(E_i - E_f) \, dE = \frac{dN}{dE}\bigg|_{E_i = E_f}.$$

---

* Defined as: $\int_b^c dx \, \delta(x - a) = 1$ for a in the interval $b \to c$; $= 0$ for a outside the interval $b \to c$; $\delta(x - a) = 0$ for $x \neq a$.

If we write an energy-momentum 4-vector

$$p = (p_1, p_2, p_3, p_4) = (\mathbf{p}, p_4) = (\mathbf{p}, iE),$$

then the density of final states per unit total energy is

$$\frac{dN}{dE_{\text{total}}} = \int \frac{\prod_1^n d^3\mathbf{p}_k \, \delta(\mathbf{p}_i - \mathbf{p}_f)}{dE} = \int \prod_1^n d^3\mathbf{p}_k \, \delta^{(4)}(p_i - p_f),$$

where $\delta^{(4)}$ indicates a δ-function conserving 4-momentum:

$$\delta^{(4)}(p_i - p_f) = \delta(p_4^2 + \mathbf{p}^2 + m^2).$$

Now, for any particle of mass m,

$$\int \delta(p_4^2 + \mathbf{p}^2 + m^2) \, dp_4 = \int \frac{1}{2p_4} \delta(p_4^2 + \mathbf{p}^2 + m^2) \, d(p_4^2)$$

$$= \left. \left| \frac{1}{2p_4} \right| \right|_{p_4^2 = -(p^2 + m^2)} = \frac{1}{2E}.$$

Hence

$$\rho_f = \frac{dN}{dE} = \frac{\int \prod_n \delta^4 p_k \, \delta(p_k^2 + m_k^2) \, \delta^{(4)}(p_i - p_f)}{\prod_{(n-1)} (2\pi)^3}. \qquad (D.5)$$

The last factor in the numerator takes care of energy-momentum conservation; the second factor ensures that the integral over the element of energy and three-momentum, $d^4 p_k$ for the kth particle, is such that $p_k^2 + m_k^2 = 0$, i.e. $p_k^2 = \mathbf{p}_k^2 - E_k^2 = -m_k^2$. The above formula, incorporating 4-vectors only, is "manifestly covariant".

G-Parity of the Pion

In Chapter 4 the G-operation was defined as

$$G = C\exp(i\pi I_2),$$

that is, a rotation through $180°$ about the y-axis in isospin space, followed by charge conjugation. For the neutral pion, $G = -1$. For charged pions, let us first find the result of the isospin rotation. From the definition of the raising and lowering operators (Appendix C),

$$I^{\pm}\psi(I, I_3) = \sqrt{I(I+1) - I_3(I_3 \pm 1)}\,\psi(I, I_3 \pm 1),$$

where

$$I^{\pm} = I_1 \pm iI_2, \qquad I_2 = -\tfrac{1}{2}i(I^+ - I^-), \qquad I_1 = \tfrac{1}{2}(I^+ + I^-).$$

From these expressions applied to pion states, one finds

$$I_2|\pi^0\rangle = -\frac{i}{\sqrt{2}}(|\pi^+\rangle - |\pi^-\rangle) = \alpha, \quad \text{say,}$$

and

$$I_2|\pi^{\pm}\rangle = \pm\frac{i}{\sqrt{2}}|\pi^0\rangle = \pm\beta.$$

Applying the operator I_2 repeatedly gives

$$I_2|\pi^0\rangle = (I_2)^3|\pi^0\rangle = \cdots = (I_2)^{2n+1}|\pi^0\rangle = \alpha,$$

$$(I_2)^2|\pi^0\rangle = \cdots = (I_2)^{2n}|\pi^0\rangle = |\pi^0\rangle,$$

$$I_2|\pi^{\pm}\rangle = (I_2)^3|\pi^{\pm}\rangle = \cdots = (I_2)^{2n+1}|\pi^{\pm}\rangle = \pm\beta,$$

$$(I_2)^2|\pi^{\pm}\rangle = \cdots = (I_2)^{2n}|\pi^{\pm}\rangle = \pm\frac{i}{\sqrt{2}}\alpha = \pm\tfrac{1}{2}(|\pi^+\rangle - |\pi^-\rangle).$$

Then, expanding the operator $R = \exp(i\pi I_2)$ by a power series,

$$\exp(i\pi I_2)\,|\pi^+\rangle = \left(1 + i\pi I_2 - \frac{\pi^2}{2!}(I_2)^2 + \cdots\right)|\pi^+\rangle$$

$$= \frac{|\pi^+\rangle + |\pi^-\rangle}{2} + \frac{|\pi^+\rangle - |\pi^-\rangle}{2}\cos\pi - \frac{|\pi^0\rangle}{\sqrt{2}}\sin\pi.$$

$$= +|\pi^-\rangle. \tag{E.1}$$

Similarly one finds that

$$\exp(i\pi I_2)\,|\pi^-\rangle = +|\pi^+\rangle \quad\text{and}\quad \exp(i\pi I_2)\,|\pi^0\rangle = -|\pi^0\rangle.$$

Since

$$C|\pi^0\rangle = +|\pi^0\rangle,$$

if follows that if we define

$$C|\pi^\pm\rangle = -|\pi^\mp\rangle,$$

then all pion states have the same G-parity:

$$G|\pi^{+,-,0}\rangle = -|\pi^{+,-,0}\rangle.$$

Some texts define the rotation R about the x-axis. Using similar methods to those above, it is then found that

$$\exp(i\pi I_1)\,|\pi^{+,-,0}\rangle = -|\pi^{+,-,0}\rangle$$

with the same phase for all charge states. In this case, one must define

$$C|\pi^\pm\rangle = +|\pi^\mp\rangle$$

if one requires the same G-parity for all pion states.

A much quicker and neater method of obtaining the result (E.1) is to make use of the fact that the isospin functions $\psi(I, I_3)$ have the same properties under rotations in "isospin space" as the spherical harmonics $Y_l^m(\theta, \phi)$ have in real space. Setting $I = l$ and $I_3 = m$, and referring to Fig. 3.1 and the spherical-harmonic tables on p. 398, we note that the rotation R corresponds to the substitution $\theta \to \pi - \theta, \phi \to \pi - \phi$. Since

$$Y_1^1(\theta, \phi) = -\sqrt{\frac{3}{8\pi}}\sin\theta\, e^{i\phi},$$

then

$$Y_1^1(\theta, \phi) \xrightarrow{R} +\sqrt{\frac{3}{8\pi}}\sin\theta\, e^{-i\phi} = Y_1^{-1}(\theta, \phi).$$

Thus

$$\psi(1, 1) \xrightarrow{R} +\psi(1, -1).$$

The Dirac Equation

An account of the Dirac equation can be found in most standard texts on quantum mechanics. We append a brief account here for convenience only. We start off with the de Broglie prescription of a plane wave representing a free particle. In units $h = c = 1$ this is

$$\psi = e^{i(\mathbf{k}\cdot\mathbf{r} - \omega t)}, \tag{F.1}$$

with $\mathbf{k}$ and ω the propagation vector and angular frequency respectively, which in our units are numerically equal to the momentum $\mathbf{p}$ and energy E. The wavefunction (F.1) is a solution of the *Klein-Gordon equation*:

$$\frac{\partial^2 \psi}{\partial t^2} = \frac{\partial^2 \psi}{\partial x^2} + \frac{\partial^2 \psi}{\partial y^2} + \frac{\partial^2 \psi}{\partial z^2} - m^2 \psi$$

$$= (\nabla^2 - m^2)\psi, \tag{F.2}$$

suitable for describing free relativistic spinless particles. If we insert (F.1) in (F.2) we get the familiar relation

$$E^2 = p^2 + m^2. \tag{F.3}$$

In the nonrelativistic case, expanding (F.3) in powers of p/m gives to first order $E = m + p^2/2m$. In this approximation (F.1) becomes, on dividing by a factor e^{-imt},

$$\phi = e^{i[\mathbf{p}\cdot\mathbf{r} - (p^2/2m)t]},$$

which satisfies the *Schrödinger equation* describing nonrelativistic particles:

$$\frac{\partial \phi}{\partial t} - \frac{i}{2m} \nabla^2 \phi = 0. \tag{F.4}$$

Dirac formulated his relativistic wave equation under the assumption that derivatives of *both* time and space coordinates should occur to *first order*. It turns out that such an equation describes particles of spin $\frac{1}{2}$. The simplest form that we could write would be the two *Weyl equations*

$$\frac{\partial \psi}{\partial t} = \pm \left(\sigma_x \frac{\partial \psi}{\partial x} + \sigma_y \frac{\partial \psi}{\partial y} + \sigma_z \frac{\partial \psi}{\partial z} \right) = \pm \, \boldsymbol{\sigma} \cdot \frac{\partial}{\partial \mathbf{r}} \psi, \tag{F.5}$$

where the σ's are constants. The condition for this to satisfy (F.2), so that (F.3) may hold, is found by squaring (F.5) and equating coefficients, whence we obtain

$$\sigma_x^2 = \sigma_y^2 = \sigma_z^2 = 1,$$

$$\sigma_x \sigma_y + \sigma_y \sigma_x = 0, \quad \text{etc.,} \tag{F.6}$$

$$m = 0.$$

Thus the Weyl equations describe massless particles (in fact, neutrinos). The σ's cannot be numbers, since they do not commute. They can, however, be represented by *matrices*. The 2×2 Pauli spin matrices are a suitable choice (not the only one). They are

$$\sigma_x = \begin{bmatrix} 0 & 1 \\ 1 & 0 \end{bmatrix}, \qquad \sigma_y = \begin{bmatrix} 0 & -i \\ i & 0 \end{bmatrix}, \qquad \sigma_z = \begin{bmatrix} 1 & 0 \\ 0 & -1 \end{bmatrix}, \tag{F.7}$$

where

$$\sigma_x^2 = \sigma_y^2 = \sigma_z^2 = \begin{bmatrix} 1 & 0 \\ 0 & 1 \end{bmatrix},$$

as required. The wave function ψ must now have two components, since it is operated on by 2×2 matrices. Thus

$$\psi = \begin{bmatrix} \psi_1 \\ \psi_2 \end{bmatrix}.$$

Now we wish to include a mass term, and therefore a further matrix. The simplest set of four anticommuting matrices are 4×4, and ψ is then a four-component spinor. First we write down the space-time coordinates as components of a 4-vector (i.e. in covariant notation),

$$x_1 = x, \qquad x_2 = y, \qquad x_3 = z, \qquad x_4 = it.$$

Generally any component is denoted by x_μ, where $\mu = 1, \ldots, 4$, and the relativistically invariant scalar product of two 4-vectors is $x_\mu X_\mu$, where a repeated index implies a summation over μ. Now we write down the *Dirac equation* as

$$\left(\gamma_\mu \frac{\partial}{\partial x_\mu} + m \right) \psi = 0, \tag{F.8}$$

which is just generalizing (F.5) by inserting a mass term. The γ's are 4×4 matrices to be determined. We can do this by requiring (F.8) to satisfy the Klein-Gordon equation (F.2), which we can rewrite in covariant form as

$$\left(\frac{\partial^2}{\partial x_\mu \partial x_\mu} - m^2\right)\psi = 0. \tag{F.9}$$

Multiplying (F.8) by $[\gamma_v(\partial/\partial x_v) - m]$, we get

$$\left(\gamma_v \frac{\partial}{\partial x_v} - m\right)\left(\gamma_\mu \frac{\partial}{\partial x_\mu} + m\right)\psi = \left(\gamma_v \gamma_\mu \frac{\partial^2}{\partial x_v \partial x_\mu} - m^2\right)\psi = 0, \tag{F.10}$$

a sum over μ, v being again understood. Equation (F.10) can be verified by writing out the individual components $\mu, v = 1, \ldots, 4$ and proving that the cross terms cancel. Comparing (F.10) with (F.9), we therefore require the following conditions on the γ's:

$$\gamma_v \gamma_\mu + \gamma_\mu \gamma_v = 2\delta_{\mu v}, \qquad \text{where} \quad \delta_{\mu v} = \begin{cases} 1, & v = \mu, \\ 0, & v \neq \mu. \end{cases} \tag{F.11}$$

The usual representation of the γ-matrices obeying these commutation relations is

$$\gamma_k = \begin{bmatrix} 0 & -i\sigma_k \\ i\sigma_k & 0 \end{bmatrix}, \qquad k = 1, 2, 3,$$
$$\gamma_4 = \begin{bmatrix} 1 & 0 \\ 0 & -1 \end{bmatrix}, \tag{F.12}$$

where each element stands for a 2×2 matrix, and the σ_k are defined in (F.7). For example,

$$\gamma_1 = \begin{bmatrix} 0 & 0 & 0 & -i \\ 0 & 0 & -i & 0 \\ 0 & i & 0 & 0 \\ i & 0 & 0 & 0 \end{bmatrix}, \qquad \gamma_4 = \begin{bmatrix} 1 & 0 & 0 & 0 \\ 0 & 1 & 0 & 0 \\ 0 & 0 & -1 & 0 \\ 0 & 0 & 0 & -1 \end{bmatrix},$$

and so on. It is also useful to define the product matrix

$$\gamma_5 = \gamma_1 \gamma_2 \gamma_3 \gamma_4 = \begin{bmatrix} 0 & -1 \\ -1 & 0 \end{bmatrix},$$

$$\text{with} \quad \gamma_5^2 = 1, \quad \gamma_5 \gamma_\mu + \gamma_\mu \gamma_5 = 0, \quad \mu = 1, \ldots, 4. \tag{F.13}$$

In order to make quite clear what is happening, let us write out the Dirac equation for free particles in full, for the individual components of

$$\psi = \begin{bmatrix} \psi_1 \\ \psi_2 \\ \psi_3 \\ \psi_4 \end{bmatrix}.$$

We obtain the four simultaneous equations

$$-i\frac{\partial \psi_4}{\partial x_1} - \frac{\partial \psi_4}{\partial x_2} - i\frac{\partial \psi_3}{\partial x_3} + \frac{\partial \psi_1}{\partial x_4} + m\psi_1 = 0,$$

$$-i\frac{\partial \psi_3}{\partial x_1} + \frac{\partial \psi_3}{\partial x_2} + i\frac{\partial \psi_4}{\partial x_3} + \frac{\partial \psi_2}{\partial x_4} + m\psi_2 = 0,$$

$$+i\frac{\partial \psi_2}{\partial x_1} + \frac{\partial \psi_2}{\partial x_2} + i\frac{\partial \psi_1}{\partial x_3} - \frac{\partial \psi_3}{\partial x_4} + m\psi_3 = 0,$$

$$+i\frac{\partial \psi_1}{\partial x_1} - \frac{\partial \psi_1}{\partial x_2} - i\frac{\partial \psi_2}{\partial x_3} - \frac{\partial \psi_4}{\partial x_4} + m\psi_4 = 0.$$

(F.14)

Writing the plane-wave solution as

$$\psi(\mathbf{r}, t) = u_j e^{i p_\mu x_\mu}, \qquad j = 1, \ldots, 4,$$ (F.15)

where the u_j are now four-component spinors, and taking the simplest case $\mathbf{p} = p_3$, we get from (F.14) four simultaneous equations:

$$(-E + m)u_1 + p u_3 = 0,$$ (F.16a)

$$\cdot (-E + m)u_2 - p u_4 = 0,$$ (F.16b)

$$(E + m)u_3 - p u_1 = 0,$$ (F.16c)

$$(E + m)u_4 + p u_2 = 0.$$ (F.16d)

One possible solution from (F.16c) is $u_1 = 1$, $u_3 = p/(E + m)$, $u_2 = u_4 = 0$. Or, from (F.16d), we could set $u_2 = 1$, $u_4 = -p/(E + m)$, $u_3 = u_1 = 0$. So two of the four solutions can be denoted

$$u_{++} = \begin{bmatrix} 1 \\ 0 \\ p/(E + m) \\ 0 \end{bmatrix} \quad \text{and} \quad u_{+-} = \begin{bmatrix} 0 \\ 1 \\ 0 \\ -p/(E + m) \end{bmatrix}.$$ (F.17a)

In these expressions, an overall normalization factor $\sqrt{(E + m)/2m}$ has been omitted. Equations (F.16a, b) give the same solutions, with label interchange $1 \leftrightarrow 4$ and $3 \leftrightarrow 2$, and the sign of E reversed. However, an equally good plane wave (F.15) can be written

$$\psi = u_j e^{-i p_\mu x_\mu},$$

i.e. with momentum $-\mathbf{p}$ and energy $-E$. This is clearly allowed, since $E = \sqrt{p^2 + m^2}$ does not determine the sign of E. So the other two of the four solutions are really the *negative-energy* solutions:

$$u_{-+} = \begin{bmatrix} -p/(|E| + m) \\ 0 \\ 1 \\ 0 \end{bmatrix} \quad \text{and} \quad u_{--} = \begin{bmatrix} 0 \\ p/(|E| + m) \\ 0 \\ 1 \end{bmatrix}. \quad \text{(F.17b)}$$

Combinations of any of these solutions are also permitted. To see what this all means, let us go to the particle rest frame, where $\mathbf{p} = 0$. Each pair of u's can then be written as two-component spinors; for example,

$$u_{++} = \begin{bmatrix} 1 \\ 0 \end{bmatrix}, \quad u_{+-} = \begin{bmatrix} 0 \\ 1 \end{bmatrix},$$

identical with the two-component Pauli spinors of the nonrelativistic theory. These are interpreted as positive-energy electron states with two possible *spin* directions, up and down — thus for electrons with half-integral spin. u_{-+} and u_{--} are the two negative-energy spin states. Dirac supposed that there was a completely filled sea of negative-energy states; a "hole" in this sea of electrons was interpreted as a *positron*, which would be produced whenever a negative energy electron received an energy $\geqslant 2mc^2$ and was kicked into a positive-energy state ($e^{\pm}$ pair creation, see Fig. F.1).

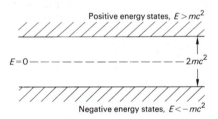

Fig. F.1

A word should be said here about the complex conjugate Dirac wave functions. Normally we define particle density by

$$\rho = \psi^*\psi,$$

where the asterisk represents complex conjugation.

If ψ is a column matrix (4-spinor), then by the rules for multiplying any matrices A and B, the number of columns of A and rows of B must be equal. We must therefore represent ψ^* by a row matrix. Thus

$$\psi = \begin{bmatrix} u_1 \\ u_2 \\ u_3 \\ u_4 \end{bmatrix} \quad \text{and} \quad \psi^* = [u_1^* \quad u_2^* \quad u_3^* \quad u_4^*].$$

Then

$$\rho = \psi^*\psi = \sum_1^4 u_i^* u_i.$$

It is useful to define the spinor

$$\bar{\psi} = \psi^*\gamma_4,$$

so that, from (F.12),

$$\bar{\psi}\gamma_4 = \psi^*$$

and

$$\rho = \bar{\psi}\gamma_4\psi. \qquad (F.18)$$

From the foregoing relations, it is easy to show that $\bar{\psi}$ satisfies the Dirac equation

$$\frac{\partial}{\partial x_\mu}\bar{\psi}\gamma_\mu - m\bar{\psi} = 0.$$

F.1   PARITY OF PARTICLE AND ANTIPARTICLE

Suppose $\psi(\mathbf{r}, t)$ satisfies the Dirac equation (F.8). Spatial inversion of this state, simply by replacing $\mathbf{r}$ by $-\mathbf{r}$, does not satisfy the Dirac equation, since it is first-order in the space coordinates. $\psi(-\mathbf{r}, t) = \psi(-x_k, x_4)$ satisfies

$$\left(\gamma_4\frac{\partial}{\partial x_4} - \gamma_k\frac{\partial}{\partial x_k} + m\right)\psi(-\mathbf{r}, t) = 0, \qquad k = 1, 2, 3.$$

On multiplying through from the left by γ_4, and using the rule $\gamma_4\gamma_k + \gamma_k\gamma_4 = 0$, we get

$$\left(\gamma_\mu\frac{\partial}{\partial x_\mu} + m\right)\gamma_4\psi(-\mathbf{r}, t) = 0.$$

Thus $\gamma_4\psi(-\mathbf{r}, t)$ is the inversion of the original state $\psi(\mathbf{r}, t)$, and γ_4 is the parity operator for the Dirac wavefunction. Now consider a spin-up positive-energy state (in the particle rest frame, $p = 0$), Eq. (F.17a):

$$u_{++} = \begin{bmatrix} 1 \\ 0 \\ 0 \\ 0 \end{bmatrix}, \qquad \gamma_4 u_{++} = \begin{bmatrix} 1 \\ 0 \\ 0 \\ 0 \end{bmatrix}.$$

For a spin-up negative energy state (F.17b),

$$u_{-+} = \begin{bmatrix} 0 \\ 0 \\ 1 \\ 0 \end{bmatrix}, \qquad \gamma_4 u_{-+} = -\begin{bmatrix} 0 \\ 0 \\ 1 \\ 0 \end{bmatrix}.$$

Thus, if the positive energy state is assigned an even intrinsic parity, then the negative-energy state, or, equivalently, the antiparticle, has odd intrinsic parity.

F.2   APPLICATION OF DIRAC THEORY TO β-DECAY

The phenomenon of β-decay is described in terms of the interaction of four particles at a point. For example, neutron decay is written $n \rightarrow p + e^- + \nu^-$. If we were dealing with scalar particles, we should write for the matrix element for $A + B \rightarrow C + D$

$$M = \text{const } \psi_C^*(\mathbf{r})\psi_D^*(\mathbf{r})\psi_A(\mathbf{r})\psi_B(\mathbf{r}).$$

For spin-$\frac{1}{2}$ particles, we deal with four-component wave functions (spinors) and the interaction would be written

$$M = \text{const } \sum_{ijkl} g_{ijkl}\psi_{C_i}^*\psi_{D_j}^*\psi_{A_k}\psi_{B_l}, \tag{F.19}$$

$i, j, k,$ and l being the spinor indices. In principle, the interaction can therefore involve $4^4 = 256$ arbitrary constants g_{ijkl}. This number can be reduced drastically by insisting that the laws of β-decay should be Lorentz-invariant, which implies that the possible interaction forms should have well-defined properties, i.e. be covariant under Lorentz transformations. It is then found that, using the Dirac matrices γ and the spinors $\bar{\psi}$ and ψ, one can construct just five basic combinations, which are named after their Lorentz transformation properties. They are:

		No. of components
$\bar{\psi}\psi \; (= \psi^*\gamma_4\psi)$	scalar S	1
$\bar{\psi}\gamma_\mu\psi$	4-vector V	4
$i\bar{\psi}\gamma_\mu\gamma_\nu\psi$	6-vector or tensor T	6
$i\bar{\psi}\gamma_5\gamma_\mu\psi$	axial 4-vector A	4
$\bar{\psi}\gamma_5\psi$	pseudoscalar P	1

By way of example, consider the inversion $\psi(-\mathbf{r}, t)$ of the state $\psi(\mathbf{r}, t)$. As mentioned above, $\psi(-\mathbf{r}, t)$ does not satisfy the Dirac equation, but $\gamma_4\psi(-\mathbf{r}, t)$, or $\bar{\psi}(-\mathbf{r}, t)\gamma_4$, does. Thus under inversion

$$\psi\bar{\psi} \rightarrow \bar{\psi}\gamma_4^2\psi = \bar{\psi}\psi$$

and is therefore a scalar, while

$$\bar{\psi}\gamma_5\psi \rightarrow \bar{\psi}\gamma_4\gamma_5\gamma_4\psi = -\bar{\psi}\gamma_5\psi$$

changes sign, and is therefore a pseudoscalar. It can be shown that both $\bar{\psi}\gamma_\mu\psi$ and $i\bar{\psi}\gamma_5\gamma_\mu\psi$ behave under rotations like 4-vectors; however, the first changes sign under inversions, while the second does not, and is therefore an axial vector.

Returning to the case of neutron decay, we may write it in the equivalent form

$$v + n \rightarrow e^- + p,$$

in which a neutrino and neutron are transformed into electron and proton. We denote the wave functions by

ψ_v to represent destruction of a neutrino (or creation of antineutrino),

$\bar{\psi}_e$ creation of an electron (or destruction of positron),

ψ_n destruction of a neutron (or creation of antineutron),

$\bar{\psi}_p$ creation of proton (or destruction of antiproton).

We see that this nomenclature is consistent with the physical process of β-decay as in the diagram. The matrix element (interaction energy density) can then be written in the bilinear form

$$M = G(\bar{\psi}_p O_i \psi_n) \times (\bar{\psi}_e O_i' \psi_v), \qquad (F.20)$$

$$\underbrace{\qquad\qquad}_{\text{nucleon bracket}} \quad \underbrace{\qquad\qquad}_{\text{lepton bracket}}$$

where O_i stands for one of the five forms written down above. In general, one expects a linear combination of the O_i ($i = 1, \ldots, 5$) for the matrix elements. The above form was first written down by Fermi in 1934 with $O_i = \gamma_\mu$ (pure vector interaction). G is a suitable constant specifying the strength of the interaction. If the weak interaction is parity-conserving, we require that M be a scalar quantity under space inversion. Since, experimentally, parity is not conserved in β-decay, M should contain *both* scalar and pseudoscalar parts. We can achieve this by including a factor γ_5 in one of the brackets — conventionally the lepton bracket. Thus the interaction has the general form

$$(\bar{\psi}_p O_i \psi_n)[\bar{\psi}_e O_i(C_i + \gamma_5 C_i')\psi_v], \qquad (F.21)$$

where C and C' are constants, and O_i represents the S, V, T, A, and P interactions. We can rewrite the lepton bracket

$$\bar{\psi}_e O_i(C_i + \gamma_5 C_i')\psi_v = (C_i + C_i')\bar{\psi}_e O_i(1 + \gamma_5)\psi_v/2$$

$$+ (C_i - C_i')\bar{\psi}_e O_i(1 - \gamma_5)\psi_v/2. \qquad (F.22)$$

To find out the meaning of the operator $1 \pm \gamma_5$ on ψ_v, we note that the Dirac equation (F.8) for a massless neutrino has the form

$$\gamma_\mu \frac{\partial \psi}{\partial x_\mu} = 0, \quad \text{or} \quad \gamma_4 \frac{\partial \psi}{\partial x_4} = -\gamma_k \frac{\partial \psi}{\partial x_k}, \quad k = 1, 2, 3.$$

We set

$$\gamma_k = i\gamma_4 \gamma_5 \hat{\sigma}_k,$$

where $\hat{\sigma}_k$ are a set of 4×4 matrices formed from the Pauli spinors (F.7):

$$\hat{\sigma}_k = \begin{bmatrix} \sigma_k & 0 \\ 0 & \sigma_k \end{bmatrix}, \tag{F.23}$$

whence

$$\frac{\partial \psi}{\partial x_4} = - i\gamma_5 \hat{\sigma}_k \frac{\partial \psi}{\partial x_k} = - i\hat{\sigma}_k \gamma_5 \frac{\partial \psi}{\partial x_k},$$

and

$$\frac{\partial \gamma_5 \psi}{\partial x_4} = - i\hat{\sigma}_k \frac{\partial \psi}{\partial x_k},$$

using the fact that $\gamma_5^2 = 1$ and $\gamma_5 \hat{\sigma}_k = \hat{\sigma}_k \gamma_5$.

Adding and subtracting the last two equations, setting $x_4 = ict$, and multiplying through by $i\hbar$ to make things more recognizable, we obtain

$$i\hbar \frac{\partial}{\partial t}[(1 + \gamma_5)\psi_v] = \hat{\sigma}_k ci\hbar \frac{\partial}{\partial x_k}[(1 + \gamma_5)\psi_v], \tag{F.24a}$$

$$i\hbar \frac{\partial}{\partial t}[(1 - \gamma_5)\psi_v] = - \hat{\sigma}_k ci\hbar \frac{\partial}{\partial x_k}[(1 - \gamma_5)\psi_v]. \tag{F.24b}$$

Thus, $(1 \pm \gamma_5)\psi_v$ are plane-wave eigenstates of the energy operator, $i\hbar \partial/\partial t$, with eigenvalues

$$E = - \boldsymbol{\sigma} \cdot \mathbf{p}c \quad \text{for} \quad \phi = (1 + \gamma_5)\psi_v \tag{F.25a}$$

$$E = + \boldsymbol{\sigma} \cdot \mathbf{p}c \quad \text{for} \quad \chi = (1 - \gamma_5)\psi_v \tag{F.25b}$$

where $\boldsymbol{\sigma}$ is the spin vector, with $\boldsymbol{\sigma}^2 = \sum \hat{\sigma}_k^2 = 1$. Since $E = pc$ for neutrinos, it follows that the state $\phi = (1 + \gamma_5)\psi_v$ represents a particle of helicity

$$H = \frac{\boldsymbol{\sigma} \cdot \mathbf{p}}{p} = - 1,$$

and $\chi = (1 - \gamma_5)\psi_v$ represents one of helicity $+ 1$. ψ_v in (F.20) defines the neutrino state, since it transforms to a negative electron. Experimentally it is found to have *negative* helicity. Thus, we need the term ϕ, which is associated with left-handed neutrinos (and right-handed antineutrinos), and have to discard the term χ. So we see that this amounts to keeping just the first of the two Weyl equations—in fact (F.24) *are* the Weyl equations [compare (F.5)]. The Dirac equation, describing particles with mass by four-component wavefunctions, breaks down for massless particles into two decoupled equations, each with two-component wavefunctions as solutions.

Thus we require $C_i = C_i'$ in (F.21), which thereby becomes

$$\bar{\psi}_e O_i (1 + \gamma_5)\psi_v = \begin{cases} \bar{\psi}_e (1 + \gamma_5) O_i \psi_v & \text{for} \quad i = S, T, P, \\ \bar{\psi}_e (1 - \gamma_5) O_i \psi_v & \text{for} \quad i = V, A, \end{cases}$$

as may be established from the commutation rules. By extending the argument used for massless particles, it may be shown that the term $\bar{\psi}_e(1 - \gamma_5)$ corresponds to the generation of left-handed electrons with intensity $1 - \boldsymbol{\sigma} \cdot \mathbf{p}/E$, i.e. a helicity $-\boldsymbol{\sigma} \cdot \mathbf{p}/E$, and $\bar{\psi}_e(1 + \gamma_5)$ to right-handed electrons (and conversely for positrons). Experimentally, it is found that electrons have negative helicity, $-\boldsymbol{\sigma} \cdot \mathbf{p}/E$. Thus, the experiments show that the β-decay interactions are V, A and not S, T, or P. The matrix element may now be written

$$M = G\{C_V(\bar{\psi}_p\gamma_\mu\psi_n)[\bar{\psi}_e\gamma_\mu(1 + \gamma_5)\psi_\nu] + i^2C_A(\bar{\psi}_p\gamma_\mu\gamma_5\psi_n)[\bar{\psi}_e\gamma_\mu\gamma_5(1 + \gamma_5)\psi_\nu]\}$$

$$= G\{[\bar{\psi}_p\gamma_\mu(C_V - \gamma_5 C_A)\psi_n][\bar{\psi}_e\gamma_\mu(1 + \gamma_5)\psi_\nu]\},$$

using $\gamma_5^2 = 1$. We see that, if $C_V = -C_A$, i.e. the axial-vector and vector couplings are equal in magnitude but opposite in sign, we obtain the famous V-A *interaction*. The nucleon bracket can then also be written in the form $\bar{\psi}_p(1 - \gamma_5)\gamma_\mu(1 + \gamma_5)\psi_n$, since

$$(1 + \gamma_5) = \tfrac{1}{2}(1 + \gamma_5)^2$$

and thus

$$\gamma_\mu(1 + \gamma_5) = \tfrac{1}{2}(1 - \gamma_5)\gamma_\mu(1 + \gamma_5).$$

In this theory, nucleons as well as leptons are left-handed, antinucleons right-handed. Since experimentally, in nucleon decay, $C_A \simeq -1.25C_V$, the degree of longitudinal polarization is in fact only $\simeq 0.9v/c$ instead of v/c. If we write the combinations $\bar{\psi}(1 - \gamma_5) = \bar{\phi}$, $(1 + \gamma_5)\psi = \phi$, the exact V-A hypothesis of Feynman and Gell-Mann gives

$$M = \text{const}\,(\bar{\phi}_p\gamma_\mu\phi_n)(\bar{\phi}_e\gamma_\mu\phi_\nu),$$

which is to be contrasted with the original Fermi theory, some 23 years earlier, in which the interaction was pure vector:

$$M = \text{const}\,(\bar{\psi}_p\gamma_\mu\psi_n)(\bar{\psi}_e\gamma_\mu\psi_\nu).$$

Color Forces between Quarks and Antiquarks

G.1 GLUON COUPLING COEFFICIENTS

The color force between quarks is mediated by a color-anticolor octet of vector gluons as in (8.54), which was written down in analogy with the meson flavor-antiflavor octet states of Table 5.2. Denoting the three colors by r, b, g (for red, blue and green), we have

$$r\bar{b} \tag{G.1a}$$

$$r\bar{g} \tag{G.1b}$$

$$b\bar{g} \tag{G.1c}$$

$$b\bar{r} \tag{G.1d}$$

$$g\bar{b} \tag{G.1e}$$

$$g\bar{r} \tag{G.1f}$$

$$(r\bar{r} + g\bar{g} - 2b\bar{b})/\sqrt{6} \tag{G.1g}$$

$$(r\bar{r} - g\bar{g})/\sqrt{2}. \tag{G.1h}$$

The coupling between a pair of like quarks, such as $rr \to rr$ in Fig. G.1, is mediated by exchange of states (G.1g) and (G.1h). The first gives a factor $\beta^2(1/\sqrt{6})(1/\sqrt{6})$, and the second, $\beta^2(1/\sqrt{2})(1/\sqrt{2})$, for a total coupling of $+ 2\beta^2/3$, where β represents the color charge. For $rb \to rb$ as in Fig. G.2, only the gluon state (G.1g) can be exchanged, giving a factor $- \beta^2(2/\sqrt{6})(1/\sqrt{6}) = - \beta^2/3$. For the interaction $rb \to br$ as in Fig. G.3, the gluon state (G.1a) is involved, yielding a factor $+ \beta^2$.

For antiquarks, we assume the coupling to gluons is given by $- \beta$, in analogy with the electrical case. So $r\bar{r} \to r\bar{r}$ involves a gluon coupling coefficient of $- 2\beta^2/3$, as in Fig. G.4; $r\bar{r} \to b\bar{b}$ as in Fig. G.5 involves the gluon state (G.1a), just as in Fig. G.3, and so has a coupling factor $- \beta^2$. Finally $r\bar{b} \to r\bar{b}$ of Fig. G.6 is the same as for Fig. G.2 with the sign reversed, that is, $+ \beta^2/3$.

In summary, the coupling coefficients for the gluon octet to pairs of quarks or antiquarks will be

$$\langle rr|G|rr\rangle = +\tfrac{2}{3}\beta^2, \qquad \langle r\bar{r}|G|r\bar{r}\rangle = -\tfrac{2}{3}\beta^2,$$

$$\langle rb|G|rb\rangle = -\tfrac{1}{3}\beta^2, \qquad \langle r\bar{b}|G|r\bar{b}\rangle = +\tfrac{1}{3}\beta^2, \qquad (\text{G.2})$$

$$\langle br|G|rb\rangle = +\beta^2, \qquad \langle r\bar{r}|G|b\bar{b}\rangle = -\beta^2.$$

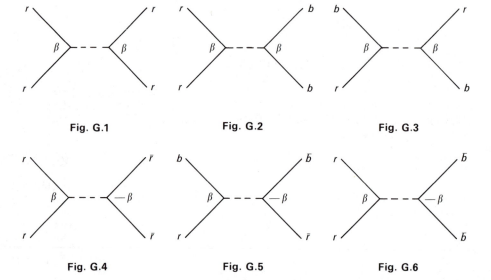

Fig. G.1   Fig. G.2   Fig. G.3

Fig. G.4   Fig. G.5   Fig. G.6

G.2 COUPLING FOR THE BARYON COLOR SINGLET

We now compute the color coupling between a pair of quarks in a color singlet, that is, a color-antisymmetric baryon state. In analogy with the flavor singlet (5.4), the baryon color wavefunction must be of the form

$$(QQQ)_{\text{singlet}} = \frac{1}{\sqrt{6}}[(rb - br)g + (bg - gb)r + (gr - rg)b]. \qquad (\text{G.3})$$

Looking at the interaction between two of the three quarks, we see that the first term in round brackets will involve $rb \to rb$ with a coefficient of $\tfrac{1}{6}$, and $rb \to br$ with $-\tfrac{1}{6}$. Including the gluon coupling factors (G.2), these two contributions together give for the rb term

$$\frac{1-\beta^2}{6} \cdot 3 + \frac{-1}{6}\beta^2 = -\frac{2\beta^2}{9}. \qquad (\text{G.4})$$

Obviously the br term in (G.3) gives the same number, as do the bg, gb, gr, and rg

states. So we need to multiply (G.4) by a factor 6 to obtain the total potential summed over all colors, for a pair of quarks in a three-quark color singlet of

$$V_{(QQ \text{ in } QQQ \text{ singlet})} = -\frac{4\beta^2}{3}\frac{1}{r}. \tag{G.5}$$

The negative sign indicates that the quark-quark interaction is attractive, and the $1/r$ dependence at short range indicates that the color field has the Coulombic form in (5.47).

G.3 COUPLING FOR THE MESON COLOR SINGLET

A quark-antiquark color-singlet wavefunction has the form, in analogy with the flavor singlet in Table 5.2,

$$(Q\bar{Q})_{\text{singlet}} = \frac{1}{\sqrt{3}}(r\bar{r} + b\bar{b} + g\bar{g}). \tag{G.6}$$

The contribution $r\bar{r} \to r\bar{r}$ in (G.6), using (G.2), will be $(1/\sqrt{3})(1/\sqrt{3})(-2\beta^2/3)$ $= -2\beta^2/9$. There are three terms of this type in (G.6), giving together $-2\beta^2/3$. The interactions $r\bar{r} \to b\bar{b}$ and $r\bar{r} \to g\bar{g}$ each give $-\beta^2/3$. Again, multiplying by 3 for all such terms in (G.6) gives $-2\beta^2$. Hence the total potential for the color-singlet meson state will be

$$V_{(Q\bar{Q} \text{ singlet})} = \frac{-2\beta^2 - 2\beta^2/3}{r} = -\frac{8\beta^2}{3}\frac{1}{r}. \tag{G.7}$$

G.4 COUPLINGS FOR COLOR NONSINGLETS

The couplings for nonsinglet color combinations of QQ and $Q\bar{Q}$ can be obtained in a similar fashion. As an example, consider a "6" (sextet) representation of color SU(3). This will have the form

$$(QQ)_{\text{sextet}} = \frac{1}{\sqrt{6}}\left[rr + bb + gg + \frac{1}{\sqrt{2}}(rb + br) \right.$$
$$\left. + \frac{1}{\sqrt{2}}(rg + gr) + \frac{1}{\sqrt{2}}(gb + bg)\right]. \tag{G.8}$$

It is left as an exercise to show that

$$V_{(QQ \text{ sextet})} = +\frac{2\beta^2}{3}\frac{1}{r}. \tag{G.9}$$

As another example we can consider a $Q\bar{Q}$ color octet. The color wavefunction of the $Q\bar{Q}$ pair is the same as that of the gluon octet given in (G.1a)–(G.1h). It is straightforward to show that

$$V_{(Q\bar{Q} \text{ octet})} = +\frac{\beta^2}{3}\frac{1}{r}. \tag{G.10}$$

We observe that both the potentials (G.9) and (G.10), which involve quarks and antiquarks with net color, are repulsive. Only the color-singlet combinations (G.5) and (G.7) have positive binding.

The foregoing results can be obtained from more general formulae, or with the help of Young diagrams; the interested reader is referred to the review lectures by Feynman (1973) and by Rosner (1980).

The above expressions for the QQ and $Q\bar{Q}$ color-singlet potentials (G.5) and (G.7) lead to the formulae for the hyperfine splitting in hadron multiplets in Section 5.8. The strong coupling constant is defined as $\alpha_s = 2\beta^2$. The values of ΔE in the quark case are obtained from the expression (5.33) in the electrical case by the substitutions $e_i e_j = 4\pi\alpha_s/3$ and $2\pi\alpha_s/3$ in (5.34) and (5.35) respectively.

TABLE I Atomic constants

Avogadro's number	N_0	$6.02205(\pm 3) \times 10^{23}\,\text{mole}^{-1}$
Velocity of light *in vacuo*	c	$2.99792458 \times 10^{10}\,\text{cm sec}^{-1}$
		$= 2.99792458 \times 10^{8}\,\text{m sec}^{-1}$
Charge on electron	e	$4.80325(\pm 2) \times 10^{-10}\,\text{esu}$
		$= 1.602189(\pm 5) \times 10^{-19}$ coulomb
Planck's constant reduced	$h/2\pi = \hbar$	$6.58217(\pm 2) \times 10^{-22}\,\text{MeV sec}$
		$= 1.054589(\pm 6) \times 10^{-27}\,\text{erg sec}$
Fine structure constant	$\alpha = e^2/4\pi\hbar c$	$1/137.0360(\pm 1)$
Mass of electron	$m_e c^2$	$9.10953(\pm 5) \times 10^{-28}\,\text{g}$
		$= 0.511003(\pm 1)\,\text{MeV}$
Mass of proton	$m_p c^2$	$938.280(\pm 3)\,\text{MeV}$
		$= 1836.152 m_e$
Classical radius of electron	$r_e = e^2/4\pi m_e c^2$	$2.81794(\pm 1) \times 10^{-13}\,\text{cm}$
Compton wavelength of electron	$\hbar/m_e c = r_e/\alpha$	$3.86159(\pm 1) \times 10^{-11}\,\text{cm}$
First Bohr radius for infinitely heavy nucleus	$a_\infty = 4\pi\hbar^2/e^2 m_e = r_e/\alpha^2$	$0.5291771(\pm 4) \times 10^{-8}\,\text{cm}$
Thomson cross section	$\frac{8}{3}\pi r_e^2$	$665.245(\pm 3)\,\text{mb}$
		$(\text{or} \times 10^{-31}\,\text{m}^2)$
Bohr magneton	$\mu_B = e\hbar/2m_e c$	$0.578838(\pm 1) \times 10^{-14}$ MeV G^{-1}
Nuclear magneton	$\mu_n = e\hbar/2m_p c$	$3.15245(\pm 1) \times 10^{-18}$ MeV G^{-1}
Unit of precession frequency	$\begin{cases} \omega = \dfrac{\mu_B}{\hbar} = \dfrac{e}{2m_e c} \\[2ex] \omega = \dfrac{\mu_n}{\hbar} = \dfrac{e}{2m_p c} \end{cases}$	$8.79401(\pm 3)$ $\times 10^6\,\text{rad sec}^{-1}\,\text{G}^{-1}$ $4.78948(\pm 3)$ $\times 10^3\,\text{rad sec}^{-1}\,\text{G}^{-1}$

Units

1 MeV $= 1.602189(\pm 5) \times 10^{-6}\,\text{erg}$
1 fermi $= 1\,\text{fm} = 10^{-13}\,\text{cm}$
1 barn $= 1\,\text{b} = 10^3\,\text{mb} = 10^6\,\mu\text{b} = 10^{-24}\,\text{cm}^2 = 10^{-28}\,\text{m}^2.$

Useful approximations

$$\hbar c \simeq 200\,\text{MeV fm}$$

Classical radius of electron:

$$e^2/4\pi m_e c^2 \simeq 2.8\,\text{fm} \simeq 2 \times \hbar/m_\pi c$$
$$\simeq 2 \times (\text{range of nuclear forces})$$

Radius of curvature for momentum p:

$$pc \simeq 0.03 H\rho \quad (pc \text{ in GeV}, H \text{ in kG}, \rho \text{ in m})$$

TABLE II   Spherical harmonics

$$Y_l^m(\theta, \phi) = \sqrt{\frac{(2l+1)(l-m)!}{4\pi(l+m)!}}\, P_l^m(\cos\theta)e^{im\phi}$$

$$P_l^m(\cos\theta) = (-1)^m \sin^m\theta\left[\left(\frac{d}{d(\cos\theta)}\right)^m P_l(\cos\theta)\right] \qquad (m \leqslant l)$$

$$P_l(\cos\theta) = \frac{1}{2^l l!}\left[\left(\frac{d}{d(\cos\theta)}\right)^l (-\sin^2\theta)^l\right]$$

$$Y_l^{-m}(\theta, \phi) = (-1)^m [Y_l^m(\theta, \phi)]^*$$

$$P_l^{-m}(\cos\theta) = (-1)^m \frac{(l-m)!}{(l+m)!}P_l^m(\cos\theta)$$

$l = 0$ $\qquad Y_0^0 = \dfrac{1}{\sqrt{4\pi}}$

$l = 1$ $\qquad Y_1^0 = \sqrt{\dfrac{3}{4\pi}}\cos\theta$ $\qquad\qquad Y_1^1 = -\sqrt{\dfrac{3}{8\pi}}\sin\theta\, e^{i\phi}$

$l = 2$ $\qquad Y_2^0 = \sqrt{\dfrac{5}{16\pi}}(3\cos^2\theta - 1)$ $\qquad Y_2^1 = -\sqrt{\dfrac{15}{8\pi}}\sin\theta\cos\theta\, e^{i\phi}:$

$\qquad\qquad Y_2^2 = \sqrt{\dfrac{15}{32\pi}}\sin^2\theta\, e^{2i\phi}$

$l = 3$ $\qquad Y_3^0 = \sqrt{\dfrac{7}{16\pi}}(5\cos^3\theta - 3\cos\theta)$ $\qquad Y_3^1 = -\sqrt{\dfrac{21}{64\pi}}\sin\theta\,(5\cos^2\theta - 1)e^{i\phi}$

$\qquad\qquad Y_3^2 = \sqrt{\dfrac{105}{32\pi}}\sin^2\theta\cos\theta\, e^{2i\phi}$ $\qquad Y_3^3 = -\sqrt{\dfrac{35}{64\pi}}\sin^3\theta\, e^{3i\phi}$

TABLE III Clebsch-Gordan coefficients

As an example of the use of this table, take the case of combining two angular momenta $j_1 = 1, m_1 = 1$ and $j_2 = 1, m_2 = -1$. We look up the entry for combining 1×1, and the fourth line gives for the coefficients in Eq. (C1) of Appendix C

$$\phi_1(1, 1)\phi_2(1, -1) = \sqrt{\tfrac{1}{6}}\psi(2, 0) + \sqrt{\tfrac{1}{2}}\psi(1, 0) + \sqrt{\tfrac{1}{3}}\psi(0, 0).$$

This tells us how two particles of angular momentum (or isospin) unity combine to form states of angular momentum $j = 0, 1$, or 2. Alternatively, a state of particular (j, m) can be decomposed into constituents. Thus $j = 2, m = 0$ can be decomposed into products of states of $j = 1$, with of course $m_1 + m_2 = m = 0$. The fourth column of the 1×1 table gives

$$\psi(2, 0) = \sqrt{\tfrac{1}{6}}\phi_1(1, 1)\phi_2(1, -1) + \sqrt{\tfrac{2}{3}}\phi_1(1, 0)\phi_2(1, 0) + \sqrt{\tfrac{1}{6}}\phi_1(1, -1)\phi_2(1, 1).$$

The sign convention in the table follows that of Condon and Shortley (1951).

$\tfrac{1}{2} \times \tfrac{1}{2}$

	$J:$	1	1	0	1
	$M:$	$+1$	0	0	-1
m_1 m_2					
$+\tfrac{1}{2}$ $+\tfrac{1}{2}$		1			
$+\tfrac{1}{2}$ $-\tfrac{1}{2}$			$\sqrt{\tfrac{1}{2}}$	$\sqrt{\tfrac{1}{2}}$	
$-\tfrac{1}{2}$ $+\tfrac{1}{2}$			$\sqrt{\tfrac{1}{2}}$	$-\sqrt{\tfrac{1}{2}}$	
$-\tfrac{1}{2}$ $-\tfrac{1}{2}$					1

$1 \times \tfrac{1}{2}$

	$J:$	$\tfrac{3}{2}$	$\tfrac{3}{2}$	$\tfrac{1}{2}$	$\tfrac{3}{2}$	$\tfrac{1}{2}$	$\tfrac{3}{2}$
	$M:$	$+\tfrac{3}{2}$	$+\tfrac{1}{2}$	$+\tfrac{1}{2}$	$-\tfrac{1}{2}$	$-\tfrac{1}{2}$	$-\tfrac{3}{2}$
m_1 m_2							
$+1$ $+\tfrac{1}{2}$		1					
$+1$ $-\tfrac{1}{2}$			$\sqrt{\tfrac{1}{3}}$	$\sqrt{\tfrac{2}{3}}$			
0 $+\tfrac{1}{2}$			$\sqrt{\tfrac{2}{3}}$	$-\sqrt{\tfrac{1}{3}}$			
0 $-\tfrac{1}{2}$					$\sqrt{\tfrac{2}{3}}$	$\sqrt{\tfrac{1}{3}}$	
-1 $+\tfrac{1}{2}$					$\sqrt{\tfrac{1}{3}}$	$-\sqrt{\tfrac{2}{3}}$	
-1 $-\tfrac{1}{2}$							1

Clebsch-Gordan coefficients (cont'd)

$\frac{3}{2} \times \frac{1}{2}$

		J: 2	2	1	2	1	2	1	2
		M: $+2$	$+1$	$+1$	0	0	-1	-1	-2
m_1	m_2								
$+\frac{3}{2}$	$+\frac{1}{2}$	1							
$+\frac{3}{2}$	$-\frac{1}{2}$		$\sqrt{\frac{1}{4}}$	$\sqrt{\frac{3}{4}}$					
$+\frac{1}{2}$	$+\frac{1}{2}$		$\sqrt{\frac{3}{4}}$	$-\sqrt{\frac{1}{4}}$					
$+\frac{1}{2}$	$-\frac{1}{2}$				$\sqrt{\frac{1}{2}}$	$\sqrt{\frac{1}{2}}$			
$-\frac{1}{2}$	$+\frac{1}{2}$				$\sqrt{\frac{1}{2}}$	$-\sqrt{\frac{1}{2}}$			
$-\frac{1}{2}$	$-\frac{1}{2}$						$\sqrt{\frac{3}{4}}$	$\sqrt{\frac{1}{4}}$	
$-\frac{3}{2}$	$+\frac{1}{2}$						$\sqrt{\frac{1}{4}}$	$-\sqrt{\frac{3}{4}}$	
$-\frac{3}{2}$	$-\frac{1}{2}$								1

$2 \times \frac{1}{2}$

		J: $\frac{5}{2}$	$\frac{5}{2}$	$\frac{3}{2}$	$\frac{5}{2}$	$\frac{3}{2}$	$\frac{5}{2}$	$\frac{3}{2}$	$\frac{5}{2}$	$\frac{3}{2}$	$\frac{5}{2}$
		M: $+\frac{5}{2}$	$\frac{3}{2}$	$+\frac{3}{2}$	$+\frac{1}{2}$	$+\frac{1}{2}$	$-\frac{1}{2}$	$-\frac{1}{2}$	$-\frac{3}{2}$	$-\frac{3}{2}$	$-\frac{5}{2}$
m_1	m_2										
$+2$	$\frac{1}{2}$	1									
$+2$	$-\frac{1}{2}$		$\sqrt{\frac{1}{5}}$	$\sqrt{\frac{4}{5}}$							
$+1$	$+\frac{1}{2}$		$\sqrt{\frac{4}{5}}$	$-\sqrt{\frac{1}{5}}$							
$+1$	$-\frac{1}{2}$				$\sqrt{\frac{2}{5}}$	$\sqrt{\frac{3}{5}}$					
0	$+\frac{1}{2}$				$\sqrt{\frac{3}{5}}$	$-\sqrt{\frac{2}{5}}$					
0	$-\frac{1}{2}$						$\sqrt{\frac{3}{5}}$	$\sqrt{\frac{2}{5}}$			
-1	$+\frac{1}{2}$						$\sqrt{\frac{2}{5}}$	$-\sqrt{\frac{3}{5}}$			
-1	$-\frac{1}{2}$								$\sqrt{\frac{4}{5}}$	$\sqrt{\frac{1}{5}}$	
-2	$+\frac{1}{2}$								$\sqrt{\frac{1}{5}}$	$-\sqrt{\frac{4}{5}}$	
-2	$-\frac{1}{2}$										1

Clebsch-Gordan coefficients (cont'd)

1×1

	$J:$	2	2	1	2	1	0	2	1	2
	$M:$	$+2$	$+1$	$+1$	0	0	0	-1	-1	-2
m_1	m_2									
$+1$	$+1$	1								
$+1$	0		$\sqrt{\frac{1}{2}}$	$\sqrt{\frac{1}{2}}$						
0	$+1$		$\sqrt{\frac{1}{2}}$	$-\sqrt{\frac{1}{2}}$						
$+1$	-1				$\sqrt{\frac{1}{6}}$	$\sqrt{\frac{1}{2}}$	$\sqrt{\frac{1}{3}}$			
0	0				$\sqrt{\frac{2}{3}}$	0	$-\sqrt{\frac{1}{3}}$			
-1	$+1$				$\sqrt{\frac{1}{6}}$	$-\sqrt{\frac{1}{2}}$	$\sqrt{\frac{1}{3}}$			
0	-1							$\sqrt{\frac{1}{2}}$	$\sqrt{\frac{1}{2}}$	
-1	0							$\sqrt{\frac{1}{2}}$	$-\sqrt{\frac{1}{2}}$	
-1	-1									1

$\frac{3}{2} \times 1$

	$J:$	$\frac{5}{2}$	$\frac{5}{2}$	$\frac{3}{2}$	$\frac{5}{2}$	$\frac{3}{2}$	$\frac{1}{2}$	$\frac{5}{2}$	$\frac{3}{2}$	$\frac{1}{2}$	$\frac{5}{2}$	$\frac{3}{2}$	$\frac{5}{2}$
	$M:$	$+\frac{5}{2}$	$+\frac{3}{2}$	$+\frac{3}{2}$	$+\frac{1}{2}$	$+\frac{1}{2}$	$+\frac{1}{2}$	$-\frac{1}{2}$	$-\frac{1}{2}$	$-\frac{1}{2}$	$-\frac{3}{2}$	$-\frac{3}{2}$	$-\frac{5}{2}$
m_1	m_2												
$+\frac{3}{2}$	$+1$	1											
$+\frac{3}{2}$	0		$\sqrt{\frac{2}{5}}$	$\sqrt{\frac{3}{5}}$									
$+\frac{1}{2}$	$+1$		$\sqrt{\frac{3}{5}}$	$-\sqrt{\frac{2}{5}}$									
$+\frac{3}{2}$	-1				$\sqrt{\frac{1}{10}}$	$\sqrt{\frac{2}{5}}$	$\sqrt{\frac{1}{2}}$						
$+\frac{1}{2}$	0				$\sqrt{\frac{3}{5}}$	$\sqrt{\frac{1}{15}}$	$-\sqrt{\frac{1}{3}}$						
$-\frac{1}{2}$	$+1$				$\sqrt{\frac{3}{10}}$	$-\sqrt{\frac{8}{15}}$	$\sqrt{\frac{1}{6}}$						
$+\frac{1}{2}$	-1							$\sqrt{\frac{3}{10}}$	$\sqrt{\frac{8}{15}}$	$\sqrt{\frac{1}{6}}$			
$-\frac{1}{2}$	0							$\sqrt{\frac{3}{5}}$	$-\sqrt{\frac{1}{15}}$	$-\sqrt{\frac{1}{3}}$			
$-\frac{3}{2}$	$+1$							$\sqrt{\frac{1}{10}}$	$-\sqrt{\frac{2}{5}}$	$\sqrt{\frac{1}{2}}$			
$-\frac{1}{2}$	-1										$\sqrt{\frac{3}{5}}$	$\sqrt{\frac{2}{5}}$	
$-\frac{3}{2}$	$+0$										$\sqrt{\frac{2}{5}}$	$-\sqrt{\frac{3}{5}}$	
$-\frac{3}{2}$	-1												1

TABLE IV Elementary particles

Long-Lived particles (stable or decaying by weak or electromagnetic transitions)[a]

Particle	J^P[b]	I^G[c]	Mass,[d] MeV	Mean life,[d] sec	Decay Mode	Decay Fraction	p_{max}, MeV/c[e]
γ	1^-	—	0	stable	—	—	—
Leptons							
ν_e	$\frac{1}{2}$	—	0(< 50 eV)	stable			
ν_μ	$\frac{1}{2}$	—	0(< 0.57)				
ν_τ	$\frac{1}{2}$	—	0(< 250)				
$e^{\pm}$	$\frac{1}{2}$	—	0.511003($\pm$ 1)	stable	—	—	—
$\mu^{\pm}$	$\frac{1}{2}$	—	105.6595($\pm$ 2)	2.1971($\pm$ 1) $\times 10^{-6}$	$e\nu\bar{\nu}$	100%	53
$\tau^{\pm}$	$\frac{1}{2}$	—	1784($\pm$ 4)	$< 2 \times 10^{-12}$	$\mu\nu\bar{\nu}$	18%	889
					$e\nu\bar{\nu}$	18%	892
					$\pi\nu$	8%	887
					$\rho\nu$	22%	723
					hadrons $+ \nu$	34%	—

Elementary particles

Particle	J^P	I^G	Mass, MeV	Mean life, sec	Decay Mode	Fraction	p_{max}, MeV/c
Nonstrange mesons							
$\pi^{\pm}$	0^-	1^-	$139.567(\pm 1)$	$2.603(\pm 2) \times 10^{-8}$	$\mu\nu$	$\simeq 100\%$	30
					$e\nu$	1.27×10^{-4}	70
					$\mu\nu\gamma$	1.24×10^{-4}	30
					$\pi^0 e\nu$	1.02×10^{-8}	5
					$e\nu\gamma$	6×10^{-8}	70
π^0	0^-	1^-	$134.963(\pm 4)$	$0.83(\pm 6) \times 10^{-16}$	$\gamma\gamma$	98.8%	67
					$\gamma e^+ e^-$	1.17%	67
η	0^-	0^+	$548.8(\pm 6)$	(Width = 0.9 keV)	$\gamma\gamma$	39%	274
					$\pi^0\gamma\gamma$	3%	258
					$3\pi^0$	30%	179
					$\pi^+\pi^-\pi^0$	23%	174
					$\pi^+\pi^-\gamma$	5%	236
η'	0^-	0^+	$958(\pm 1)$	(Width = 0.3 MeV)	$\eta\pi\pi$	66%	–
					$\rho^0\gamma$	30%	–
					$\gamma\gamma$	$\simeq 2\%$	–

Elementary particles

Particle	J^P	I^G	Mass, MeV	Mean life, sec	Mode	Fraction	p_{max}, MeV/c
Strange mesons							
$K^\pm$	0^-	$\frac{1}{2}$	$493.67(\pm2)$	$1.237(\pm3)\times10^{-8}$	$\mu^\pm\nu$	$63.5\%\,(K_{\mu2})$	236
					$\pi^\pm\pi^0$	$21.2\%\,(K_{\pi2})$	205
					$\pi^\pm\pi^+\pi^-$	$5.6\%\,(K_{\pi3})$	126
					$\pi^\pm\pi^0\pi^0$	1.7%	133
					$\mu^\pm\pi^0\nu$	$3.2\%\,(K_{\mu3})$	215
					$e^\pm\pi^0\nu$	$4.9\%\,(K_{e3})$	228
					$e^\pm\nu$	$1.5\times10^{-5}\,(K_{e2})$	247
						and others	
$K^0, \bar{K}^0$	0^-	$\frac{1}{2}$	$497.7(\pm1)$	50% K_S, 50% K_L			
K_S				$0.892(\pm2)\times10^{-10}$	$\pi^+\pi^-$	68.6%	206
					$\pi^0\pi^0$	31.3%	209
			$m_{K_L}-m_{K_S}=\dfrac{0.477}{\tau_S}$				
K_L				$5.18(\pm4)\times10^{-8}$	$\pi^+\pi^-?$	3×10^{-3}	206
					$\pi^0\pi^0\pi^0$	21.5%	139
					$\pi^+\pi^-\pi^0$	12.6%	133
					$\pi\mu\nu$	26.8%	216
					$\pi e\nu$	38.8%	229
					$\pi^+\pi^-$	2.0×10^{-3}	206
						and others	

Elementary particles

Particle	J^P	I^G	Mass, MeV	Mean life, sec	Decay Mode	Fraction	p_{max}, MeV/c
Charmed mesons							
$D^\pm$	0^-	$\frac{1}{2}$	1868(± 1)	$\simeq 2 \times 10^{-13}$	D^+ $\begin{cases} K^-\pi^+\pi^+ \\ K^- + \text{any} \\ K^+ + \text{any} \\ e^\pm + \text{any} \\ \bar{K}^0 + \text{any} \end{cases}$	4% 10% <6% 8% 40%	845 — — — —
$D^0, \bar{D}^0$	0^-	$\frac{1}{2}$	1863(± 1)	$\simeq 4 \times 10^{-13}$	D^0 $\begin{cases} K^-\pi^+ \\ K^-\pi^+\pi^0 \\ K^-\pi^+\pi^+\pi^- \\ K^- + \text{any} \\ \bar{K}^0 + \text{any} \\ e^\pm + \text{any} \end{cases}$	1.8% 12% 3.5% 35% 60% 8%	860 843 812 — — —

Elementary particles

Particle	S	J^P	I	Mass, MeV	Mean life, sec	Decay Mode	Decay Fraction	p_{max} MeV/c
Baryons								
p	0	$\frac{1}{2}^+$	$\frac{1}{2}$	938.280($\pm$ 3)	stable			
n	0	$\frac{1}{2}^+$	$\frac{1}{2}$	939.573($\pm$ 3)	917 $\pm$ 14	$pe^-\nu$	100%	1
Λ	-1	$\frac{1}{2}^+$	0	1115.60($\pm$ 5)	2.63($\pm$ 2) $\times 10^{-10}$	$p\pi^-$	65%	100
						$n\pi^0$	35%	104
						$pe\nu$	0.81×10^{-3}	163
						$p\mu\nu$	1.6×10^{-4}	131
Σ^+	-1	$\frac{1}{2}^+$	1	1189.4($\pm$ 1)	0.800($\pm$ 4) $\times 10^{-10}$	$p\pi^0$	52%	189
						$n\pi^+$	48%	185
						$\Lambda e^+\nu$	2×10^{-5}	72
						$p\gamma$	1.2×10^{-3}	225
						$n\pi^+\gamma$	9×10^{-4}	185
Σ^0	-1	$\frac{1}{2}^+$	1	1192.5($\pm$ 1)	6×10^{-20}	$\Lambda\gamma$	100%	75
Σ^-	-1	$\frac{1}{2}^+$	1	1197.3($\pm$ 1)	1.48($\pm$ 1) $\times 10^{-10}$	$n\pi^-$	100%	193
						$\Lambda e^-\nu$	6×10^{-5}	79
						$ne^-\nu$	1.1×10^{-3}	230
						$n\mu^-\nu$	0.5×10^{-3}	210
						$n\pi^-\gamma$	10^{-4}	193

Elementary particles

Particle	S	J^P	I	Mass, MeV	Mean life, sec	Decay		
						Mode	Fraction	p_{max}, MeV/c
Ξ^0	-2	$\frac{1}{2}^+$	$\frac{1}{2}$	$1314.7(\pm 7)$	$3.0(\pm 2) \times 10^{-10}$	$\Lambda\pi^0$	100%	135
Ξ^-	-2	$\frac{1}{2}^+$	$\frac{1}{2}$	$1321.3(\pm 2)$	$1.64(\pm 2) \times 10^{-10}$	$\Lambda\pi^-$	100%	139
Ω^-	-3	$\frac{3}{2}^+$	0	$1672.2(\pm 3)$	$0.82(\pm 3) \times 10^{-10}$	$\Xi^0\pi^-$	23%	293
						$\Xi^-\pi^0$	8%	289
						ΛK^-	69%	210
Λ_c ($C = +1$)	0	$\frac{1}{2}^+$	0	$2273(\pm 6)$	$\sim 7 \times 10^{-13}$	$pK^-\pi^+$	$\simeq 3\%$ and others	814

Elementary particles

Particle	J^P	I^G	Mass, MeV	Width Γ (MeV)	Decay Mode	Fraction
Meson resonances (nonstrange)						
$\rho(770)$	1^-	1^+	776 ± 3	158 ± 5	$\pi\pi$	100%
					e^+e^-	6×10^{-5}
					$\mu^+\mu^-$	6×10^{-5}
$\omega(783)$	1^-	0^-	782.4 ± 2	10.1 ± 0.3	$\pi^+\pi^-\pi^0$	90%
					$\pi^0\gamma$	9%
					$\pi^+\pi^-$	1.4%
					e^+e^-	7×10^{-5}
$\delta(980)$	0^+	1^-	981 ± 3	52 ± 8	$\eta\pi, K\bar{K}$	
$S^*(980)$	0^+	0^+	$\simeq 980$	$\simeq 40$	$\pi\pi, K\bar{K}$	
$\phi(1020)$	1^-	0^-	1019.6 ± 1	4.1 ± 0.2	K^+K^-	49%
					$K_L K_S$	36%
					$\pi^+\pi^-\pi^0$	15%
					e^+e^-	3×10^{-4}
					$\mu^+\mu^-$	3×10^{-4}
$A1$	1^+	1^-	1100–1300	$\simeq 300$	$\rho\pi$	100%
$B(1235)$	1^+	1^+	1231 ± 10	130 ± 10	$\omega\pi$	100%
$f(1270)$	2^+	0^+	1273 ± 5	178 ± 20	$\pi\pi$	$\simeq 80\%$ and others
D	1^+	0^+	1288 ± 7	27 ± 10	$K\bar{K}\pi, \eta\pi\pi$	–
$A2(1310)$	2^+	1^-	1317 ± 5	102 ± 5	$\rho\pi$	40%
					$K\bar{K}$	$\simeq 5\%$
					$\eta\pi$	15%
					$\omega\pi\pi$	11%
$E(1420)$	1^+	0^+	1418 ± 10	50 ± 10	$K\bar{K}\pi$	
$f'(1515)$	2^+	0^+	1516 ± 12	67 ± 10	$K\bar{K}, \pi\pi$	
$\rho'(1600)$	1^-	1^+	$\simeq 1600$	$\simeq 300$	4π	85%
					2π	15%
$A3(1660)$	2^-	1^-	1660 ± 10	200 ± 50	$f\pi$	60%
					$\rho\pi$	30%

Elementary particles

Particle	J^P	I^G	Mass, MeV	Width Γ (MeV)	Decay Mode	Decay Fraction
$\omega(1670)$	3^-	0^-	1666 ± 5	166 ± 15	$\rho\pi$	
$J/\psi(3100)$	1^-	0^-	3097 ± 1	0.063 ± 0.009	e^+e^- $\mu^+\mu^-$ hadrons $\pi^0\pi^+\pi^+\pi^-\pi^-$ $3(\pi^+\pi^-)\pi^0$ and others	7% 7% 86% 3.7% 3%
$\chi(3415)$	(0^+)	0^+	3414 ± 4	$-$	$2(\pi^+\pi^-)$ $\pi^+\pi^- K^+K^-$ $\gamma\psi(3100)$ $3(\pi^+\pi^-)$ and others	5% 4% 3% 2%
$\chi(3510)$	$-$	0^+	3507 ± 4	$-$	$\gamma\psi(3100)$ $2(\pi^+\pi^-)$ $3(\pi^+\pi^-)$ and others	30% 2% 3%
$\chi(3550)$	$-$	0^+	3551 ± 5	$-$	$\gamma\psi(3100)$ $2(\pi^+\pi^-)$ $\pi^+\pi^- K^+K^-$ and others	15% 3% 2%
$\psi(3685)$	1^-	0^-	3685 ± 1	0.22	e^+e^- $\mu^+\mu^-$ $\gamma\chi$ $\pi^+\pi^-\psi(3100)$ $2\pi^0\psi(3100)$ and others	1% 1% 20% 33% 17%
$\psi(3770)$	1^-	$-$	3768 ± 3	25 ± 3	$D\bar{D}$	100%
$\Upsilon(9460)$	1^-	$-$	9458 ± 6	0.06	l^+l^-	5%
$\Upsilon(10020)$	1^-	$-$	10016 ± 14	<10	l^+l^-	

Elementary particles

Particle	J^P	I^G	Mass, MeV	Width Γ (MeV)	Decay Mode	Fraction
Meson resonances (strange)						
$K^*(890)$	1^-	$\frac{1}{2}$	$891.8(\pm 4)$	50 ± 1	$K\pi$	~100%
					$K\pi\pi$	0.2%
$K^*(1420)$	2^+	$\frac{1}{2}$	1434 ± 5	100 ± 10	$K\pi$	50%
					$K^*(890)\pi$	27%
					$K\rho$	7%
					$K\omega$	4%
					$K\eta$	2%
$K(1240$ $-1420)$	1^+	$\frac{1}{2}$	Probably 2 resonances		$K\pi\pi$ $[K^*(890)\pi$ or $K\rho]$	
Meson resonances (charmed)						
$D^{*+}(2010)$	1^-	$\frac{1}{2}$	2009 ± 1	< 2	$D^0\pi^+$	65%
					$D^+\pi^0$	29%
					$D^+\gamma$	8%
$D^{*0}(2010)$	1^-	$\frac{1}{2}$	2006 ± 2	< 5	$D^0\pi^0$	60%
					$D^0\gamma$	40%

Elementary particles

Resonance	J^P	I	Mass, (MeV)	Width Γ, MeV	Decay Mode	Fraction, %
Baryon resonances (nonstrange)						
$N(1470)$	$\frac{1}{2}^+$	$\frac{1}{2}$	1435–1505	200–400	$N\pi$	60
					$N\pi\pi$	40
$N(1520)$	$\frac{3}{2}^-$	$\frac{1}{2}$	1510–1540	100–150	$N\pi$	50
					$N\pi\pi$	50
$N(1535)$	$\frac{1}{2}^-$	$\frac{1}{2}$	1500–1600	50–160	$N\pi$	34
					$N\eta$	66
$N(1650)$	$\frac{1}{2}^-$	$\frac{1}{2}$	1655–1680	105–175	$N\pi$	60
					$N\pi\pi$	30
$N(1670)$	$\frac{5}{2}^-$	$\frac{1}{2}$	1660–1690	120–180	$N\pi$	40
					$N\pi\pi$	60
$N(1688)$	$\frac{5}{2}^+$	$\frac{1}{2}$	1680–1692	105–180	$N\pi$	60
					$N\pi\pi$	40
$N(1700)$	$\frac{3}{2}^-$	$\frac{1}{2}$	1665–1730	70–120	$N\pi$	10
					$N\pi\pi$	90
$N(1710)$	$\frac{1}{2}^+$	$\frac{1}{2}$	1650–1750	100–150	$N\pi$	20
					$N\pi\pi$	> 50
$N(1810)$	$\frac{3}{2}^+$	$\frac{1}{2}$	1700–1810	150–250	$N\pi$	17
					$N\pi\pi$	70
$N(1990)$	$\frac{7}{2}^+$	$\frac{1}{2}$	1950–2050	100–400	$N\pi, N\pi\pi$	
$\Delta(1236)$	$\frac{3}{2}^+$	$\frac{3}{2}$	1230–1236	115 ± 5	$N\pi$	99.4
					$N\gamma$	0.6
$\Delta(1650)$	$\frac{1}{2}^-$	$\frac{3}{2}$	1620–1695	130–200	$N\pi$	27
					$N\pi\pi$	73
$\Delta(1670)$	$\frac{3}{2}^-$	$\frac{3}{2}$	1650–1690	175–300	$N\pi, N\pi\pi$	
$\Delta(1690)$	$\frac{3}{2}^+$	$\frac{3}{2}$	1500–1900	150–350	$N\pi$	20
					$N\pi\pi$	80
$\Delta(1890)$	$\frac{5}{2}^+$	$\frac{3}{2}$	1890–1930	250–400	$N\pi, N\pi\pi$	
$\Delta(1910)$	$\frac{1}{2}^+$	$\frac{3}{2}$	1850–1950	200–330	$N\pi, N\pi\pi$	
$\Delta(1950)$	$\frac{7}{2}^+$	$\frac{3}{2}$	1935–1980	140–220	$N\pi$	45
					$\Delta(1236)\pi$	50
					ΣK	2
					and others	
$\Delta(1960)$	$\frac{5}{2}^-$	$\frac{3}{2}$	1910–1950	150–300	$N\pi$	
$\Delta(2420)$	$\frac{11}{2}^+$	$\frac{3}{2}$	2380–2450	300–500	$N\pi$	

Elementary particles

Resonance	J^P	I	Mass, (MeV)	Width Γ, MeV	Decay Mode	Fraction, %
Baryon resonances ($S = -1$)						
$\Lambda(1405)$	$\frac{1}{2}^-$	0	1405 ± 5	40 ± 10	$\Sigma\pi$	100
$\Lambda(1520)$	$\frac{3}{2}^-$	0	1518 ± 2	16 ± 2	$N\bar{K}$	46
					$\Sigma\pi$	41
					$\Lambda\pi\pi$	10
					and others	
$\Lambda(1670)$	$\frac{1}{2}^-$	0	1670	$\simeq 30$	$N\bar{K}$	15
					$\Lambda\eta$	35
					$\Sigma\pi$	50
$\Lambda(1690)$	$\frac{3}{2}^-$	0	1690	27–85	$N\bar{K}$	20
					$\Sigma\pi$	55
					$\Lambda\pi\pi$	15
					$\Sigma\pi\pi$	10
$\Lambda(1800)$	$\frac{1}{2}^-$	0	1700–1850	200–400	$N\bar{K}$	40
					$\Sigma\pi$ etc.	
$\Lambda(1815)$	$\frac{5}{2}^+$	0	1815 ± 5	75 ± 10	$N\bar{K}$	65
					$\Sigma\pi$	11
					$\Sigma(1385)\pi$	17
$\Lambda(1830)$	$\frac{5}{2}^-$	0	$\simeq 1835$	66–145	$N\bar{K}$	10
					$\Sigma\pi$	30
$\Lambda(1860)$	$\frac{3}{2}^+$	0	$\simeq 1890$	60–200	$N\bar{K}$	30
					$\Sigma\pi$ etc.	
$\Lambda(2100)$	$\frac{7}{2}^-$	0	2100	40–145	$N\bar{K}$	25
					and others	
$\Lambda(2110)$	$\frac{5}{2}^+$	0	$\simeq 2100$	150–250	$N\bar{K}$	–
$\Sigma(1385)$	$\frac{3}{2}^+$	1	1385 ± 1	36 ± 3	$\Lambda\pi$	90
					$\Sigma\pi$	10
$\Sigma(1660)$	$\frac{1}{2}^+$	1	1580–1690	30–200	$N\bar{K}, \Sigma\pi, \Lambda\pi$	
$\Sigma(1670)$	$\frac{3}{2}^-$	1	1670	50	$\Sigma\pi$	$\simeq 50$
					$\Lambda\pi$	$\simeq 30$
					and others	
$\Sigma(1750)$	$\frac{1}{2}^-$	1	1750	80	$N\bar{K}$	15
					$\Lambda\pi$	
$\Sigma(1765)$	$\frac{5}{2}^-$	1	1774 ± 5	100–150	$N\bar{K}$	45
					$\Lambda\pi$	15
					$\Lambda(1520)\pi$	15
					$\Sigma(1385)\pi$	13
$\Sigma(1915)$	$\frac{5}{2}^+$	1	1910	$\simeq 70$	$N\bar{K}$	~ 10
					$\Lambda\pi$	~ 5
$\Sigma(2030)$	$\frac{7}{2}^+$	1	2030	80–170	$N\bar{K}$	~ 10
					$\Lambda\pi$	~ 35
					$\Sigma\pi$	~ 5

Elementary particles

Resonance	J^P	I	Mass, (MeV)	Width Γ, MeV	Decay Mode	Fraction, %
Baryon resonances ($S = -2$)						
$\Xi(1530)$	$\frac{3}{2}^+$	$\frac{1}{2}$	1532 ± 0.3	9 ± 0.5	$\Xi\pi$	100
$\Xi(1820)$	$\frac{3}{2}$	$\frac{1}{2}$	1823	$\simeq 30$	$\Lambda\bar{K}$	30
					$\Xi\pi$	10
					$\Xi(1530)\pi$	30
					$\Sigma\bar{K}$	30

[a] Adapted from "Review of particle properties" of Particle Data Group, R. L. Kelly *et al.*, *Reviews of Modern Physics* **52**, S1 (Apr. 1980).

[b] J^P, spin and parity of particle. ( ), spin-parity assignment still in some doubt.

[c] I^G, isospin and G-parity.

[d] Errors given in a bracket [thus (± 2)] refer to the last decimal place.

[e] p_{max}, momentum of each secondary in two-body decay, or maximum momentum of any secondary in three-body decay.

Answers to Problems

CHAPTER 1

1.1 $\lambda_c = 0.67 \times 10^6 \text{ sec}^{-1}$; $\Lambda = 26.5$ cm; $G_F \simeq 10^{-7}$.

1.2 $E = \frac{1}{2}E_\pi(1 + \beta \cos \theta)$; $D = (1 + \beta \cos \theta)/(1 - \beta \cos \theta)$.

1.3   Binding of $(H_2\mu)^+$ larger than $(H_2e)^+$. Reduced mass $\mu_H = m_\mu/(1 + m_\mu/M_H)$ $< \mu_D = m_\mu/(1 + m_\mu/M_D)$. Internuclear distance $\simeq 3 \times 10^{-11}$ cm. HD $\rightarrow$ He^3 + μ + 5.4 MeV. [References: L. Alvarez *et al.*, *Phys. Rev.* **105**, 1127 (1957); G. Feinberg and L. Lederman, *Ann. Rev. Nucl. Science* **13**, 431 (1963).]

1.5 $g/G \sim 10^{-6}$.

1.6   (i): Allowed. (ii): Forbidden for free protons, allowed for nuclei when *p-n* binding-energy difference is sufficient. (iii): Forbidden by conservation of muon number. (iv): Forbidden by strangeness conservation.

1.7 $|\Delta e/e| > (KM^2/e^2)^{1/2} = 10^{-18}$.

CHAPTER 2

2.1   (a) 0.2, (b) 0.07, (c) 0.02.

2.2 $y_{\text{rms}} = (21/p\beta\sqrt{2})(S^{3/2}/\sqrt{3})$ radiation lengths. (a) 4.9 cm; (b) 1.65 cm.

2.3   1.6 km; 4.95 m.

2.5 $E = 7\,mc^2$.

2.6   (a) $1 - \dfrac{p_f}{M}$,   (b) $1 + \dfrac{p_f}{M}$,   (c) 1.

2.7   106 MeV; 6×10^{-3}.

2.8   TLI = 310 g cm^{-2}; 60,000 photons; 5%.

CHAPTER 3

3.1 (a) $J_z = \pm \frac{1}{2}$. (d) Same method as (b) and (c): S-state capture ensures $J_z(\Sigma) = \pm \frac{1}{2}$.

3.2 $M(\text{even}) = \boldsymbol{\sigma} \cdot (\mathbf{k} \times \boldsymbol{\varepsilon})$, even relative parity, $M1$ photon; $M(\text{odd}) = \boldsymbol{\sigma} \cdot \boldsymbol{\varepsilon}$, odd relative parity, $E1$ photon. Invariant mass of $e^+ e^-$ is $M(e^+ e^-) = \sqrt{(E_+ + E_-)^2 - k^2}$ · Large $M(e^+ e^-)$ implies small k and a smaller matrix element for even relative parity. Thus $f(M(e^+ e^-))$ drops off more steeply in even-parity case.

3.4 Two Λ's in ^1S state if $P(\Xi) = +1$, ^3P state if $P(\Xi) = -1$.

3.8 (a) J even.   (b) None (parity is not conserved in weak decay).

CHAPTER 4

4.1 As for $\pi^- p \to n\pi^0$, $\pi^- p \to p\pi^-$, $\pi^+ p \to \pi^+ p$.

4.2 $1:2$.

4.3 $2:1$.

4.4 (a) No: $I = 0$ or 2 only. (b) Yes. (c) No: $I \geqslant 2$, since $I_3 = 2$. (d) No: $I = 0$ or 2 only. (e) Yes.

4.5 (a) $I = 0$, 1, 2, or 3. (b) $I = 1$ or 3.

4.6 $I = 0$ or 1. $\sigma_{(a)}/\sigma_{(b)} = 1$ if $I = 0$ only; $= 0$ if $I = 1$ only.

4.7 (a) $+1$; $I = 0$: $^3P_0$, $^3P_2$, $^3F_0$, $^3F_2, \ldots$; $I = 1$: $^3S_1$, $^3D_1$, $^3D_3, \ldots$. (b) $+1$; $I = 0$: $^3P_0$, $^3P_2$, $^3F_0$, $^3F_2, \ldots$, (c) -1; $I = 1$: $^1S_0$, $^1D_2, \ldots$, $^3P_1$, $^3P_2, \ldots$. $p\bar{p} \nrightarrow 2\pi^0$ implies annihilation from S-states only.

4.8 Ratio $= 1$.

4.9 ρ-ω interference in $\pi^+\pi^-$ mode, with amplitude α, typical of G-violating electromagnetic interactions. A narrow dip (or peak) will occur in $\pi^+\pi^-$ mass spectrum in ω-region.

4.12 $2\sqrt{E_1^2 - m^2}\sqrt{E_2^2 - m^2} = \pm [M^2 + 2E_1 E_2 - 2M(E_1 + E_2) + m^2]$.

4.14 $61°$, $1.06\,\text{GeV}/c$; $0°$, $9.25\,\text{GeV}/c$; $16\,\text{GeV}^2$.

4.15 $(p/p_0)^{2l+1}(E_0/E)$.

CHAPTER 5

5.1 $\mu_p = 3$; $\mu_n = -2$; $\mu_{\Xi^0} - \mu_{\Xi^-} = -1$; $\mu_{\Sigma^+} - \mu_{\Sigma^-} = 4$.

CHAPTER 6

6.1 $\sim 10^{-8}$.

6.2 (a) $E_\nu = E_\nu(\text{max})/(1 + \gamma^2 \theta^2)$, where $\gamma = E_{\pi,K}/m_{\pi,K}$. (b) $E_\nu(\text{max}) = 0.42 E_\pi$ or $0.96 E_K$. (c) 40 GeV. (d) 2.8×10^8. (e) 0.03. (f) To range out muons.

6.4 $\sim 10^{-4}$ (experiment: 0.7×10^{-4}).

6.5   (a) $2:1$; (b) $2:1$.

6.6   (a) $1:2$; (b) $2:1$.

6.8 4×10^{-15}.

6.10   0.17/day. No difference.

CHAPTER 7

7.2 $m^3 \, d\sigma/dm = (8\pi\alpha^2/9)AB(1 - 5\tau)\Sigma e_i^2/20$.

7.3 $\overline{Q}/Q = 0.20$; $\langle y_\nu \rangle = 0.484$; $\langle y_{\bar{\nu}} \rangle = 0.345$.

7.5 $L = 1.6 \times 10^{28} \text{ cm}^{-2} \text{ sec}^{-1}$; 90 hr^{-1}.

7.6 $W = 2.1$ GeV.

7.8   (a) 1.22×10^5; (b) 1460.

CHAPTER 8

8.1 $\sin^2 \theta_w = \frac{3}{8}; \frac{1}{8}; \frac{1}{4}$.

8.3 $\overline{R} = \frac{20}{9}x^2 - x + \frac{1}{2}$, $R = \frac{20}{27}x^2 - x + \frac{1}{2}$, where $x = \sin^2 \theta_w$.

8.4 $\tau_p > 10^{16}$ yr.

Worked Solutions to Selected Problems

CHAPTER 1

1.4   If m denotes the neutrino mass, the muon energy in pion decay at rest is

$$E_\mu = (m_\pi^2 + m_\mu^2 - m^2)/2m_\pi.$$

If the neutrino masses are m_1 and m_2 respectively, the difference in energy of the muon in the two cases is

$$\delta E_\mu = (m_1^2 - m_2^2)/2m_\pi.$$

The irreducible error in the measurement of E_μ due to the finite lifetime, and hence the uncertainty in mass of the pion, is

$$\Delta E_\mu = \frac{\hbar}{2\tau_\pi}\left(1 - \frac{m_\mu^2}{m_\pi^2}\right).$$

So $\delta E_\mu > \Delta E_\mu$ if

$$m_1^2 - m_2^2 > m_\pi\hbar\frac{1 - m_\mu^2/m_\pi^2}{\tau_\pi} = 1.2\,\text{eV}^2.$$

CHAPTER 2
2.3

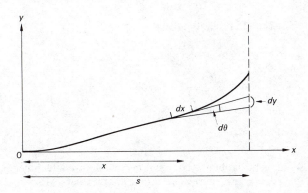

The mean squared lateral deflection, projected in the xy-plane, due to scattering element dx radiation lengths is

$$dy^2 = (s - x)^2 \, d\theta^2 = \frac{1}{2}\left(\frac{21}{p\beta c}\right)^2 (s - x)^2 \, dx \qquad (p\beta c \text{ in MeV}).$$

Integrated over the whole track length s, assuming $p\beta c = $ constant, the rms deflection is

$$y_{rms} = \left(\frac{21}{\sqrt{2}\,p\beta c}\right)\frac{s^{3/2}}{\sqrt{3}}.$$

In the question, $p\beta c$ is not constant. Approximating $\beta = 1$, $pc = E$, we may write, in obvious notation,

$$E(x) = E_0 - \alpha x = \alpha(s - x),$$

where $\alpha = dE/dx = $ constant. Then

$$dy^2 = \frac{1}{2}\frac{(21)^2 \, dx(s - x)^2}{\alpha^2(s - x)^2},$$

or

$$y_{rms} = \left(\frac{21}{\sqrt{2}\,E_0}\right)s^{3/2}.$$

The radial spread is

$$r_{rms} = \sqrt{2}\,y_{rms} = \frac{21}{E_0}s^{3/2}.$$

The mean range of the muons will be

$$R = \frac{E_0}{dE/dx} = 0.5 \times 10^6 \text{ g} = \underline{1.6 \text{ km rock}}.$$

The range measured in radiation lengths is

$$s = R/x_0 = 2 \times 10^4.$$

Then

$$r_{rms} = 59.5 \text{ radiation lengths}$$

$$= \underline{4.95 \text{ m}}.$$

2.5   The threshold energy is obtained by requiring that the available kinetic energy in the CMS be $2m$, in order to create a pair. This means the total CMS energy must be $4m$, corresponding to three electrons and one positron mutually at rest in the CMS frame. If E, p are the energy and momentum of the incident electron in the laboratory frame, the total (4-momentum)^2 of the colliding particles is

$$(E + m)^2 - (p + 0)^2 = 2m^2 + 2mE.$$

This must be equal to the (total energy)^2 in the CMS frame, where the total momentum is zero. Therefore

$$(4m)^2 = 2m^2 + 2mE,$$

or

$$E^{\text{threshold}} = 7m.$$

2.7 The equation for the maximum transferable energy is

$$E'_{\text{max}} = 2m\beta^2\gamma^2 c^2,$$

where mc^2 is the electron rest energy, and γ is the Lorentz factor of the primary particle. To identify a particle as a pion, it is necessary that $E' > (E'_{\text{max}})_K$. For kaons of 5 GeV/c, $\gamma = 10.2$, $\beta = 1$, so that $(E'_{\text{max}})_K = 104 \times 1.02 = 106\,\text{MeV}$. For a pion, $(E'_{\text{max}})_\pi = 1320$ MeV.

The probability of observing such a δ-ray in liquid hydrogen is, with $\beta \simeq 1$,

$$P(> E') = 2\pi\left(\frac{e^2}{mc^2}\right)^2 \times mc^2 \times N_0 \frac{1}{E'}\left[1 - \frac{E'}{E'_{\text{max}}}\left(1 + \beta^2 \log_e \frac{E'_{\text{max}}}{E'}\right)\right]$$

per g/cm^2 traversed. Inserting the numerical values, and with $E' = 106$ MeV, $E'_{\text{max}} = 1320$ MeV, one finds

$$P(> 106\text{ MeV}) = 1.04 \times 10^{-3}\,\text{g}^{-1}\,\text{cm}^2.$$

Therefore in 1 m path, the probability is 6.2×10^{-3}.

CHAPTER 3

3.6 Since $J_D = 1$ and S-state capture is involved, the reaction would involve $J = 1$ in initial and final states. Since the Q-value is only 0.5 MeV, the final state $nn\pi^0$ must be an S-state. Then the two neutrons must be in a triplet spin state, forbidden by the Pauli principle.

CHAPTER 4

4.10 The $p\bar{p}$ system has C-parity $(-1)^{l+S}$, where l is the relative orbital angular momentum, and S is the total spin (0 or 1).

The space parity of the $p\bar{p}$ system is $P = (-1)^{l+1}$, since particle and antiparticle have opposite intrinsic parity. Thus, the initial state has

$$(CP)_{p\bar{p}} = (-1)^{2l+S+1} = (-1)^{S+1} \qquad \text{for all } l.$$

Now let the total angular momentum of the two K^0's be J, where

$$|l + S| \geqslant J \geqslant |l - S|.$$

Measured in the K^0 rest frame, the K_1^0 and K_2^0 mesons have CP-eigenvalues $+1$ and -1 respectively. Thus, the final state has

$$CP = \begin{cases} (\pm 1)(\pm 1)(-1)^J = (-1)^J & \text{for } 2K_1^0 \text{ or } 2K_2^0, \\ (+1)(-1)(-1)^J = (-1)^{J+1} & \text{for } K_1^0 K_2^0, \end{cases}$$

$l = 0$: For annihilation from an S-state, $J = S$, so that the initial state has

$$(CP)_{p\bar{p}} = (-1)^{J+1}, \qquad J = 0, 1.$$

Thus, the $K_1^0 K_2^0$ final state is allowed, and $2K_1^0$ or $2K_2^0$ is forbidden.

$l = 1$: For the triplet state ($S = 1$), $J = 0, 1$, or 2. $(CP)_{p\bar{p}}$ is even, so that $J = 0, 2$ allows $2K_1^0$ or $2K_2^0$ in the final state, while $J = 1$ gives $K_1^0 K_2^0$ only. For the singlet state ($S = 0$), $J = 1$, $(CP)_{p\bar{p}}$ is odd, and only the states $2K_1^0$ or $2K_2^0$ are allowed.

Experimentally, it is found that for annihilations at rest, only $K_1^0 K_2^0$ is observed, proving that annihilation takes place from an $l = 0$ state. On the contrary, annihilation of antiprotons in flight takes place from p-states also, and the modes $K_1^0 K_2^0$, $2K_1^0$, and $2K_2^0$ all appear.

4.11 If we let E, T represent total and kinetic energies of the pions respectively, and Q the total kinetic energy available in the decay, it is readily verified that the conditions are

$$(E_1 E_2 E_3) = \text{maximum} \qquad \text{when} \quad T_1 = T_2 = T_3 = Q/3,$$

$$(E_1 E_2 E_3) = \text{minimum} \qquad \text{when} \quad T_1 = 0, \quad T_2 = T_3 = Q/2.$$

With $x = Q/m$, where m is the pion mass, then

$$(E_1 E_2 E_3)_{\text{max}} = m^3 \left[1 + x + \frac{x^2}{3} + \frac{x^3}{27} \right],$$

$$(E_1 E_2 E_3)_{\text{min}} = m^3 \left[1 + x + \frac{x^2}{4} \right],$$

so that

$$\varepsilon = \frac{(E_1 E_2 E_3)_{\text{max}} - (E_1 E_2 E_3)_{\text{min}}}{(E_1 E_2 E_3)_{\text{max}}} \simeq \frac{x^2}{12(1 + x)}, \qquad \text{where} \quad x \simeq \frac{75}{140} \simeq \frac{1}{2}.$$

Thus $\varepsilon \simeq 0.014$.

4.15 The Lorentz-invariant phase-space factor (Appendix D) for a two-body final state is proportional to

$$p^2 \frac{dp}{dE} \times \frac{1}{E_1 E_2},$$

where E_1 and E_2 are the total energies of the particles and p their momentum, in the CMS. The total energy $E = E_1 + E_2$. Writing

$$E = \sqrt{p^2 + m_1^2} + \sqrt{p^2 + m_2^2},$$

it is found that

$$\frac{dp}{dE} = \frac{E_1 E_2}{p(E_1 + E_2)}.$$

Thus the factor

$$\frac{p^2 \, dp}{E_1 E_2 \, dE} = \frac{p}{E}$$

enters into the width Γ.

The treatment here refers to an S-wave resonance. For one decaying to two (spinless) particles of orbital angular momentum l, it is necessary also to include a centrifugal barrier factor, familiar from reaction theory in nuclear physics (see, for example, J. M. Blatt and V. F. Weisskopf, *Theoretical Nuclear Physics*, John Wiley, 1952, pp. 320 *et seq.* and Appendix A). This factor is associated with the behavior of partial-wave amplitudes near the origin (in the interaction region), which have a radial dependence of approximately $(kr)^l$ when $kr < l$. The width Γ then includes a factor $(pR_0)^{2l}$, where R_0 is some (unknown) range parameter (of order 1 fm). Then it is plausible to write

$$\frac{\Gamma(E)}{\Gamma(E_0)} = \left(\frac{p}{p_0}\right)^{2l+1} \times \left(\frac{E_0}{E}\right),$$

where E_0 and p_0 refer to values at the resonance peak. For decay into particles with spin, the formula is more complicated [see, for example, J. D. Jackson, *Nuovo Cimento* **34**, 1644 (1964)].

CHAPTER 6

6.10 The reaction rate is given by

$$R = \sigma\phi N,$$

where σ is the cross-section per nucleus for neutrino absorption, N is the total number of nuclei in the detector, and ϕ is the neutrino flux in $\sec^{-1} \mathrm{cm}^{-2}$.

We take 164 as the molecular weight of C_2Cl_4, and the total mass of liquid as 6×10^8 g; the number of ^{37}CL nuclei is $N = 2.2 \times 10^{30}$. The solar heat flux is 2 cal $\mathrm{cm}^{-2} \min^{-1}$, or 8.8×10^{11} MeV $\mathrm{cm}^{-2} \sec^{-1}$. Of this, 10% appears as neutrinos, of mean energy 1 MeV, and 1% of the neutrinos are supposed sufficiently energetic to produce a reaction, so that $\phi = 8.8 \times 10^8$ cm^{-2} $\sec^{-1}$. Thus

$$R = \sigma\phi N = (10^{-45})(8.8 \times 10^8)(2.2 \times 10^{30}) = 1.9 \times 10^{-6} \sec^{-1}$$

$$= 0.17 \text{ day}^{-1}.$$

6.11 We apply the $\Delta I = 1$ rule by combining the baryon ($I = \frac{1}{2}$) with a "spurion" of $I = 1$, to give a final hadronic state of $I = \frac{3}{2}$ and $I_3 = \frac{3}{2}$ or $\frac{1}{2}$.

Referring to Table III of Clebsch-Gordan coefficients, and using the $I = 1$ and $I = \frac{1}{2}$ combination, we may write for reaction (i)

$$\phi(1, 1)\phi(\tfrac{1}{2}, \tfrac{1}{2}) = \psi(\tfrac{3}{2}, \tfrac{3}{2}),$$

$$\uparrow \qquad \uparrow$$
$$\text{spurion} \quad \text{nucleon}$$

and for (ii)

$$\phi(1, 1)\phi(\tfrac{1}{2}, -\tfrac{1}{2}) = \sqrt{\tfrac{1}{3}}\,\psi(\tfrac{3}{2}, \tfrac{1}{2}) + \sqrt{\tfrac{2}{3}}\,\psi(\tfrac{1}{2}, \tfrac{1}{2}).$$

If the pion-nucleon system is in a pure $I = \frac{3}{2}$ state, the cross-section ratio obtained by squaring the above amplitudes is $\sigma_{(i)}/\sigma_{(ii)} = 3/1$. For a $\Delta I = 2$ transition, we use $I = 2$ and $I = \frac{1}{2}$ entry of the table, and find for reaction (i)

$$\phi(2, 1)\phi(\tfrac{1}{2}, \tfrac{1}{2}) = -\sqrt{\tfrac{1}{5}}\,\psi(\tfrac{3}{2}, \tfrac{3}{2}) + \sqrt{\tfrac{4}{5}}\,\psi(\tfrac{5}{2}, \tfrac{3}{2}),$$

and for (ii)

$$\phi(2, 1)\phi(\tfrac{1}{2}, -\tfrac{1}{2}) = \sqrt{\tfrac{3}{5}}\,\psi(\tfrac{3}{2}, \tfrac{1}{2}) + \sqrt{\tfrac{2}{5}}\,\psi(\tfrac{5}{2}, \tfrac{1}{2}).$$

For a final state of $I = \frac{3}{2}$ only, the ratio is then $\sigma_{(i)}/\sigma_{(ii)} = \frac{1}{3}$.

6.12 We assume that the three pions are in a relative S-state. Then, by Bose symmetry, any pair must be in a symmetric isospin state, i.e. $I = 0$ or $I = 2$. Call the amplitudes for an $I = 2$ and an $I = 0$ dipion state A and B, respectively. From the $\Delta I = \frac{1}{2}$ rule, we add a "spurion" of $\Delta I = \frac{1}{2}$ to the kaon, of $\Delta I = \frac{1}{2}$, to form states of $I = 0$ or 1. $I = 0$ is forbidden for the three-pion state, which is formed from a dipion of $I = 0$ or 2 and a third pion of $I = 1$. So we consider a three-pion state of $I = 1$, obtained by adding together $I = 1$ with $I = 0$ or 2. Referring to Table III of coefficients we find, in self-evident notation,

Charged kaon:

$$\psi(1,1) = A[\sqrt{\tfrac{3}{5}}\phi(2,2)\phi(1,-1) - \sqrt{\tfrac{3}{10}}\phi(2,1)\phi(1,0) + \sqrt{\tfrac{1}{10}}\phi(2,0)\phi(1,1)]$$
$$+ B[\phi(0,0)\phi(1,1)], \tag{a}$$

Neutral kaon:

$$\psi(1,0) = A[\sqrt{\tfrac{3}{10}}\phi(2,1)\phi(1,-1) - \sqrt{\tfrac{2}{5}}\phi(2,0)\phi(1,0) + \sqrt{\tfrac{3}{10}}\phi(2,-1)\phi(1,1)]$$
$$+ B[\phi(0,0)\phi(1,0)]. \tag{b}$$

The next step is to express the various pion combinations in terms of the isospin functions ϕ. The three-pion wave function must be completely symmetric under pion label interchange, as required for identical bosons. So we write the $\pi^+\pi^+\pi^-$ combination as

$$(+ + -) = \sqrt{\tfrac{1}{6}}(\pi_1^+\pi_2^+\pi_3^- + \pi_2^+\pi_1^+\pi_3^- + \pi_3^-\pi_2^+\pi_1^+$$
$$+ \pi_3^-\pi_1^+\pi_2^+ + \pi_2^+\pi_3^-\pi_1^+ + \pi_1^+\pi_3^-\pi_2^+), \tag{c}$$

the factor $\sqrt{\tfrac{1}{6}}$ being to normalize the amplitude to unity. Referring to the 1×1 entry in Table III, treating the first two pions as the "pair" gives

$$(\pi^+\pi^+\pi^-) = \sqrt{\tfrac{3}{5}}A; \qquad (\pi^+\pi^-\pi^+) = (\pi^-\pi^+\pi^+) = \sqrt{\tfrac{1}{60}}A + \sqrt{\tfrac{1}{3}}B.$$

The second result, for example, follows from the fact that the coefficient for combining $I = 1, I_3 = +1$ and $I = 1, I_3 = -1$ to give $I = 2, I_3 = 0$ is $\sqrt{\tfrac{1}{6}}$; and to give $I = 0, I_3 = 0$ the coefficient is $\sqrt{\tfrac{1}{3}}$. These factors are then multiplied into the appropriate terms in (a), in order to find $\langle\psi(1,1)|\pi^+\pi^+\pi^-\rangle$, etc. Adding together all terms in (c) gives us

$$\langle\psi(1,1)|++-\rangle = 2\sqrt{\tfrac{2}{3}}C, \qquad \text{where} \quad C = \sqrt{\tfrac{4}{15}}A + \sqrt{\tfrac{1}{3}}B.$$

Similarly, one finds for the other charge combinations

$$\langle\psi(1,1)|+00\rangle = -\sqrt{\tfrac{2}{3}}C,$$
$$\langle\psi(1,0)|000\rangle = -C,$$
$$\langle\psi(1,0)|+-0\rangle = \sqrt{\tfrac{2}{3}}C.$$

Squaring these amplitudes, we obtain the ratios of the decay rates

$$\Gamma(K_L \to 3\pi^0) = C^2 = \tfrac{3}{2}\Gamma(K_L \to \pi^+\pi^-\pi^0),$$
$$\Gamma(K^+ \to \pi^+\pi^+\pi^-) = \tfrac{8}{3}C^2 = 4\Gamma(K^+ \to \pi^+\pi^0\pi^0).$$

In order to compare the neutral $(K_L \to \pi^+\pi^-\pi^0)$ with the charged $(K^+ \to \pi^+\pi^0\pi^0)$ decay rate, we need a result from Chapter 6 [Eq. (6.79)]. We have actually calculated the transition for $K^0 \to \pi^+\pi^-\pi^0$. Actually the weak conservation laws only allow half the K^0's to decay in this mode, called K_2^0 or K_L, which has CP eigenvalue -1. The other half is called K_1^0 or K_S, has $CP = +1$, and does not decay to three pions. Thus

$$\langle K^0|T|\pi^+\pi^-\pi^0\rangle = \frac{1}{\sqrt{2}}\langle K_L|T|\pi^+\pi^-\pi^0\rangle.$$

Using this result, one obtains

$$\Gamma(K_L \to \pi^+\pi^-\pi^0) = 2\Gamma(K_0 \to \pi^+\pi^-\pi^0) = 2\Gamma(K^+ \to \pi^+\pi^0\pi^0).$$

Experimentally, there is a small deviation from this prediction, indicating that $\Delta I = \frac{3}{2}$ as well as $\Delta I = \frac{1}{2}$ transitions are involved.

For a more complete discussion of the $K \to 3\pi$ decay modes, the reader is referred to G. Källèn, *Elementary Particle Physics*, Addison-Wesley, 1964, Ch. 16; and for a more general treatment of three-pion decays, to the classic paper by C. Zemach, *Phys. Rev.* **133**, B1201 (1964).

CHAPTER 7

7.5   A current of 10 mA of relativistic particles in a ring of radius 10 m corresponds to a circulating charge $q = (2\pi r/c)i$, or, inserting appropriate numbers, $N = 1.3 \times 10^{10}$ circulating electrons or positrons. If the cross-sectional area of the beam is A, the particle density transverse to the beam will be N/A. The reaction rate will therefore be

$$R = \sigma \left(\frac{N}{A}\right)^2 \times Afn,$$

where f is the revolution frequency, and the bunches meet n times per revolution. With $n = 2$, $f = c/2\pi r$, $A = 0.1 \text{ cm}^2$, and $\sigma = 1.5 \times 10^{-30} \text{ cm}^2$, this formula gives $L = R/\sigma = 1.6 \times 10^{28} \text{ cm}^{-2}\text{sec}^{-1}$ and $R = 90 \text{ hr}^{-1}$.

References

Abashian, A., R. J. Abrams, D. W. Carpenter, G. P. Fisher, B. M. Nefkens, and J. H. Smith, *Phys. Rev. Lett.* **13**, 243 (1964).

Abrams, G. S., *et al.*, *Phys. Rev. Lett.* **33**, 1452 (1974).

Adair, R. K., *Phys. Rev.* **100**, 1540 (1955).

Alff, C., *et al.*, *Phys. Rev. Lett.* **9**, 325 (1962).

Alston, M., L. Alvarez, P. Eberhard, M. Good, W. Graziano, H. Ticho, and S. Wojcicki, *Phys. Rev. Lett.* **5**, 520 (1960).

Altarelli, G., and G. Parisi, *Nucl. Phys.* **B126**, 298 (1977).

Anderson, C. D., *Phys. Rev.* **43**, 491 (1933).

Anderson, C. D., and S. Neddermeyer, *Phys. Rev.* **51**, 884 (1937); **54**, 88 (1938).

Anderson, H., T. Fujii, R. Miller, and L. Tau, *Phys. Rev.* **119**, 2050 (1960).

Anderson, K. J., *et al.*, Proc. 19th Int. Conf. on HEP, Tokyo, 1978.

Aubert, J. J., *et al.*, *Phys. Rev. Lett.* **33**, 1404 (1974).

Augustin, J., J. Bizot, J. Buon, J. Haissinski, D. Labanne, P. Martin, H. Nguyen Ngoc, J. Perez-y-Jorba, F. Rumpf, E. Silva, and S. Tavernier, *Phys. Lett.* **28B**, 508 (1969).

Augustin, J. E., *et al.*, *Phys. Rev. Lett.* **33**, 1406 (1974).

Ayed, R., P. Bareyre, and G. Villet, *Phys. Lett.* **31B**, 598 (1970).

Bacino, W., *et al.*, *Phys. Rev. Lett.* **41**, 13 (1978).

Bailey, J., *et al.*, *Phys. Lett.* **55B**, 420 (1977).

Barnes, V., *et al.*, *Phys. Rev. Lett.* **12**, 204 (1964).

Barnes, W. S., *et al.*, *Phys. Rev.* **117**, 226 (1960).

Bartel, W., B. Dudelzak, H. Krehbiel, J. McElroy, U. Meyer-Berkheut, W. Schmidt, V. Walther, and G. Weber, *Phys. Lett.* **28B**, 148 (1968).

Bathow, G., *et al.*, *Nucl. Phys.* **B20**, 592 (1970).

Baton, J., B. Berthelot, B. Deler, O. Goussu, M. Neveu-Rene, A. Rogozinski, F. Shively, V. Alles-Borelli, E. Benedetti, R. Gasseroli, and P. Waloschek, *Nuov. Cim.* **35**, 713 (1965).

Bergkvist, K. E., *Nucl. Phys.* **B39**, 319 (1972).

Bethe, H. A., and J. Ashkin, "Passage of radiations through matter," *Exptl. Nucl. Phys.* **1**, 166 (1953).

Bjorken, J. D., *Phys. Rev.* **163**, 1767 (1967).

Bjorklund, R., W. E. Crandall, B. J. Moyer, and H. F. York, *Phys. Rev.* **77**, 213 (1950).

Blackett, P. M. S., and G. P. Occhialini, *Proc. Roy. Soc.* **A139**, 699 (1933).

Brabant, J., B. Cork, N. Horowitz, B. Moyer, J. Murray, R. Wallace, and W. Wenzel, *Phys. Rev.* **101**, 498 (1956).

Cabibbo, N., *Phys. Rev. Lett.* **10**, 531 (1963).

Callan, C. G., and D. G. Gross, *Phys. Rev. Lett.* **21**, 311 (1968); **22**, 156 (1969).

Carlson, A. G., J. E. Hooper, and D. T. King, *Phil. Mag.* **41**, 701 (1950).

Cartwright, W. F., C. Richman, M. Whitehead, and H. Wilcox, *Phys. Rev.* **91**, 677 (1953).

Cavanagh, P., J. Turner, C. Coleman, G. Gard, and B. Ridley, *Phil. Mag.* **2**, 1105 (1957).

Chamberlain, O., E. Segre, C. Wiegand, and T. Ypsilantis, *Phys. Rev.* **100**, 947 (1955).

Charpak, G., Bouclier, T. Bressani, J. Favier, and C. Zupancic, *Nucl. Instr. Methods* **62**, 262 (1968).

Charpak, G., *et al.*, *Nucl. Inst. Meth.* **80**, 13 (1970).

Chew, G. F., and F. E. Low, *Phys. Rev.* **101**, 1570 (1956).

Chew, G., S. Frautschi, and S. Mandelstam, *Phys. Rev.* **126**, 1202 (1962).

Christenson, J. H., J. Cronin, V. Fitch, and R. Turlay, *Phys. Rev. Lett.* **13**, 138 (1964).

Clark, D. L., A. Roberts, and R. Wilson, *Phys. Rev.* **83**, 649 (1951); **85**, 523 (1952).

Coffin, C., N. Dikmen, L. Ettlinger, D. Meyer, A. Saulys, K. Terwilliger, and D. Williams, *Phys. Rev. Lett.* **17**, 458 (1966).

Condon, E. U., and G. H. Shortley, *The Theory of Atomic Spectra*, Cambridge University Press, 1951.

Conversi, M., E. Pancini, and O. Piccioni, *Phys. Rev.* **71**, 209 (1947).

Courant, E. D., M. S. Livingston, and H. S. Snyder, *Phys. Rev.* **88**, 1190 (1952).

Dalitz, R. H., *Phil. Mag.* **44**, 1068 (1953).

Davis, R., *et al.*, *Phys. Rev. Lett.* **20**, 1205 (1968).

De Shalit, A., S. Cuperman, H. Lipkin, and T. Rothem, *Phys. Rev.* **107**, 1459 (1957).

Deutsch, M., *Prog. Nucl. Phys.* **3**, 131 (1953).

Dirac, P. A. M., *Proc. Roy. Soc.* **A117**, 610 (1928); also *The Principles of Quantum Mechanics*, Oxford University Press, 1947.

Drell, S. D., and T. M. Yan, *Phys. Rev. Lett.* **24**, 181 (1970); *Ann. Phys.* (*N.Y.*) **66**, 595 (1971).

Dress, W. B., J. K. Baird, P. D. Miller, and N. F. Ramsey, *Phys. Rev.* **170**, 1200 (1968).

Durbin, R., H. Loar, and J. Steinberger, *Phys. Rev.* **84**, 581 (1951).

Erwin, A., R. March, W. Walker, and E. West, *Phys. Rev. Lett.* **6**, 628 (1961).

Fabri, E., *Nuovo Cim.* **11**, 479 (1954).

Fermi, E., *Z. Physik* **88**, 161 (1934).

Feynman, R. P., *Phys. Rev. Lett.* **23**, 1415 (1969).

Feynman, R. P., and M. Gell-Mann, *Phys. Rev.* **109**, 193 (1958).

Feynman, R. P., Proc. 5th Hawaii Topical Conf. in Particle Phys. Hawaii University Press, 1973.

Fortson, E. N., and L. Wilets, *Adv. in Atomic and Mol. Phys.* July (1981).

Fox, D. J., *et al.*, *Phys. Rev. Lett.* **33**, 1504 (1974).

Frauenfelder, H., A. Hanson, N. Levine, A. Rossi, and G. De Pasquali, *Phys. Rev.* **107**, 643 (1957).

Friedman, J. T., and H. W. Kendall, *Ann. Rev. Nucl. Science* **22**, 203 (1972).

Gallinaro, G., *et al.*, *Phys. Rev. Lett.* **38**, 1255 (1977).

Gell-Mann, M., *Phys. Rev.* **92**, 833 (1953).

Gell-Mann, M., *Phys. Lett.* **8**, 214 (1964).

Gell-Mann, M., and F. E. Low, *Phys. Rev.* **95**, 1300 (1954).

Gell-Mann, M., and A. Pais, *Phys. Rev.* **97**, 1387 (1955).

Georgi, H., and S. L. Glashow, *Phys. Rev. Lett.*, **32**, 438 (1974).

Geweniger, C., *et al.*, *Phys. Lett.* **B48**, 487 (1974).

Glashow, S. L., J. Iliopoulos, and L. Maiani, *Phys. Rev. D* **2**, 1285 (1970).

Glück, M., and E. Reya, *Nucl. Phys.* **B156**, 456 (1979).

Goldhaber, A. S., and M. M. Nieto, *Rev. Mod. Phys.* **43**, 277 (1971).

Goldhaber, M., L. Grodzins, and A. Sunyar, *Phys. Rev.* **109**, 1015 (1958).

de Groot, J. G. H., *et al.*, *Z. Physik* **C1**, 143 (1979).

Gross, D. J., and F. Wilczek, *Phys. Rev. D* **8**, 3633 (1973); **9**, 980 (1974).

Hasert, F. J., *et al.*, *Phys. Lett.* **46B**, 138 (1973), *Nucl. Phys.* **B73**, 1 (1974).

Heisenberg, W., *Z. Physik* **77**, 1 (1932).

Herb, S. W., *et al.*, *Phys. Rev. Lett.* **39**, 252 (1977).

Hofstadter, R., *Rev. Mod. Phys.* **28**, 214 (1956).

't Hooft, G., *Phys. Lett.* **37B**, 195 (1971).

Innes, W. R., *et al.*, *Phys. Rev. Lett.* **39**, 1240 (1977).

Jackson, J. D., *Rev. Mod. Phys.* **37**, 484 (1965).

Jones, L. W., *Rev. Mod. Phys.* **49**, 717 (1977).

Kim, Y. S., *Contemp. Phys.* **14**, 289 (1973).

Kobayashi, M., and K. Maskawa, *Prog. Theor. Phys.* **49**, 282 (1972).

Konopinski, E., *Ann. Rev. Nucl. Science* **9**, 99 (1959).

Konopinski, E. J., and A. M. Mahmoud, *Phys. Rev.* **92**, 1045 (1953).

Landau, L., and I. Pomerancuk, *Dokl. Akad. Nauk. SSSR* **92**, 535, 735 (1953).

Langer, L., and R. Moffat, *Phys. Rev.* **88**, 689 (1952).

LaRue, G. S., *et al.*, *Phys. Rev. Lett.* **38**, 1011 (1977); **46**, 967 (1981).

Lattes, C. M. G., H. Muirhead, C. F. Powell, and G. P. Occhialini, *Nature* **159**, 694 (1947).

Lautrup, B. E., A. Peterman, and E. De Rafael, *Phys. Rep.* **3C**, 193 (1972).

Lee, T. D., and C. N. Yang, *Phys. Rev.* **98**, 1501 (1955).

Lee, T. D., and C. N. Yang, *Phys. Rev.* **104**, 254 (1956).

Lehraus, I., *et al.*, *Nucl. Inst. Meth.* **153**, 347 (1978).

Liede, I., and M. Roos, *Nucl. Phys.* **B167**, 397 (1980); see also J. E. Kim, *et al.*, *Rev. Mod. Phys.* **53**, 211 (1981).

Llewellyn-Smith, C. H., and J. F. Wheater, *Phys. Lett.* **105B**, 486 (1981).

Lobashov, V. M., *et al.*, *Nucl. Phys.* **A197**, 241 (1972).

Lyubimov, V. A., E. G. Novikov, V. Z. Nozik, E. F. Tretyakov, and V. S. Kosik, *Phys. Lett.* **94B**, 266 (1980).

Maglic, B., L. Alvarez, A. Rosenfeld, and M. Stevenson, *Phys. Rev. Lett.* **7**, 178 (1961).

Maki, Z., *et al.*, *Prog. Theor. Phys.* **28**, 870 (1962).

Marciano, W. J., and A. Sirlin, *Nucl. Phys.* **189**, 442 (1981).

Marshak, R., and H. Bethe, *Phys. Rev.* **72**, 506 (1947).

Marshak, R., and E. Sudarshan, *Phys. Rev.* **109**, 1860 (1958).

Ne'eman, Y., *Nucl. Phys.* **26**, 222 (1961).

Neubeck, N., *et al.*, *Phys. Rev. C* **10**, 320 (1974).

Nishijima, K., *Prog. Theor. Phys.* **13**, 285 (1955).

Occhialini, G. P. S., and C. F. Powell, *Nature* **159**, 186 (1947).

Orear, J., G. Harris, and S. Taylor, *Phys. Rev.* **102**, 1676 (1956).

Pais, A., *Phys. Rev.* **86**, 663 (1952).

Pais, A., and O. Piccioni, *Phys. Rev.* **100**, 1487 (1955).

Panofsky, W. (data of E. Bloom *et al.*), Int. Conf. High Energy Physics, Vienna, 1968.

Pauli, W., *Handbuch der Physik* **24**, 1, 233 (1933).

Pauli, W., *Phys. Rev.* **58**, 716 (1940).

Perkins, D. H., *Nature* **159**, 126 (1947).

Perl, M. L., *et al.*, *Phys. Rev. Lett.* **35**, 1489 (1975); *Phys. Lett.* **63B**, 366 (1976).

Plano, R., A. Prodell, N. Samios, M. Schwartz, and J. Steinberger, *Phys. Rev. Lett.* **3**, 525 (1959).

Politzer, H. D., *Phys. Rep.* **14C**, 129 (1974).

Pontecorvo, B., *Sov. Phys. JETP* **26**, 984 (1968).

Prescott, C. Y., *et al.*, *Phys. Lett.* **77B**, 347 (1978); **84B**, 524 (1979).

Primakoff, H., *Phys. Rev.* **81**, 899 (1951).

Pryce, M. H. L., and J. C. Ward, *Nature* **160**, 435 (1947).

Regge, T., *Nuovo Cim.* **14**, 951 (1959).

Reines, F., and C. Cowan, *Phys. Rev.* **113**, 273 (1959).

Rochester, G. D., and C. C. Butler, *Nature* **160**, 855 (1947).

Rosner, J., Proc. Advanced Study Inst. on Techniques and Concepts in HEP, St. Croix, USVI (ed. T. Ferbel), 1980.

Rossi, B., *High Energy Particles*, Prentice-Hall, New York, 1952.

Rossi, B., and K. Greisen, *Rev. Mod. Phys.* **13**, 240 (1941).

Sacton, J., Proc. Topical Conf. on Neutrino Physics, Oxford (Rutherford Lab. RL-78-081) (1978).

Sakurai, J. J., *Invariance Principles and Elementary Particles*, Princeton University Press, Princeton, New Jersey, 1964.

Salam, A., *Elementary Particle Theory* (ed. N. Svartholm), Almquist and Wiksells, Stockholm, 1968.

Shafer, J., J. Murray, and D. Huwe, *Phys. Rev. Lett.* **10**, 176 (1963).

Snyder, L., *et al.*, *Phys. Rev.* **63**, 440 (1948).

Sonderegger, P., J. Kirz, O. Guisan, P. Falk-Vairant, C. Bruneton, P. Borgeaud, A. Stirling, C. Caverzasio, J. Guillaud, M. Yvert, and B. Amblard, *Phys. Lett.* **20**, 75 (1966).

Sternheimer, R. M., "Interactions of radiation with matter," *Methods of Exptl. Phys.* **5A**, 1 (1961).

Stevenson, M., L. Alvarez, B. Maglic, and A. Rosenfeld, *Phys. Rev.* **125**, 687 (1962).

Street, J. C., and E. C. Stevenson, *Phys. Rev.* **52**, 1003 (1937).

Stueckelberg, E. C. G., *Helv. Phys. Acta* **11**, 225, 299 (1938).

Tadic, C., *Rep. Prog. Phys.* **43**, 67 (1980).

Tretyakov, E. F., *et al.*, Proc. Neutrino Conf. Aachen, 1976.

Tripp, R. D., "Spin and parity determination of elementary particles," *Ann. Rev. Nucl. Science* **15**, 325 (1965).

Van Dyck, R. S., R. A. Ekstrom, and H. G. Dehmelt, *Nature* **262**, 776 (1976).

Van Dyck, R. S., P. G. Schwinberg, and H. G. Dehmelt, *Phys. Rev. Lett.* **38**, 310 (1977).

Van Royen, R., and V. F. Weisskopf, *Nuovo Cim.* **50**, 617 (1967); **51**, 583 (1967).

Von Dardel, B., D. Dekkers, R. Mermod, J. D. Van Putten, M. Vivargent, G. Weber, and K. Winter, *Phys. Lett.* **4**, 51 (1963).

Von Witsch, W., A. Richter, and P. von Brentano, *Phys. Rev.* **169**, 923 (1968).

Weber, G., Proc. 1967 Int. Sym. on Electron and Photon Interactions at High Energies, Stanford, California, 1967, p. 59.

Weinberg, S., *Phys. Rev. Lett.* **19**, 1264 (1967).

Winter, K., Proc. Int. Symp. on Lepton and Photon Interactions at High Energies, Fermilab, Ill. (Ed. T. B. W. Kirk and H. D. I. Abarbanel).

Weyl, H., *Z. Physik* **56**, 330 (1929).

Wigner, E. P., *Proc. Am. Phil. Soc.* **93**, 529 (1949).

Wolfenstein, L., *Phys. Lett.* **13**, 562 (1964).

Wu, C. S., and I. Shaknov, *Phys. Rev.* **77**, 136 (1950).

Wu, C. S., E. Ambler, R. Hayward, D. Hoppes, and R. Hudson, *Phys. Rev.* **105**, 1413 (1957).

Yang, C. N., and R. L. Mills, *Phys. Rev.* **96**, 191 (1954).

Yoh, J. K., *et al.*, *Phys. Rev. Lett.* **41**, 684 (1978).

Yukawa, H., *Proc. Phys. Math. Soc. Japan* **17**, 48 (1935).

Zweig, G., CERN Report 8419/Th 412, 1964.

Index

Index

429